The study of the earth and the environment requires an
the physical processes within and at the surface of the ear
allow the student to develop a broad working knowledge of mechanics and
its application to the earth and environmental sciences, providing the background necessary to study the professional literature. The only mathematical
background required is that given by a first course in calculus; all other
mathematical concepts are introduced in the context of their application to
the earth and environmental sciences.

For each topic within mechanics, the physics is developed to give an insight
into a wide range of natural phenomena. For example, the theory of stress
and strain is applied to landslides, rock fractures, and earthquakes; the theory
of flow through porous media is applied to the movement of groundwater,
and the compaction of sediments; and the properties of Newtonian and non-
Newtonian fluids are discussed in the context of the flow of lava, ice, and
debris. The breadth of applications described and the many worked examples
will help both students and professionals to understand the physical systems
at work in the earth's crust and at its surface.

Mechanics in the Earth and Environmental Sciences

Mechanics in the Earth and Environmental Sciences

Gerard V. Middleton
McMaster University

Peter R. Wilcock
The Johns Hopkins University

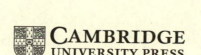

CAMBRIDGE
UNIVERSITY PRESS

Published by the Press Syndicate of the University of Cambridge
The Pitt Building, Trumpington Street, Cambridge CB2 1RP
40 West 20th Street, New York, NY 10011-4211, USA
10 Stamford Road, Oakleigh, Melbourne 3166, Australia

© Cambridge University Press 1994

First published 1994

Printed in Great Britain at the University Press, Cambridge

A catalogue record of this book is available from the British Library

Library of Congress cataloguing in publication data

Middleton, Gerard V.
Mechanics in the earth and environmental sciences / Gerard V. Middleton, Peter R. Wilcock.
p.   cm.
Includes bibliographical references and index.
ISBN 0 521 44124 2. – ISBN 0 521 44669 4 (pbk)
1. Mechanics, Analytic.   2. Geodynamics.   I. Wilcock, Peter R.
II. Title.
QA808.M49   1994
550'.1'531–dc20   93-49455   CIP

ISBN 0 521 44124 2 hardback
ISBN 0 521 44669 4 paperback

TAG

Contents

Preface

This book is the product of our conviction that the earth and environmental science curriculum, as generally taught in departments of geology, geography, and environmental science, must provide better training in the mathematical and physical aspects of the earth sciences. A quantitative approach using models is increasingly important for all research in these disciplines, even in supposedly "qualitative" fields, such as stratigraphy or environmental design. More and more, a working knowledge of physical processes is needed to understand the earth's surface, or near-surface, environment. Environmental scientists need to understand the flow of fluids over or within surface materials, the stability of slopes, and the effects of storms, earthquakes, and volcanic eruptions. A good grasp of physical processes is also increasingly important for traditional applications in the mineral and petroleum industries.

The training necessary to understand natural physical processes is not covered in the freshman mathematics and physics that is traditionally required of graduates in earth and environmental science programs. Indeed, it is difficult to find this material in an accessible and relevant form in any upper level classes. In contrast, most earth science curricula require at least one formal course in chemistry, and further study of chemical principles is an essential part of the training provided by most courses in crystallography, mineralogy, petrology, and environmental processes. The student who wishes to pursue such applications further will find many suitable courses offered by departments of chemistry or environmental engineering. The earth science student who needs to understand more mathematics or physics is not generally so lucky. Courses in linear algebra (or vectors) and ordinary differential equations are required for most introductory courses in mechanics, but even a first course in mechanics does not cover the topics that are most important for the earth sciences, namely the mechanics of continua, including aspects of elasticity, fluid mechanics, and rheology. It is true that these are large fields that cannot be mastered in a single course – but it is better that a student know something about them than nothing at all. In this book, we aim to provide a basic working knowledge of the principles underlying these fields, without attempting a thorough survey of any of them.

The range of knowledge touched on in this book is vast, and extends far beyond the expertise of the two authors, who are a sedimentologist and a geomorphologist respectively. We have choosen appropriate topics and examples and have deliberately avoided confining these to our own fields of specialist knowledge. We have tried to write a book that will be useful as a text for an upper level undergraduate or beginning graduate course in a department of geology, geography, or environmental science. The course can be taught by anyone with a good background in the physical aspects of one of these disciplines. We have provided carefully selected references partly because we feel an obligation to give sources in areas where we go beyond our own expertise. We hope, however, that other instructors will take up the challenge of teaching similar material, and that they and at least some students will make use of the references provided in order to pursue topics that particularly interest them.

The main aims of the book are to introduce some topics in the mechanics of continua, and their applications in the earth sciences. To do this, however, we need to teach some mathematical material – vectors, tensors, and some aspects of ordinary and partial differential equations. These are all topics that have extensive applications in other areas of physics: in gravitation and in electricity and magnetism, for example. Such mathematics is generally considered to be relatively advanced, yet it is our strong belief that earth scientists need to have at least some acquaintance with these topics, even though they may never have the time or opportunity to take rigorous courses in them. We have introduced all mathematical concepts using no more knowledge than is learned in a freshman course in calculus.

In this book, breadth is stressed at the expense of depth. Yet we do strive for as complete an understanding of the basic physics as is possible under the circumstances. Breadth can be achieved because the underlying mechanics is common to many different natural processes. For example, the concepts of stress and strain find applications to rocks, soils, groundwater, rivers, oceans, and atmosphere. Diffusion is a process operating at many scales, from the molecular diffusion of momentum or chemical species to large-scale turbulent diffusion of heat or sediment in the ocean or atmosphere. Stress, deformation, flow, diffusion are examples of the broad concepts whose understanding is critical for all of the geophysical sciences. We hope to give some understanding of these ideas, and the notation and methods that make use of them – to the point where the student will be able to read original papers in, for example, the *Journal of Geophysical Research*, without simply skipping all of the equations.

We believe that, depending on the students' background, this book contains more material than can be taught in a single semester. It would be easy to expand the subject matter to a year-long treatment by including more thorough discussion of examples. In order to shorten the course some topics can fairly readily be omitted: they include some of the later parts of Chapter 2, some topics in Chapters 5 and 6, and all of Chapter 12. Our views on the main topics that must be developed by the instructor and understood by the student are explained more fully in Chapter 1.

The beginnings of this book can be traced to 1979, when Gerry Middleton and Paul Clifford introduced a course like this in the Department of Geology at McMaster University. Over several years, Middleton developed a set of course notes, which served as a text for that course and its successors. In 1987, Peter Wilcock joined the Department of Geography and Environmental Engineering at The Johns Hopkins University and began to teach a similar course to a mixed class of students in geology and environmental science. He heard about the course at McMaster and began using a set of the course notes as a text. The book developed out of the long-distance collaboration that resulted. After accumulating four years of modifications and additions from the experience of teaching the material at both universities, the notes were completely rewritten by Middleton during a sabbatical year at the University of Washington. Large parts of the text have been added, deleted or rewritten by Wilcock, to the point where we now find it hard to remember exactly who wrote what.

We thank our universities for supporting us while we struggled with this relatively new teaching and scholarly initiative. Middleton received a McMaster Teaching and Learning grant one summer, which enabled him to experiment with the use of personal computers in this course. He is indebted to the University of Washington, which provided fine facilities and a most stimulating intellectual environment during the year of writing (1991–2): thanks go particularly to Darrel Cowen, Tom Dunne, Bernard Hallet, and Jody Bourgeois. The book has been prepared using the LATEX system developed by Donald Knuth and Leslie Lamport. The particular implementation used in the later stages has been emTEX, a Shareware program for personal computers, made available without charge by its author, Eberhard Mattes. The authors of this book express their gratitude to those who have made available such a sophisticated typesetting system, without taking any personal profit. Drafts of various chapters, and of the notes that preceded them have been read critically by Tom Dunne, Hugh Ellis, David Goodings, Monty Hampton, Jon Major, Cesar Mendoza, Paul Myrow, Chris Paola, and John Southard. Section 10.5 benefited from discussion of debris flows with Dick Iverson. We thank our copy-editor, Susan Parkinson, whose contributions to this book went far beyond the normal work of a copy-editor. Last, but certainly not least, we thank the students who took our courses over the years.

A set of computer programs designed to accompany this text, together with a brief set of instructions for their use, may be obtained at a nominal cost from the Computer Oriented Geological Society, P.O. Box 371046, Denver CO 80237-0246, USA. Tel. (303) 751-8553, Email: cogs@mines.colorado.edu. This set of programs includes both source code (in Turbo Pascal) and executable code and should run on most IBM-compatible computers. Computer programs referred to in the text (e.g., EJECTA, LAPLACE, EHEAT) are all part of this set.

1
Introduction

1.1 What this book is about

This book is about applying the concepts of mechanics to the earth and environmental sciences. Mechanics is an old and well-developed branch of physics, which has been applied successfully to many problems in the earth sciences. It is certainly not the only branch of science that has been applied to such problems, but learning how mechanics can be applied to geological problems is probably the best way for a student to learn how to apply other branches of physics to the earth sciences. In the following discussion, we find it useful to make a distinction (not followed by all authorities) between the natural sciences, such as geology or biology, and the physical sciences, such as physics and chemistry.

The natural sciences have made much use of methods developed in mathematics and in the physical sciences. Nevertheless, they have also developed their own unique methods. In geology, for example, students learn to observe in the field and in the laboratory, and to use their observations to give historical explanations of natural phenomena. Mathematics and physics certainly cannot teach us to do this, because they are not much interested in descriptive observation or historical explanation.

Further, all natural sciences are necessarily concerned with particular circumstances, that is, with combinations of materials (e.g., minerals, rocks, plant and animal species) and processes (e.g., the accumulation and flow of ice in continental glaciers, the settling of crystals in magma chambers) actually observed or inferred to be present in the natural environment. Because of the complexity of natural phenomena, the natural sciences have a strong empirical component, and are much concerned with observation and classification. The methodology required to organize this information takes an important and central role in the natural sciences.

1

The role of the physical sciences, including mechanics, in the earth and environmental sciences can be viewed from two complementary perspectives.

(1) The natural sciences may be seen as an obvious and useful source of applications of basic physical principles. This view fits this book well. The basis of its organization is a (more-or-less) systematic introduction to continuum mechanics. The examples used illustrate the application of these principles to the earth and environmental sciences.

(2) Another view is that mechanics provides a means of explaining many natural physical processes and, more important, a means of generalizing observations from one location to another. Thus, one might attempt to show how the earth's surface can be explained using mechanics. This is not the basis we used to organize the book. Although we have tried to provide examples from a variety of natural environments, we have not attempted to cover all the physical processes acting on, or near, the earth's surface.

Nonetheless, this second view of the role of mechanics in earth and environmental science represents the essential reason why both of the authors, as practitioners in the earth and environmental sciences, have an interest in mechanics and believe that a firm grasp of the subject is necessary for understanding how the earth works and for conducting research that finds utility beyond the time and place of a particular study.

The mechanics of a natural process can be assumed to be universal, so it is not necessary to completely re-explain a physical phenomenon (whether it is river flow, a volcanic eruption, or a landslide) every time it is encountered. This does not mean that to understand one flood, or eruption, or landslide, is to understand them all, because every natural physical process has a setting that is unique. If the essential mechanics is known, however, the basic features of any particular natural event may be more readily understood, unusual aspects may be more easily discovered, and, most important, the nature of the process in the past and future may be assessed.

Much of classical mathematics and physics was developed in order to describe and explain events taking place in three-dimensional space: events such as the motion of projectiles and planets, the bending and breaking of beams and plates, the flow of fluids like air and water, and the slow deformation of crystalline materials like ice. Obviously, these things are interesting not just because they are aspects of how matter behaves, but also because they are a large part of how the earth behaves. To explain any aspect of the earth we need to understand the processes that produced the effects we now see.

Understanding of processes is particularly important in the environmental applications of the earth sciences. In the past geologists were more concerned with understanding history than with making predictions: a geologist spent more time trying to understand how a particular volcano developed over the past few million years, than he did trying to predict how it would behave over the next hundred years. Understanding history is important in the search for minerals and fuels, and it is certainly relevant to the environmental sciences too – but it does not go very far in answering the types of practical questions that the environmental sciences must deal with, such as the following. Will this volcano erupt in the next few years? Will this slope fail? Will the groundwater supply be contaminated by waste disposed at this site?

The physical sciences provide the basic knowledge and techniques needed to answer questions of this type. Generally this type of knowledge has been mostly applied in engineering design: in learning how to construct buildings and bridges that do not fall down, or machinery that operates for some years without serious wear or failure. The environmental sciences are now learning how to combine this type of knowledge about processes with an understanding of historical development in order to make assessments of how the natural environment is likely to change in the future.

This is a challenging and difficult task – and we do not pretend that it can be taught in a single course or learned by reading a single book. But those hoping to practise environmental or earth sciences should at least begin to understand the basic principles, if only so that they can communicate better with experts in the various disciplines involved. Luckily, there is now much that is common to all the major physical sciences, and this core of common knowledge and techniques can readily be applied to the earth. The principles of stress and strain, to which substantial sections of this book are devoted, apply to the deformation of continuous materials, be they fluid or solid, in any environment or application (e.g., the minerals and rocks of the solid earth, natural soils and fluids, the oceans and atmospheres). The laws governing the flow of heat apply to solid rock masses, magma chambers, and lakes, and also have much in common with the laws governing the flow of groundwater. The basic principles of fluid mechanics apply to fluids in all natural environments, and also have close analogues in electrical and magnetic phenomena.

Most of the practical applications of mechanics are based on the assumption that the materials involved constitute a *continuum* in which properties such as mass and strength vary smoothly in space. We discuss this assumption in the next section.

In the remaining sections of this chapter, we discuss four aspects of physical theory that will form themes elaborated in the rest of the book.

(1) In most branches of physical science it is possible to set up *governing equations* which explain and predict the phenomena, at least in principle. Newton's second law relates the change in momentum of a body to the net force acting on it. Though applied by Newton to bodies whose mass could be considered to be concentrated at a point (the "centre of mass"), it took more than a hundred years of research to understand how it could be applied to extended solid or fluid bodies in general. To understand the motion of such materials we need very general formulations not only of Newton's second law (the *equations of motion*) but also of the conservation of mass (the *continuity equations*), and the conservation of angular momentum. We also need to understand how particular materials respond internally to the application of force, a behaviour generally expressed in the *constitutive equations* of that particular material.

(2) Almost all the phenomena of real interest to the earth sciences involve motion in three-dimensional space. To describe such motion, and the forces that produce and resist it, we need appropriate mathematics: it is provided by coordinate geometry, and by vector and tensor analysis.

(3) Once in possession of an appropriate set of governing equations, we need to understand how it can be applied to the problem in hand: this means knowing how to set the problem up by a statement of the *initial and boundary conditions* and how to solve it for those conditions. As the governing equations are generally a set of differential equations (in three dimensions, *partial differential equations*) we need to understand how to integrate these equations. The integrations are generally carried out by a combination of analytical and numerical techniques. Though we cannot explore this aspect of the problem very far, we need to understand the general principles involved.

(4) Besides the science, solving real problems generally involves a good deal of art – something that is generally described as the "art of modeling". This can really only be learned by example and practice, so this book includes many examples of simple applications of mechanics to different branches of the earth and environmental sciences. The examples discussed in the book are necessarily chosen to be as simple as possible, and they may strike some readers as unrealistic. References to more realistic (and therefore more extended and complicated) applications are

given in each chapter. The reader should review at least some of these to get a better sense of the current "state of the art".

1.2 Definition of a continuum

From the point of view of classical physics, matter is generally assumed to take one of three forms: a point mass, a rigid body, or a continuum. All three forms represent an idealization of real matter but the level of idealization decreases as we move from a point mass to a continuum. Though the concept of a point mass is an extreme idealization of real bodies of matter, it is a simplification of the real world that works very well in some applications. It was Newton himself who first showed that we do not need to know the exact distribution of mass within the earth in order to apply the universal law of gravitation. For such purposes, we can assume that all the mass of the earth is concentrated at a point, its centre of mass.

For many other problems concerned with solid bodies, we must be concerned with the size and shape of the material, as well as its mass. For example, the force needed to roll a boulder in flowing water or air depends on the size and shape of the boulder, as well its mass. In this case, we still can simplify the problem somewhat by assuming that the solid body (the boulder) is rigid, even though we know real materials are not perfectly so. Often, the deformation of the solid is very small and if we neglect it we can still obtain interesting and useful results.

The case is not so straightforward when we consider the air or water flowing around the boulder. In doing so, the fluid clearly deforms. That is, there is a change in the position of fluid particles relative to each other as they move around the obstacle. The forces involved in this deformation depend on a variety of factors, including the mass of the fluid and the resistance it offers to deformation. Sometimes we can neglect these properties of the fluid and the associated forces, as we do in Chapter 2, where we first consider the motion of a volcanic bomb without accounting for the drag force imparted to the bomb by the air. Later in that chapter, we add the fluid drag to our discussion and find that the bomb trajectory is considerably different. The difference is caused by forces exerted on the bomb by fluid deformations produced by its movement.

When solids or fluids deform in response to applied forces, we must be concerned with the distribution of material properties (e.g., mass, resistance to deformation) within the material. We still need to make an important simplifying assumption, namely that the distribution of these properties is continuous in space. This is the *continuum hypothesis*. Strictly speaking, this hypothesis cannot be true. We know that the distribution of properties, such as mass, of real materials is decidedly discontinuous at the molecular scale, so we must limit the problems we consider to those with a characteristic length scale several orders of magnitude larger than that scale.

Another problem arises at much larger scales (those typical of many geophysical applications of mechanics) because the properties of a material vary in space. This variation is central to most of the practical problems of concern to us here because it is the gradient, or variation in space, of mass or force that produces motions and deformation. As a result, the useful range in length scale for the continuum hypothesis has two limits: the lower limit is defined by an element of volume that is much bigger than a molecule, and the upper limit is defined by an element of volume that is smaller than any important spatial variation in material properties.

It turns out that such a middle ground exists for most problems. There are of the order of 10^7 molecules in $\approx 10^{-9}$ mm^3 of air. This is both a huge number of molecules (so we need not worry about the exact number of molecules in an elemental volume) and a very small elemental volume (much smaller than the physical dimensions of any practical problem). A limiting volume of 10^{-9} mm^3 is applicable to all liquids and to gases at atmospheric pressure.

Further, consider the density of a fluid. In the natural sciences, we rarely consider discrete "bodies" of fluid, so that it is usually difficult to define what is meant by the total fluid mass in a particular case. It is easier to measure the mass contained within a given volume of the fluid. Then we can define the mass density (or simply, density) ρ, as the measured mass divided by the measured volume. Mass is a property of points (or bodies considered as point masses); density is the corresponding property of a continuum. To make the concept more precise we reduce the measured volume to "elemental" size. As long as our elemental volume of fluid is much larger than the lower scale limit, so that it contains a huge number of molecules, we need not be concerned with exactly how many molecules, or even which molecules, are in the volume and we can define the density with reasonable accuracy. As long as the elemental volume is also much smaller than the upper scale limit over which fluid density varies appreciably, the concept of density can be usefully applied to the real world.

The minimum length scale for continuum analyses is not necessarily the molecular scale. For example, consider the density and strength of a pile of cobbles the size of baseballs. For a test volume of the order of a cubic millimetre, the density would vary wildly between that of air and that of the rock cobbles, depending on where the volume was located. On the other hand, the density within a volume of many cubic metres would show only negligible variation with the exact placement of the test volume. Similarly, if we wished to measure the aggregate strength of the cobbles, it is reasonable that the scale of the testing apparatus would have to be at least two orders of magnitude larger than an individual cobble to avoid the variation that would result from placing different numbers and arrangements of cobbles in the test volume. These issues of a minimum length scale apply not only to solids such as soils and rock, but also to mixtures of fluids of different composition (e.g., oil and water), and mixtures of fluids and solid particles (e.g., slurries or suspensions of sediment in water).

The reason we use the continuum hypothesis is simply that it allows us to use differential calculus to analyze the properties of a material and its motion and deformation. Continuous changes, or *gradients*, in physical properties and forces define mechanical problems, and differential calculus is the mathematics that treats such gradients precisely and efficiently. In using the continuum hypothesis, we are not limited to cases without abrupt boundaries between different materials. In these cases, we have to define the appropriate *boundary conditions* and then we can use continuum mechanics to describe what happens within those boundaries.

1.3 Governing equations

In Chapter 2, we shall briefly review the mechanics of points and extended bodies. We show how Newton's second law can be regarded as the governing equation for the motion of point masses. It relates the observed change in momentum $d(mv)/dt$ to the sum of all the forces $\sum F_i$, acting on a point mass, m:

$$\frac{d(mv)}{dt} = \sum F_i \tag{1.1}$$

Equations like this, generally written out for each of the three coordinate directions separately, are called the *equations of motion*. They are found not only in the classical mechanics of point masses, but also in the mechanics of rigid bodies, and in the mechanics of continua. In the mechanics of rigid bodies, applied forces not only produce accelerating motion in the direction of the mean force ("translation"), but may also produce accelerating rotations. In the mechanics of continua, applied forces may also produce distortion, that is, elongation or contraction, and shear.

For the motion of point masses, the governing equations consist of Newton's three laws, together with the (generally implicit) assumption of conservation of mass – that is, the mass does not change with time. But for the dynamics of rigid bodies we need to add further concepts, those of torque, or the moment of a force, and angular momentum (the "moment of momentum") which were not stated by Newton, and indeed were not explicitly formulated until almost 80 years later by Euler (see Chapter 2).

For continua, the situation is much more complex. To start with, we have to consider force in a different way. In studying the application of a force to a point mass, we are not faced with the problem of how the force varies over the surface of the mass. But in studying the deformation of continua, it is not only the total applied force that matters but also how it is distributed over the surface to which it is applied. We are concerned, therefore, with forces per unit area, or stress, and how the stress changes across a small volume of the medium. Thus, it is the gradient of the stress that matters, and so the equations of motion are generally written in units of force per unit volume, rather than force as in Equation (1.1). (The derivative of a stress, that is a force per unit area, with respect to distance has the units of force per volume.)

Even after converting from forces to stress gradients, the governing equations for a continuum involve more than the equations of motion and continuity. These equations tell us how the stresses act on the continuum, but they tell us nothing about how a deformable continuum responds to the applied stresses. Consider small cubes of rubber, clay, and motor oil. These different materials will respond quite differently to the same applied stresses. Indeed, one fundamental difference is immediately evident: we need a container to keep oil in the shape of a cube whereas the rubber and clay are able to stand freely. The equations of motion do not explain the various responses of different deformable materials to applied stresses; another set of equations, each specific to a particular type of material, is needed. These are the *constitutive equations* for that particular continuum.

Fortunately, it turns out that three simple constitutive equations do a good job of describing the response of many natural materials to applied stresses. Rubber, clay, and motor oil are examples of materials represented by each type of equation. Each constitutive equation is also conventionally represented by a simple mechanical model.

(1) Rubber is an example of the first type of material: it is a *linearly elastic* solid. When a stress is applied, its deformation is finite and directly proportional to the applied stress. The deformation is essentially instantaneous and is completely recoverable when the stress is removed. The constitutive equation for an ideal elastic substance is a linear relation between stress and strain. After the mathematics of material deformation

(strain) are presented in Chapter 7, the behavior of linearly elastic solids is discussed in Chapter 8. The conventional mechanical model is a spring that is fixed at one end. When a force is applied to the other end, the elongation of the spring is directly proportional to the force. When the force is removed, the spring recoils to its original position.

(2) Clay represents a *plastic* material. If the applied stress is too small, the clay does not deform at all. Once the level of applied stress reaches some threshold strength, however, the clay deforms permanently. For an ideal plastic, the applied stress cannot be increased beyond this threshold value, because it is immediately relieved by flow. The constitutive equation for an ideal plastic is very simple and states that the stress within the body is less than or equal to the threshold strength. If a stress larger than this is applied, it is immediately reduced to the threshold stress level by flow. Plastic behavior and the idea of material strength are covered in Chapter 4, once the essentials of stress have been presented. The conventional mechanical model is a block resting on a flat, rough surface. The block does not move until the force applied is sufficient to overcome the friction between the block and the surface below. The deformation in a plastic material is permanent, unlike that in an elastic solid. When the force is removed, the block does not return to its original position.

(3) Motor oil represents a *viscous* substance. Viscous materials strain indefinitely in response to an applied stress and, for an ideal viscous fluid, the constitutive equation is a linear relation between the applied stress and the rate of strain. The behavior of linear, or Newtonian, viscous fluids is covered in Chapter 9. The conventional mechanical model is a *dashpot*, which is a small cylinder filled with oil. The cylinder contains a loose-fitting piston that can be pushed back and forth within the cylinder. As the piston is moved, the fluid moves from one side of the piston to the other at a rate that is proportional to the force applied to the piston. In contrast to an elastic substance, but similar to a plastic material, a fluid deforms permanently: when the force is removed from the piston, the fluid and piston remain in place. Unlike a plastic substance, but similarly to an ideal elastic material, a viscous fluid will deform at any stress, regardless of its magnitude.

Ideal elastic and viscous materials have linear relations between stress and strain or strain rate. The stress applied to an ideal plastic can be no larger than the threshold strength. Although the behavior of many natural materials can be approximated using these three simple models, many others show more complex behavior. Still, relatively simple modifications to the basic constitutive equations can be made to approximate the stress–strain behavior of most natural materials. These include nonlinear elastic or viscous constitutive equations and simple, additive combinations of the basic relations for elastic, plastic, and viscous behavior. We will consider some of these in Chapter 10.

1.4 Vectors and tensors

In elementary mechanics, a familiar concept is that of a *vector*, which can represent a quantity like force, velocity or acceleration, that has both magnitude and direction. But a force per unit area, called a *stress*, can only be defined by specifying a magnitude and *two directions*, the orientation of the force itself, and of the surface on which it acts. As continua transmit stresses to all parts of the body, a stress applied to one surface can be felt as a stress on any other surface within the body. So if we want to express the complete *state of stress* at a point we must generally specify the stresses, not just on a single surface but on three surfaces normal to the three coordinate axes. We will see in Chapter 4, that stress must therefore be described by an array of six components, not just the three components that are sufficient for force.

The simplest stress field can be described by a single scalar quantity, pressure. Some of the applications of pressure are, however, far from simple, and in Chapter 4 we introduce the important concepts of buoyancy and effective stress. The consideration of pressure in pore spaces leads naturally to the consideration of flow through porous media, in Chapter 6. An understanding of pore pressures and the movement of pore fluids is important, not only for practical applications in soils, geohydrology, and petroleum geology, but also in understanding the stress, strength and deformation (e.g., compaction) of porous soils and rocks.

Stress is by no means the only quantity that is more complicated than a vector. When we study how materials deform (Chapter 7), we will see that to describe deformation we need a quantity, called *strain*, which is at least as complicated as stress. And the physical properties of solid continua, such as the way they transmit fluids (their *permeability*, Chapter 6) or respond to stresses (their elastic or viscous moduli) may also be complicated quantities. Just as a special notation and algebra, *vector algebra*, was developed to describe vectors, so a special notation and algebra, called *tensor* algebra, has been developed to describe tensor quantities such as stress or strain. In this book we do not attempt to provide a complete discussion of either vector or tensor algebra, but we do introduce the notation (Chapters 2 and 4), and describe a few of its uses. Vector operators (grad, div, curl), essential for the description of deformation (Chapters 7 and 8) and flow (Chapter 9), are introduced in Chapter 6, in the simpler context of flow through porous media.

One of the great virtues of tensor notation is that by using it, we can write complicated results very compactly – so it is widely used in the professional literature, and without some understanding of the notation it is impossible to read the journals that describe the applications of mechanics to problems in the earth sciences. We also come to see that vectors can be regarded as a special kind of tensor: a "first-order" tensor, that needs only a single row or column of numbers, unlike the "second-order" tensors generally met in mechanics, which require square arrays of numbers for their description. Ordinary scalar quantities, like mass, that require only a single number, can in turn be regarded as "zeroth-order" tensors (Chapter 4).

The concept of stress, though it now seems to us a rather obvious extension of the idea of a force, was the product of almost a century and a half of thought by the best mathematicians and physicists of the time, and was first clearly formulated by Cauchy in 1822. It is not surprising, therefore, that most of the progress in formulating the governing equations of continuum mechanics took place in the nineteenth century: Newton himself had only a very hazy idea of what constituted a Newtonian fluid.

1.5 Solving the equations

Though Newton formulated the basic equations of motion, and derived many interesting results from them, he did not introduce the formal techniques now used to apply the governing equations to particular phenomena. This was done over the following 50–100 years by a group of Swiss and French scientists and mathematicians. The pattern now in almost universal use is to express the governing equations as sets of partial differential equations that state how the equations of motion, the continuity equation, and the constitutive equations apply to each infinitesimal "element" of matter. The problem is then to integrate these equations (generally more than once) taking into account the particular configuration of interest. So, for example, if we want to know how water flows through a channel, we need to know the *initial conditions*, that is, the depth and velocity of the water at the first time that interests us, and the *boundary conditions*, for example, the geometry of the channel, and the specifications that the velocity must fall to zero at the solid boundaries of the channel and that the shear stress is zero at the free surface. Only then can we hope to discover, for example, how the stresses vary along the channel floor, or how the velocity is distributed within the flow.

Integrating the governing equations is generally possible only for a few simple initial or boundary conditions. This does not mean the equations are useless: even if they cannot be integrated, study of the equations often reveals many interesting general properties of the phenomena – and in particular, it tells us how they can be applied at different geometric and temporal scales (Chapter 3). In fact, the scaling properties of general physical laws, first discussed thoroughly only in the mid-nineteenth century, has proved to be a very powerful general method applied more recently to many complex problems in nuclear and solid state physics, as well as to problems such as the turbulent flow of fluids (Chapters 11 and 12).

Since the development of digital computers, enormous progress has been made in the numerical solution of differential equations. The subject is a large one, but simple examples are certainly not beyond the capacity of the personal computers now readily available to students. We have prepared a set of computer programs that supplement the material presented in this book, and illustrate a few of the techniques that are now routinely used in almost all applications of mechanics to real scientific and engineering problems.

1.6 The art of modeling

In Chapter 3, we discuss at some length just what is meant by a model in a modern science such as mechanics. Models may be actual physical models of phenomenon, generally built at a reduced scale, or they may be conceptual or mathematical constructs. Constructing models generally means greatly simplifying the real world, but it also means making assumptions that bridge over the gaps in our knowledge. It is useful as an intellectual exercise, since it forces us to confront both our ignorance and also how we can apply our limited knowledge to a real situation; and it is useful for making predictions.

Almost all environmental questions demand the construction of models. Will burning fossil fuels cause global warming? We need a model of the fate of carbon dioxide released into the atmosphere by burning fuel, and another model of how this will affect the global climate. Yet further models are needed to predict how climate change would affect sea level, or energy consumption, or the growth of crops. It is impossible to predict exactly what will happen – the natural systems involved are much too complicated for any model to give a complete description of what is likely to happen. A model attempts to include what is known about all the major factors and to assess their relative importance. It is indeed helpful to discover and make use of the historical background – we could make little progress in understanding the present climate if we did not know from geological history that we have only recently emerged from a major ice age – but history is not enough. We must also try to understand the physical processes that govern natural phenomena, and how (or even if) these can be reliably applied to predict what will happen next.

2

Review of elementary mechanics

2.1 Introduction

This chapter is devoted mainly to a review of the mechanics of point masses and rigid bodies. Though the most useful branch of mechanics for the earth sciences is continuum mechanics, there are also important applications of more elementary mechanics. As examples we will consider the motion of sediment grains (sand, pebbles or boulders), volcanic ejecta (tephra), and meteorites. We introduce the concept of fluid drag forces so that we can provide a relatively realistic treatment of these phenomena.

In the course of the chapter, we will treat some topics that the student may not have studied in introductory physics or mathematics. The main one is vectors, and their notation, algebra, and statistics. Vectors are widely used to describe motions of masses and the forces that give rise to them. We pay particular attention to the rotation of vectors and reference axes, because these ideas are a useful starting point for understanding tensors, which are introduced in Chapter 4 and used throughout the rest of the book. So if some aspects of our discussion of vectors seem unnecessarily pedantic, we ask the reader to believe that the ideas will become more useful later.

We will show by example that it is useful to write out Newton's laws formally as "equations of motion". If we can obtain an algebraic expression for the integral of these equations, then we claim that we have an "analytical solution" of the equations. The types of problems generally considered in introductory physics are carefully selected so that they can be solved analytically. In many real-world applications, analytical solutions cannot be found, and we must resort to numerical solutions, which can be worked out for particular conditions using computers. In this chapter, we introduce one simple technique for obtaining such numerical solutions.

The Austrian physicist Ernst Mach (1838–1916) defined *mechanics* as

14

"concerned with the motions and the equilibrium of masses" (Mach, 1960). The study of motion, exclusive of masses and forces, is called *kinematics*. The study of the equilibrium of forces is *statics*, and the study of the relationship between motion and forces is *dynamics*. All three are therefore included within mechanics.

2.2 Newton's laws

The principles of statics were known in part to the Greeks (e.g., Archimedes) and were further developed by European scientists during the Renaissance (e.g., Pascal, 1623–62), but the principles of dynamics were not established until Newton (1642–1727) formulated his laws of dynamics. They were first published in Latin in 1687. The following is a modernized translation (after Mach, 1960).

Law I: Every body continues in its state of rest, or in uniform motion in a straight line, unless it is compelled to change this state by forces impressed upon it.

Law II: The rate of change of momentum is proportional to the impressed force and is made in the direction of the straight line in which the force is impressed.

Law III: To every action there is always opposed an equal reaction or, the mutual actions of two bodies upon each other are always equal in magnitude and opposite in direction.

Corollary I: A body acted on by two forces simultaneously will move along the diagonal of a parallelogram in the same time as it would move along the sides by those forces acting separately.

Force may be considered to be an undefined axiom of mechanics (Truesdell, 1968, p.323) or to be defined by Newton's first and second laws:

$$\text{force} = \text{rate of change of momentum}$$
$$\mathbf{F} = d(m\mathbf{v})/dt \tag{2.1}$$

where m is mass and $\mathbf{v}$ is velocity. In Law II and Corollary I, Newton also indicated that force is a *vector* characterized not only by magnitude but also by direction, and gives the "parallelogram" law for vector addition. Force is not a *scalar*, like mass and time, that can be defined by magnitude alone. In this text, bold-face symbols are used for vectors. Velocity and acceleration are further examples. The properties of vectors are explained more fully in the next section.

For constant mass, Newton's second law becomes

$$\mathbf{F} = m(d\mathbf{v}/dt) \tag{2.2}$$

or

$$\text{force} = \text{mass} \times \text{acceleration}$$

$$\mathbf{F} = m\mathbf{a} \tag{2.3}$$

Note that Equation (2.1) implies that if there is no external force applied to a particle, the momentum cannot change, i.e., it implies the *conservation of momentum* for a particle. The principle can be extended to the total momentum of systems of particles using Newton's third law.

One of Newton's great achievements was to give a clear quantitative meaning to the concept of force, and to distinguish between mass and weight, which is the force produced by gravity acting on mass. He assumed that the "inertial mass" that is related to force by his second law, is the same as the "gravitational mass" that enters into the law of gravity.

It is useful to recognize two different types of forces:

(1) *body forces*, which act on every element of mass equally, e.g., the force of gravity;
(2) *surface forces*, which act along or across surfaces, e.g., pressure, friction.

In this chapter we will be concerned mainly with body forces, though we will also consider some aspects of fluid resistance and friction, which are surface forces. The main consideration of surface forces is deferred to Chapter 4, where we introduce the key concept of *stress* as a way to describe surface forces. For the moment, we note simply that a surface force is generally described as a force per unit area.

Disciplines such as fluid mechanics often take a statics approach to the equations of motion. Thus we may rewrite Equation (2.3) as

$$\mathbf{F} - m\mathbf{a} = 0$$

Applying this to the motion of a projectile, this corresponds to the motion being described by an observer moving with the projectile. To such an observer, the projectile is at rest, and there is an equilibrium between the applied force $\mathbf{F}$ and an (imaginary) *inertia force* $-m\mathbf{a}$. The use of the term "inertia force" (often written "inertial force") has a long historical tradition established by the Swiss mathematician Leonhard Euler (1707–83). In 1736, Euler wrote (Dugas, 1955):

The force of inertia ... should be reckoned by the force or power that is necessary to take the body out of its state.

The term "inertia force" is used to describe not only the resistance with which a body opposes an applied force – for example, the resistance that your hand feels as you try to throw a heavy object; it is also used to describe the force that a body exerts on something (or somebody!) that opposes its motion. For example, when we consider the forces that a flowing fluid exerts on a stationary object (Sections 2.11, 3.3.3 and 9.5), we shall see that it is useful to distinguish "viscous" forces due to shearing of the fluid as it flows past the object, from the "inertia force" on the object caused by fluid decelerating where it meets the object head on.

The Newtonian concept of force is fundamental to mechanics, but it is frequently misunderstood by students. Even among experts its definition may be controversial. For further clarification at the elementary level see Holton (1973), French (1971) or Pippard (1972), and for historical insight see Westfall (1977) or Cohen (1987). We return to the topic later in this chapter, when we consider the use of coordinates to set up Newton's second law in the form of "equations of motion".

2.3 Vectors, coordinates, and components

In writing the equations above, we defined vectors as quantities that have both magnitude and direction, and we used bold face symbols to represent them. An arrow over a letter, or a "hat" (caret) over a letter is also used, particularly for representing vectors on drawings or on the blackboard.

Later in the book, we give a better definition of a vector than that it has magnitude and direction: it is defined as a quantity that obeys certain algebraic rules. At that time, we will introduce yet another quantity, called a tensor, which obeys its own set of algebraic rules. Thus scalars, vectors, and tensors are defined by the types of algebraic rules they follow. One advantage of this definition is that we will be able to show that vectors are just a special class of tensors, and scalars are a special class of vectors (therefore another special class of tensors).

For the moment, however, let us consider that a vector is characterized by having two properties, magnitude and direction, whereas a scalar has only the property of magnitude. We know that (for a given system of units) scalars, such as mass, can be represented by a single number, but it is now clear that we will need more than one number to represent a vector.

As a geological example of a vector, consider a *lineation*, which is any linear structure or texture in a rock or on a rock surface. For example, the intersection of bedding and cleavage produces a lineation in slates, and the friction of a glacier moving over a rock surface produces "striations",

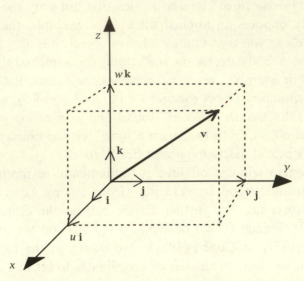

Fig. 2.1. Cartesian coordinate axes, components, and unit vectors for a vector **v**.

which are also a type of lineation. A lineation has an orientation that can be measured in the field by using the compass bearing, i.e., the *azimuth*, and the *angle of plunge* (the angle that the linear structure makes with the horizontal plane, also called the dip). It might be objected that a lineation has no magnitude, but this objection is overcome if we arbitrarily assign each measurement a magnitude of one. Such a vector might be represented by an array of three numbers:

$$\text{lineation vector} = (\text{magnitude}, \text{azimuth}, \text{plunge})$$
$$\mathbf{L} = (1, A, d)$$

where A is the azimuth, measured in the horizontal plane clockwise from north, and d is the angle of plunge (positive down, negative up).

This is a useful technique in structural geology, and can be used to represent the orientation of planes, as well as that of lineations, if we remember that the orientation of a plane may be specified by that of a line normal to the plane (the *pole* of the plane). The way in which vectors are most commonly represented is, however, by their three Cartesian components. From the rule for adding vectors (Newton's corollary I), it follows that any vector may be represented as the sum of three component vectors, each oriented parallel to a reference coordinate axis. The three axes are called "Cartesian" (after the French philosopher and mathematician René Descartes, 1596–1650) if each axis is normal to the other two (and if certain rules of vector algebra are valid). One commonly used set of reference axes is shown in Figure 2.1. In the earth sciences, however, a different set may be chosen: it is often convenient to choose the x-axis as north, the y-axis as east, and the z-axis as vertically down.

Each component in the x-, y-, and z-direction, may be considered to be the product of a scalar u, v, or w, and the unit vector in that direction. The three unit vectors are generally represented as $\mathbf{i}$, $\mathbf{j}$, and $\mathbf{k}$, or $\hat{\imath}, \hat{\jmath}$, and $\hat{k}$. Then we can represent a vector $\mathbf{v}$, for example, as

$$\mathbf{v} = u\mathbf{i} + v\mathbf{j} + w\mathbf{k} \tag{2.4}$$

Some writers call the three quantities u, v, and w, the *components* of the vector, but others reserve the term component for the three vector components $u\mathbf{i}$, $v\mathbf{j}$, $w\mathbf{k}$, and call u, v, and w the *coordinates* of the vector.

In any event, if we use the Cartesian convention, we can represent a vector by an array of three scalars (u, v, w). The particular values of these scalars change if a different set of Cartesian axes is chosen.

An algebra of vectors may then be defined in terms of these scalars. For example, we define addition of two vectors,

$$\mathbf{v}_3 = \mathbf{v}_1 + \mathbf{v}_2,$$

as meaning

$$\mathbf{v}_3 = (u_1 + u_2)\mathbf{i} + (v_1 + v_2)\mathbf{j} + (w_1 + w_2)\mathbf{k} \tag{2.5}$$

Generally, we write equations for each of the three coordinates (or components) separately, so that we can omit the unit vectors:

$$u_3 = u_1 + u_2$$
$$v_3 = v_1 + v_2 \tag{2.6}$$
$$w_3 = w_1 + w_2$$

The magnitude (or "length") of a vector is commonly written $|\mathbf{v}|$. Its value in Cartesian coordinates is

$$|\mathbf{v}| = \sqrt{(u^2 + v^2 + w^2)} \tag{2.7}$$

The square of the magnitude is the sum of the squares of the components of the vector. This type of sum recurs so often that it has been found useful to define a special notation for it in vector algebra. Thus the *dot product* or *scalar product* of two vectors $\mathbf{v}_1$ and $\mathbf{v}_2$ is defined as:

$$\mathbf{v}_1 \cdot \mathbf{v}_2 = u_1 u_2 + v_1 v_2 + w_1 w_2 \tag{2.8}$$

Another expression for the dot product, which can be derived from this definition, is

$$\mathbf{v}_1 \cdot \mathbf{v}_2 = |\mathbf{v}_1||\mathbf{v}_2| \cos \theta \qquad (2.9)$$

From this equation it is easily seen that if two vectors are perpendicular their dot product is zero, whereas if they are parallel, it is simply the product of their magnitudes. From Equation (2.8) the sum of squares of a vector's components is given by the dot product of a vector by itself. This is also called the square of the vector:

$$\mathbf{v}^2 = \mathbf{v} \cdot \mathbf{v} = u^2 + v^2 + w^2 \qquad (2.10)$$

Note that the dot product of two vectors yields a scalar, not another vector. It has an important application to the concepts of work and power, which we discuss in a later section. Also, later in this chapter, we will define a different way of taking the product of two vectors (the cross product, or vector product) that does yield a vector.

   If a situation can be fully described by the use of only one coordinate (in a suitably chosen direction) it is called *one-dimensional*. An example is the fall of a rock from rest under the action of gravity alone (all forces and motion are in a single direction). If two coordinates are needed, the situation is described as *two-dimensional*. An example is a rock sliding down a slope (we assume that the direction of the slope is constant, but the angle may vary, so that the motion remains in a single vertical plane). Another example is water flowing down a slope in a wide channel: except near the walls of the channel, the depth, velocity, etc., are uniform in the cross-stream direction. In one- or two-dimensional problems, we need write only one or two equations, respectively, to represent the vectors involved.

2.4 Position and velocity vectors

We are familar with the idea that a point in space may be represented by an array of coordinates (x, y, z). But such a point may also be considered to define a vector $\mathbf{r}$, drawn from the origin to that point (Figure 2.2). Such a vector is called a *position vector*. The reason why it is a good idea to do this only becomes apparent when we use a sequence of position vectors to trace out a path in space. Suppose that a point moves along a path from A to B (Figure 2.3). Consider the movement between two points P_1 and P_2 which are very close together. We can use two position vectors $\mathbf{r}_1$ and $\mathbf{r}_2$ to represent the two points, and the change in position from one point to the

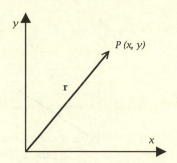

Fig. 2.2. A position vector.

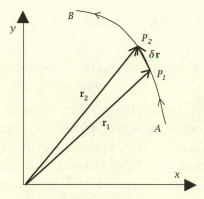

Fig. 2.3. A position vector moving along a curve.

next is represented by $\delta\mathbf{r}$, which is the vector from P_1 to P_2. By definition,

$$\mathbf{r}_2 = \mathbf{r}_1 + \delta\mathbf{r}$$

If the time taken to move this small distance $\delta\mathbf{r}$ is δt then the velocity is given approximately by

$$\mathbf{v} \approx \delta\mathbf{r}/\delta t$$

and exactly by

$$\mathbf{v} = \lim_{\delta t \to 0} \frac{\delta\mathbf{r}}{\delta t} = \frac{d\mathbf{r}}{dt} \qquad (2.11)$$

In the limit the orientation of $\mathbf{v}$ will be tangential to the path line AB at the point marked by $\mathbf{r}$. Similarly one can show that the acceleration vector $\mathbf{a}$ is the derivative with respect to time of the velocity vector $\mathbf{v}$.

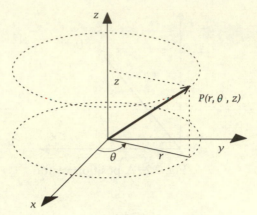

Fig. 2.4. Cylindrical coordinates of a position vector.

2.5 Cylindrical and spherical coordinates; left- and right-handed Cartesian axes

It is not always most appropriate to use Cartesian coordinates. Other reference systems are better for some applications. For example, for the flow of water or molten rock through a pipe of cylindrical cross-section, it is most convenient to make use of cylindrical coordinates (Figure 2.4). In this reference system, a position vector **r** is defined by the coordinates r, θ, and z, where:

(1) r is the distance from the origin to the projection of the vector onto the xy-plane;
(2) θ is the angle that the component r makes with the x-axis;
(3) z is the distance normal to the xy-plane.

Another coordinate system that is obviously useful in the earth sciences is the spherical coordinate system (Figure 2.5). In this case the three coordinates are:

(1) r, the distance from the origin. In the case of the Earth, this can be taken to be the mean radius (6.37×10^6 m);
(2) θ, the angle that the vector makes with the vertical axis z. For the Earth, if the z-axis is along the axis of rotation, $\theta = 90 -$ latitude (θ is called the *co-latitude*);
(3) ϕ, the angle that the projection of r onto the xy-plane makes with x. For the Earth, $\phi =$ longitude, assuming that the xz-plane passes through Greenwich.

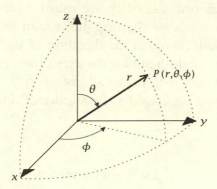

Fig. 2.5. Spherical coordinates of a position vector.

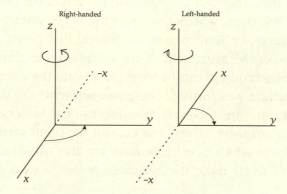

Fig. 2.6. Right- and left-handed reference frames.

Angles in cylindrical and spherical coordinates are generally measured in radians, rather than degrees. 2π radians are equivalent to 360 degrees, so one radian equals $180/\pi$ degrees. For very small angles the angle in radians is nearly equal to the tangent of the angle, which in turn is nearly equal to its sine.

It has already been emphasized that there are several different ways to represent the components of a vector. Even if we agree to use the method of Cartesian coordinates, the choice of a particular Cartesian reference frame, though it can be justified on the grounds of convenience, remains arbitrary.

One illustration of this is apparent as soon as we consider constructing a set of three axes. Figure 2.6 shows two such sets of axes: the only difference between them is the choice of the positive direction for the x-axis. Yet the two sets of axes are not equivalent, that is, we cannot change one set into the other simply by rotating it.

To distinguish the two sets, they are called *right- and left-handed axes*. If we represent the x-axis by the first finger, and the y-axis by the second finger, then the thumb points in the direction of the z-axis (verify that this works for the left and right hand for the axes shown in the figure). The usual convention in mechanics is to define all axes as right-handed sets.

Later in this chapter we consider the algebra of rotation of vectors and of coordinate axes.

2.6 Vector statistics

One common application of the algebra of vectors is to compute measures of the average and "spread" (dispersion) of vector measurements. Examples of earth science data that can be regarded as vectors (in three dimensions) include: current velocities, whether measured in air or water; the orientation of bedding, cleavage, or joint planes; structural lineations; fold axes; and paleomagnetic poles. Measurement of the orientation of sedimentary structures, such as cross-bedding, can be used to indicate the direction of flow at the time the structure was formed: these paleocurrents are directions in the horizontal plane, and are therefore two-dimensional vectors. In measuring currents in the atmosphere, rivers, and oceans, the small vertical component is often ignored – in which case, these data are also two-dimensional.

For samples of scalar data, the measures generally used are the *mean* $\bar{x}$, and *variance* s^2:

$$\bar{x} = \frac{\sum x_i}{N} \tag{2.12}$$

$$s^2 = \frac{\sum (x_i - \bar{x})^2}{N-1} \tag{2.13}$$

where $i = 1, 2, \ldots N$ (i.e., there are N measurements in the sample).

The square root of the variance, called the *standard deviation*, is often used as a measure of "spread" instead of the variance.

These measures do not work with vectors, however, as can easily be seen from the following example. Suppose we want to average two unit vectors measured in the horizontal plane: these could be two paleocurrent directions, say, indicating flow towards azimuths of 350 degrees and 50 degrees. The mean of these two azimuths is simply $(350 + 50)/2 = 200$ degrees – which is obviously not a good average of the two directions. As the average is used to calculate the variance, the variance is also inappropriate for vectors.

Instead of using the scalar mean and variance, we use the *vector mean direction* θ, calculated by summing the north–south and east–west components of the azimuths A_i, as a measure of the average direction and the *vector mean length*, L, calculated by dividing the vector magnitude by the number of measurements, as a measure of the dispersion:

$$\theta = \arctan(V/W) \tag{2.14}$$

where the east–west component is

$$V = \sum \sin A_i$$

and the north–south component is

$$W = \sum \cos A_i$$

and the vector mean length is

$$L = \frac{\sqrt{(V^2 + W^2)}}{N} \tag{2.15}$$

A mathematical theory of the statistics of vectors, analogous to that of scalars, has been worked out, and is described in texts on geological statistics (e.g., Cheeney, 1983; Fisher *et al.*, 1987). This theory was developed by the leading mathematical statistician of his day, R. A. Fisher (1890–1962), to solve problems which arose in the analysis of paleomagnetic data.

2.7 Example of the use of coordinates

A familiar problem in elementary mechanics texts is the "ballistics problem": how to predict the trajectory and speed of projectiles. In a geological context, the problem might be how far volcanic bombs can be thrown out of a volcanic vent, assuming a certain initial speed v_0. Alternatively, what was the initial speed of ejection of bombs, if we find them at a given distance x away from the vent? We will discuss this problem at some length, not so much because it is an important problem in volcanology (though it has been thoroughly discussed by Wilson, 1972) but because it can be used to illustrate the general techniques used to set up almost all mechanical problems for solution (for a more extensive discussion, see Hart and Croft, 1988). Though these general techniques are now thought to be a characteristic part of "Newtonian" physics, Truesdell (1968, Chapter II) has pointed out that, in fact, they were not developed or used by Newton himself. We owe their development to the Swiss mathematicians James Bernoulli (1655–1705), his brother John (1667–1748), John's son Daniel (1700–82), and Leonhard Euler (1707–83).

The first step towards a solution is to simplify the problem. Real volcanic bombs come in all shapes and sizes: we assume simply that a typical bomb has a mass m, which is large enough that we can neglect air resistance. At first, we will not worry about the difference in elevation between the volcanic vent and the point at which the bomb lands. It is generally best first to oversimplify a problem, and then make the solutions more realistic by introducing complexity, rather than risk starting with a problem that is too complex to be understood.

Essentially, therefore, we reduce the problem to a balance of forces acting on the bomb. This balance can be set up as an *equation of motion*, which has a general form similar to that of Equation (2.3)

$$m\mathbf{a} = \sum \mathbf{F}_i$$

where the right-hand side is the vector sum of all the real forces $\mathbf{F}_i$ acting on the bomb. In the case we are considering, the only force to be considered is the force of gravity $-m\mathbf{g}$. If the mass of the bomb is constant, we have

$$m(d\mathbf{v}/dt) = -m\mathbf{g} \qquad (2.16)$$

Note:

(1) Only one real force is acting on the bomb, namely the gravity force. Using the concepts developed at the end of Section 2.2, however, we may understand $-m(d\mathbf{v}/dt)$ as an "inertia force" resisting the action of gravity – if it did not, then application of the force would result in an infinite acceleration. If there is another real force that perfectly balances gravity, as there is when a bomb rests on the ground, then there is no acceleration, and the "inertia force" becomes zero, i.e., there is a *static* equilibrium.

(2) The force on the right-hand side is negative because it is conventional to consider that vectors are positive upward. If we decided to use the geophysical convention, that down is positive, then the left-hand side would become negative and the right-hand side positive.

Now we choose reference axes:

x-axis: horizontal, in direction of motion;
y-axis: horizontal, normal to x;
z-axis: vertical.

We can now write out the equation of motion (Equation 2.16) for each of the three components:

$$x\text{-component}: \quad m(du/dt) = 0$$
$$y\text{-component}: \quad m(dv/dt) = 0 \qquad (2.17)$$
$$z\text{-component}: \quad m(dw/dt) = -mg$$

We can neglect the y-direction because we have chosen our reference frame so that v is zero at time 0, and it follows that for all other times $v = 0$ (and so $dv/dt = 0$). The x-direction, however, cannot be neglected except for the special case where the bomb has an initial velocity only in the vertical direction so that $u_0 = 0$. In general, however, this is not so. Therefore, the problem is a two-dimensional problem. Note that we can also eliminate m, so the equations simplify to:

$$du/dt = 0 \qquad (2.18)$$
$$dw/dt = -g \qquad (2.19)$$

We have now written down the equations of motion for this special case, and we would like to "solve" these equations. To do this we

- specify the "initial" or "boundary" conditions;
- integrate the equations.

Integration without specifying limits gives formulae that have unknown "constants" of integration: their value is determined from the initial or boundary conditions. The *initial conditions* are simply those at time zero: in the ballistics example, they include the initial position and velocity of the volcanic bomb. If the initial conditions are specified, the problem is known as an "initial value problem". Often, the initial conditions are not known: the problem is to determine what they should be in order to achieve some specified result. That result constitutes the *boundary conditions*, and the problem becomes a "boundary value problem". In the example, we might know that bombs of a particular mass have been thrown to a particular place, and we want to infer what range of initial velocities would have been necessary to achieve this. The practical problem in ballistics is a boundary value problem: we know the location of the target, and we want to know what velocity (speed and direction) we should use to hit it. In general, initial value problems are easier to solve than boundary value problems.

In the example, we will consider the inital value problem: the origin of the coordinate axes is set at the vent, where the initial components of velocity have the values u_0, w_0. This means that the bomb leaves the crater at an angle to the horizontal of β, where:

$$\tan \beta = w_0/u_0$$

Equations (2.18) and (2.19) can now be integrated:

$$u = u_0 \qquad (2.20)$$
$$w = -gt + w_0 \qquad (2.21)$$

In these two equations, both u_0 and w_0 were originally constants of integration, evaluated by setting $t = 0$. Since there are no forces acting in the x-direction, the velocity component in this direction will remain constant, but the component in the z-direction will change because of the action of the gravity force.

If we suppose that w_0 is positive then w will become 0 (that is, the bomb will be at the top of its trajectory) when t is given by

$$0 = -gt + w_0$$

i.e., when

$$t = w_0/g \tag{2.22}$$

To obtain a solution for how far a bomb will travel, we need an expression in terms of x and z. We get this by rewriting Equations (2.20) and (2.21) as:

$$u = dx/dt = u_0 \tag{2.23}$$
$$w = dz/dt = -gt + w_0 \tag{2.24}$$

Integrating these equations and applying the initial conditions again (at $t = 0, x = 0$ and $z = 0$) gives

$$x = u_0 t \tag{2.25}$$
$$z = -gt^2/2 + w_0 t \tag{2.26}$$

Equation (2.22) gives the time at which the bomb rises to a maximum height h. Substituting this result into Equation (2.26) gives:

$$h = -\frac{g}{2}\left(\frac{w_0}{g}\right)^2 + w_0\left(\frac{w_0}{g}\right)$$

$$= \frac{w_0^2}{2g} \tag{2.27}$$

Note that we can also eliminate t from Equations (2.25) and (2.26) to give a single equation of the form $z = f(x)$ to describe the trajectory of the bomb:

$$z = (w_0/u_0)x - (g/2u_0^2)x^2 \tag{2.28}$$

This is the equation of a parabola.

If we know the ground elevation around the volcano, we can determine the horizontal distance travelled by the bomb, by substituting a value of z and solving Equation (2.28) for x. For example, if the bomb strikes the ground at $z = 0$, the result is $x = 2u_0 w_0/g$. It can also be shown that for a given initial speed the greatest horizontal travel is obtained by setting $u_0 = w_0$ (i.e., by ejecting the bomb at 45 degrees to the horizontal). By determining the greatest distance that large bombs are thrown, we can determine the speed of ejection from a volcanic vent, and from this data it is possible to make a rough estimate of the pressure in the magma chamber (Chapter 9, Problem 3).

2.8 Numerical solution of the equations of motion

In the example given in the previous section, it was relatively easy to obtain an analytical solution of the equations. This meant that an algebraic expression could be found for the integrals involved: to obtain a particular solution we could just put the appropriate numerical values of the various variables and constants into the equations (such as Equation 2.21) that are obtained by integration.

Unfortunately, many equations of motion, particularly those that include the various forces and complex boundary conditions of the real world, cannot be solved analytically. In other words, it is not possible to integrate the differential equations involved so as to obtain an exact algebraic solution. Even in the most complex cases, however, it is generally possible to obtain an approximate solution for a particular set of initial or boundary conditions by numerical methods. In this section, we want to illustrate the general way in which this is done by obtaining a numerical solution to Equation (2.24). Of course, we know the exact solution to this equation, so this has the further advantage that we can compare the results obtained by the two methods.

The equation we wish to integrate using approximate methods is

$$dz/dt = -gt + w_0 \qquad (2.29)$$

If the position at time t is $z(t)$, and the position a short time δt later is $z(t + \delta t)$ we can approximate the derivative dz/dt using

$$dz/dt \approx \delta z/\delta t = [z(t + \delta t) - z(t)]/\delta t \qquad (2.30)$$

Combining the last two equations, and solving for the position at time $t + \delta t$ gives:

$$z(t + \delta t) \approx z(t) + (w_0 - gt)\delta t \qquad (2.31)$$

To solve the problem numerically, therefore, we might make use of the following computational rule:

(1) Specify δt and initial numerical values of t, z, and w. Also give the numerical value of g.
(2) Specify how often we want to repeat the calculation that follows.

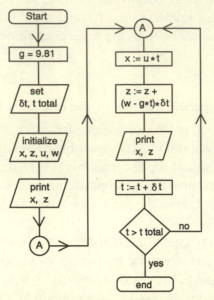

Fig. 2.7. Flow chart for numerical integration.

(3) Use Equation (2.31) to calculate the first step and print out (or plot) the value of $z(t + \delta t)$.

(4) Set this to be the new value of $z(t)$, and set $(t + \delta t)$ to be the new value of t.

(5) Repeat steps 3 and 4 until the number of repetitions specified in step 2 has been carried out.

Such a computational rule is called an *algorithm*. We have written it out in words but it could better be expressed in symbols, such as by the flow chart shown in Figure 2.7. Note that in this chart we have used the symbol := in some of the boxes. This is a symbol used in the computer language Pascal, to make clear the difference between mathematical equality or identity (indicated by =, as in $c = a + b$) and *assignment* (indicated by :=, as in $a := a + 1$). Though this is a distinction not made symbolically by all computer languages, it is an important one. For example, the expression $a = a + 1$ makes no sense mathematically, but in most computer languages it is understood to mean "replace the old value of a by a new value, $a + 1$."

Devising a suitable algorithm and expressing it as a flow chart is the first step in the construction of a computer program to solve the problem. In fact, algorithms are often written down using a computer language such as Pascal or a "pseudocode" based on this language. A program, written in

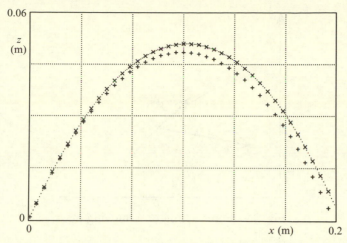

Fig. 2.8. Example of a computer plot, showing exact and numerical solutions of a projectile problem. The dotted line shows the exact solution. The upright crosses below the line and the diagonal crosses on the line are two numerical solutions (see text).

Pascal, which solves the projectile problem is used in the computer module EJECTA (see the end of the Preface for information about how to obtain these computer modules). The student is encouraged not only to experiment numerically with this and the other programs referred to in this book, but also to study the source code. The code is itself the best summary of the numerical steps necessary to solve the problem. One of the virtues of Pascal is that code written in this language is also very easy to read – even for those who have never learned to program in Pascal.

Figure 2.8 shows trajectories calculated for an initial ejection velocity of 1 m s^{-1}, at an angle of 45 degrees to the horizontal. The dotted line is the exact (analytical) solution and the upright crosses below the line are the approximation, using a value for δt of 0.005 seconds. A more accurate solution can be obtained by using a smaller time interval. We will show in Section 2.13 that a more accurate solution can also be obtained by a slight modification of the numerical method: this solution is shown in Figure 2.8 by the small diagonal crosses that lie along the dotted line.

2.9 Work, energy, and power

In Section 2.4 we saw that the velocity vector could be generated by considering the rate of change of a position vector moving along a path. Now consider that we move not simply a point, but a point with mass m. We fur-

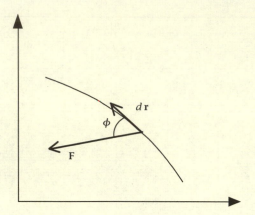

Fig. 2.9. Force acting on a moving mass.

ther consider that this point mass is being acted on by a force $\mathbf{F}$ (Figure 2.9). Let $\mathbf{F}$ make an angle ϕ with the velocity vector $\mathbf{v}$ (in other words, the force need not be applied in the same direction in which the mass is moving).

Now we define work as the product of a force times the displacement in the direction of the force. Thus, for a small displacement $d\mathbf{r}$ the work dW done by the force $\mathbf{F}$ is

$$dW = |\mathbf{F}|dr \cos \phi \qquad (2.32)$$

or, in vector notation,

$$dW = \mathbf{F} \cdot d\mathbf{r} \qquad (2.33)$$

So the total work in moving a mass from A to B is the integral of dW:

$$W = \int_A^B \mathbf{F} \cdot d\mathbf{r} \qquad (2.34)$$

As $\mathbf{v} = d\mathbf{r}/dt$, this is the equivalent of:

$$W = \int_{t_A}^{t_B} \mathbf{F} \cdot \mathbf{v}dt \qquad (2.35)$$

where it is now understood that the limits of integration are the times when the mass passes through A and B.

To evaluate W, we make the substitution, using Newton's second law (Equation 2.1):

$$\mathbf{F} \cdot \mathbf{v} = m\left(\frac{d\mathbf{v}}{dt}\right) \cdot \mathbf{v}$$

$$= \frac{d}{dt}\left(\frac{m}{2}\mathbf{v} \cdot \mathbf{v}\right)$$

$$= \frac{d}{dt}\left(\frac{m}{2}v^2\right)$$

In this equation v^2 means the square of the magnitude of **v**. So now we can write the equation for work as

$$W = \int_{t_A}^{t_B} \frac{m}{2} d(v^2) dt = \frac{m}{2}(v_B^2 - v_A^2) \tag{2.36}$$

The work that must be done in going from A to B is defined as the change in *kinetic energy* (KE). In particular, if the initial velocity is zero ($v_A = 0$), the gain in kinetic energy as the velocity increases to v is $mv^2/2$.

Equation (2.36) expresses the net work done in moving from A to B. In general, we must know the path over which the work is done to know the total amount of work done. For simple mechanical forces that do not result in the generation of frictional heat, however, the work done does not depend on the path from A to B. Such forces are called *conservative*. For work done by conservative forces, the energy at A and B depends only on conditions at these points, because it does not make any difference how one got from one point to the other. If this is the case, we should be able to define the work needed to move from A to B as a function of position alone. We define this function as the change in *potential energy* (PE) between the points A and B, and so, for conservative forces, the work done in moving from A to B is:

$$W = (PE)_A - (PE)_B \tag{2.37}$$

We now have two quantities related to work. We can equate them as:

$$W = (KE)_B - (KE)_A = (PE)_A - (PE)_B \tag{2.38}$$

or

$$(KE)_A + (PE)_A = (KE)_B + (PE)_B \tag{2.39}$$

Because there are no terms on the left-hand side related to conditions at point B, and no terms on the right-hand side related to conditions at point A, Equation (2.39) can only be true if both sides equal a constant. We define this constant as the total mechanical energy of the system. If only conservative forces are involved in a problem, Equation (2.39) tells us that the total mechanical energy will be conserved.

This is a relatively profound result. It was deduced using only the definition of work and integration of Newton's second law. It is also a limited result, because it does not include nonconservative forces, which are important in many real problems. We could have started with a general statement that energy is always conserved and that we need consider only kinetic and

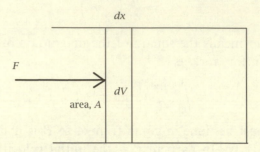

Fig. 2.10. Compression of material in a cylinder. The pressure $p = F/A$.

potential energy in problems involving only conservative forces. The statement that energy is always conserved is more general than Equation (2.39); we have not proved here that *all* energy is conserved.

Despite the similarities demonstrated here, Newton's second law and the conservation of mechanical energy provide us with two different approaches to mechanical problems. This can have important practical uses. Sometimes, either approach may be used to solve a problem. This is illustrated by Problem 4 at the end of this chapter, where the simple case of the rise of a point mass thrown exactly opposite to the direction of gravity is solved using both Newton's second law and the conservation of mechanical energy. At other times, only one or the other of these approaches can be successfully used to solve a problem, as when the amount of energy consumed in a process is desired, but unknown. In such a case, it may be possible to find a mechanical solution giving the motion, and this solution may then be used to calculate the mechanical energy at the beginning and end of the process. The difference between the two would then be equal to the energy consumed.

Kinetic energy is energy that the mass has by virtue of motion, and potential energy is energy that it has by virtue of position or configuration. We commonly think of PE as a result of the gravitational force field. But there are other types of PE, e.g., the energy contained in a compressed spring or other elastic material. Note that this form of PE (elastic strain energy) may also be converted into KE. A geological example is the conversion of energy by movement along a fault:

$$\begin{matrix} \text{strain} \\ \text{energy} \end{matrix} \rightarrow \begin{matrix} \text{seismic wave} \\ \text{motion energy} \end{matrix} \rightarrow \text{heat energy}$$

or

$$\text{PE} \rightarrow \text{KE} \rightarrow \text{heat energy}$$

Consider compression of material in a cylinder (Figure 2.10). For example, the material could be a volume of air ("confined" not by the walls of a cylinder, but by the surrounding atmosphere) and the force of compression the weight of the overlying atmosphere (atmospheric pressure). The material might also be magma in a magma chamber, and the force of compression the weight of the overlying rock. To calculate the energy stored in the material by compression, suppose that application of a normal force of magnitude F, on an area A, produces a change in length of dx (corresponding to a change in volume of dV). Then

$$dW = Fdx = pAdx = pdV \qquad (2.40)$$

where p is the pressure. We can write the work for isothermal compression more explicitly if we know the equation of state of the material in the cylinder (i.e., the equation relating pressure to volume). For example, if the cylinder contains an ideal gas, then $pV = RT$, (where T is the temperature and R is the gas constant) and the work is given by:

$$
\begin{aligned}
W &= \int_{V_1}^{V_2} pdV = \int_{V_1}^{V_2} (RT/V)dV \\
&= RT \ln V|_{V_1}^{V_2} \\
&= RT \ln(V_2/V_1) \qquad (2.41)
\end{aligned}
$$

and since V is inversely proportional to p we can also write

$$W = RT \ln(p_1/p_2) \qquad (2.42)$$

Unfortunately, it is not so easy to determine the equation of state for a magma as it is for an ideal gas! It is, however, possible to write approximations for the equation of state of air, water, and elastic materials such as rocks.

Power is defined as the time rate of doing work

$$P = dW/dt \qquad (2.43)$$

Using Equation (2.33) we have

$$P = \mathbf{F} \cdot (d\mathbf{r}/dt) = \mathbf{F} \cdot \mathbf{v} \qquad (2.44)$$

The net work done is equal to the change in mechanical energy, so the time rate of doing work is equal to the rate of energy loss in systems involving nonconservative forces.

The unit of work or energy in the SI system is the joule (kg m^2 s^{-2}), and the unit of power is the watt (kg m^2 s^{-3}). One joule (J) corresponds to 4.184 calories of heat, so it is a small unit (and remember that the calories counted by dieters are actually kilocalories – so each one corresponds to more than 4000 joules). Burning one barrel of oil (42 US gallons) releases 6×10^9 joules of heat energy.

We devote the next few sections to applying the basic concepts of work, power, and energy to some geological examples.

2.10 Application to meteorite impact

Meteorites are fragments of matter that were formed early in the history of the solar system and that approach the earth or other planets from space. The measured approach speeds vary from 11 to 72 km s^{-1}. Meteorites are made either of iron, with a density of about 8000 kg m^{-3} or of iron-rich silicates, with a density of about 3500 kg m^{-3}. Most meteorites are small, and weigh only a few grams or kilograms, but some are very large. Because of its great speed and mass, a meteorite has a very large kinetic energy, and when it strikes the earth, it first penetrates the surface, then becomes greatly compressed by the resistance of the surface rocks (and produces great pressures and compression in the rocks at the impact site). Finally, as the meteorite loses momentum, it, and the compressed rocks, experience a very rapid decompression – in effect, the meteorite and its compressed target simply explode.

The kinetic energy of a meteorite is easily computed: it is simply $mv^2/2$. Suppose the mass is one thousand metric tonnes (10^6 kg) – which would correspond to an iron meteorite only 3 m in diameter. Then if the approach speed was 11.2 km s^{-1} the kinetic energy would be 6.3×10^{12} J, or about the energy of a thousand barrels of oil, or one kilotonne of TNT (defined as 4.2×10^{12} J). Increasing the speed six times (to near the upper limit of observed approach speeds) increases the energy 36 times. It is easily seen that large meteorites release as much energy as atomic bombs: and, in fact, observations on test explosions of atomic bombs placed just below the earth's surface threw considerable light on the dynamics of meteorite crater formation (see references to Chapter 3).

Why is the minimum approach velocity of meteorites more than 11 km s^{-1}? The answer is that this is the *escape velocity* for the earth, that is the minimum speed necessary for a projectile to escape from the earth's gravitational field – and therefore also the speed produced by the earth's gravity acting on a mass at rest in space. The easiest way to compute the escape velocity is by equating the kinetic energy at the earth's surface to the work done by the earth's gravity field as it draws in a mass from far out in space:

$$\tfrac{1}{2}mv^2 = \int_R^\infty F_x dx \qquad (2.45)$$

where we assume that x is the direction of fall towards the earth's surface, and the velocity of the mass relative to the earth is zero far out in space. Impact takes place at the earth's surface, a distance of R from the centre of the earth. F_x is given by Newton's law of gravitation:

$$F_x = GmM/x^2 \qquad (2.46)$$

where m is the mass of the meteorite, M is the mass of the earth (5.975×10^{24} kg), and G is the universal gravitational constant (6.7×10^{-11} kg^{-1} m^3 s^{-2}). Evaluating the integral on the right-hand side of Equation (2.45) gives

$$\int_R^\infty \frac{GmM}{x^2} dx = GmM \left[-\frac{1}{x} \right]_R^\infty = \frac{GmM}{R} \qquad (2.47)$$

and solving for v gives 11.2 km s^{-1}.

2.11 Application to fluid drag and lift

In this section we will apply energy concepts to throw some light on the resistance offered by fluids to the passage of solid bodies through them. Some knowledge of fluid resistance is necessary for any realistic discussion of the flight of projectiles (or small meteorites) through the air, or the settling of grains in water or magma, or even the sinking of large blocks of the lithosphere through the mantle at subduction zones. Note that it is the relative motion between solid bodies and fluids that produces fluid resistance: when evaluating the magnitude of the force of resistance it does not matter if the solid moves through the fluid, or the fluid moves past the solid. Generally, we speak of the fluid exerting a drag on the solid, or the solid offering resistance to the flow. In this case, drag and resistance would be two forces, opposite in sign but equal in magnitude. Experimentally, measurements of drag are generally made in a wind or water tunnel, by measuring the force acting on a solid body held at rest, with the fluid moving past it at a constant speed (measured far upstream).

Consider the flow of a fluid of density ρ past a solid body of cross-sectional area A. As small volumes of fluid (generally called fluid particles) approach a solid obstacle, they are deflected and decelerated, as shown by the flow lines in Figure 2.11(a). One particle comes to rest at a point P on the surface of the body (called the *stagnation point* of the flow). The deceleration of fluid particles is produced by the presence of the body, and

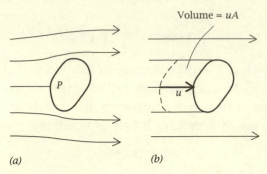

Fig. 2.11. Flow of a fluid past a solid body: (a) the real flow pattern; (b) an approximation.

requires a force exerted on the fluid by the body (the fluid resistance). In turn, we consider that an equal and opposite "inertia force" (the drag force) is exerted by the fluid on the body. Some authors call such "forces" "forces due to acceleration" (where the acceleration may, of course, be negative).

In order to make an approximate analysis let us suppose that the flow is like that shown in Figure 2.11(b). We assume that all the particles directly upstream from the body are decelerated from some initial velocity u, to zero velocity where they impact on the surface of the body, while fluid particles not directly upstream are unaffected.

In unit time, the volume of the particles coming to rest is therefore approximately uA, and the mass is $\rho u A$. Therefore the loss of kinetic energy in unit time is given by

$$\text{KE loss} = (\rho u A)u^2/2 = \rho u^3 A/2 \tag{2.48}$$

This rate of loss of KE is the power loss. For a constant velocity and force, both acting in the same direction Equation (2.44) gives

$$P = Fu$$

Combining the last two equations we can write for the drag force F_D acting on the body:

$$F_D = P/u = (\rho u^2/2)A \tag{2.49}$$

Note that the drag force is in the direction of flow relative to the solid body.

In reality, the flow is more complicated, so we introduce an empirical *drag coefficient* C_D, which must be determined by experiment:

$$F_D = C_D(\rho u^2/2)A \tag{2.50}$$

From experiment it is found that, for a given shape, the drag coefficient

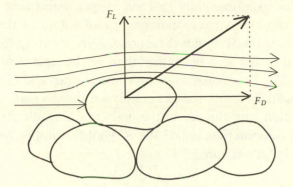

Fig. 2.12. Drag and lift force components acting on a pebble lying on the bed of a stream.

is nearly constant over a considerable range of flow conditions. Its value varies from about 0.1 for well streamlined forms, to about 1.0 for a flat plate oriented normal to the flow. A good average for subspherical bodies such as boulders or pebbles is about 0.5.

A phenomenon analogous to drag is that of fluid lift. Lift forces act where the flow around a body is asymmetrical: this may be either because of the shape of the body (as in a wing, whose shape is designed to produce maximum lift) or because the body is lying on a solid boundary, as shown in Figure 2.12. By definition the drag force acts in the direction of the mean flow (relative to the body), and the lift force acts normally to the mean flow.

By convention, the lift force F_L is expressed by an equation analogous to Equation (2.50):

$$F_L = C_L(\rho u^2/2)A \qquad (2.51)$$

As we can see from the figure, the drag and lift forces are just two components of the total force of the fluid on the body. The proportion between the coefficients of lift and drag depend on the shape of the body and several other factors: the two components may be roughly of the same order of magnitude.

Experimental studies have shown that the drag and lift coefficients are not well defined constants, even for a given shape. This is so because the analysis given above has neglected several important factors, notably the viscous components of the drag force. As we can see from Figures 2.11 and 2.12, the fluid must bend or deform in order to flow past a solid body. Viscosity is the property that relates the stress to the rate of deformation of a fluid. It is reasonable, therefore, that it should have some effect on the fluid

drag. It is found experimentally that viscosity is important at small length
scales or low velocities (corresponding to small values of the kinetic energy
of the approaching fluid): under these conditions, the drag force turns out to
be proportional to the velocity, rather than to the square of the velocity as
in Equation (2.50). The numerical values of the drag and lift coefficients are
found experimentally to depend not simply on the viscosity of the fluid, but
on its combination with the size of the body and the velocity and density of
the fluid. This combination, called the Reynolds number, will be discussed
in some detail in Subsection 3.3.3.

2.12 Stream power

The application of the power concept in the example given in the last section
is not straightforward. The straightforward application is when the force
acts over a certain distance in unit time. How can this apply to the case of
an obstacle which remains at rest in a flow ? The answer is that it is equally
valid to consider the fluid as stationary, with the obstacle moving through
it. Then if a force, F_D, is necessary to drive the obstacle at a speed u, it is
clear that the rate of doing work must be $F_D u$.

   Another application that is even less obvious is stream power. This
measures the rate at which a stream performs work on its bed, therefore
some workers believe it is one of the most important variables that controls
the nature and rate of sediment transport, and the type of bedforms (ripples,
dunes) that result from sediment movement.

   Let the force per unit area (shear stress) exerted by the water on the bed
of the river be τ_0, and let the velocity, averaged over any point on the bed,
be $\bar{u}$. Then it might be concluded that the power, per unit area of the bed,
is:

$$P = \tau_0 \bar{u} \tag{2.52}$$

This is in fact the usual formula given for *stream power*, but the derivation
given above leaves a lot to be desired. At the bed, where the shear stress
is acting, the velocity is not, in fact, $\bar{u}$ but drops to zero. Why then is the
stream power not also zero?

   A better discussion was given by Bagnold (1960), who defined stream
power as:

the time rate of conversion of potential gravity energy into kinetic turbulent eddy
energy ...

The rate at which a unit volume of water is "falling" by flowing down a channel inclined at a slope S is $\bar{u}S$, i.e., in unit time the water descends in elevation by an amount $h = \bar{u}S$. The gravitational potential energy is γh, where $\gamma = \rho g$ is the *specific weight* (i.e., weight per unit volume) of the water. Over a unit area of the bed, there are d volumes of water, where d is the depth. So we have

$$\text{Stream power} = \text{rate of loss of potential energy}$$
$$\text{per unit area of bed}$$
$$= \gamma h d$$

Substituting for h

$$P = \gamma d S \bar{u} \tag{2.53}$$

γdS is just the downslope component of the gravity force acting on the volume d of water. Because the average velocity remains constant, this force must be balanced by an equal and opposite drag force exerted on the flow by the unit area of the bed. This is the bottom shear stress τ_0. Therefore:

$$P = \tau_0 \bar{u} \tag{2.54}$$

In some cases, such as the drag acting on the bottom of a river, fluid drag can be considered to act parallel to the boundary between a fluid and a solid. It can then be considered to be a type of frictional surface force. We defer further consideration of the important topic of friction until Section 4.2. In other cases, such as the drag on an obstacle considered in Section 2.11, it can be considered to result mainly from the difference in fluid pressure between the front and back of the obstacle. We consider this aspect in Section 9.5.

2.13 The equations of motion including drag

Earlier in this chapter, we considered the problem of determining the trajectories of a volcanic bomb by integrating the equations of motion. To simplify the problem, we neglected the effect of drag forces. We now reconsider this problem, including drag.

For simplicity, we start with the assumption that the drag per unit mass on the volcanic bomb is *proportional to the speed*:

$$F_D/m = k_1 v \tag{2.55}$$

The assumption generally applies only to small particles moving slowly through viscous fluids but it has the advantage of permitting an analytical solution. In Chapter 3 (Subsection 3.3.3), we will present Stokes' law, which is the general form of Equation (2.55), and be more precise about its range of validity.

We obtain the solution by writing the equations of motion for the horizontal and vertical directions:

- **horizontal direction**, x. The equation of motion is:

$$du/dt = -k_1 u \tag{2.56}$$

or

$$du/u = -k_1 dt$$

so

$$\ln u = -k_1 t + C \tag{2.57}$$

Setting the initial condition as $u = u_0$ at $t = 0$, the constant of integration $C = \ln u_0$, so

$$u = u_0\, e^{-k_1 t} \tag{2.58}$$

This indicates that instead of remaining constant, the horizontal component of velocity decreases exponentially towards zero.

- **vertical direction**, z. The equation of motion is

$$dw/dt = -g - k_1 w \tag{2.59}$$

The integration is a little more tricky, but possible using no more than basic calculus, and gives the result

$$w = -\frac{g}{k_1} + \left(w_0 + \frac{g}{k_1}\right) e^{-k_1 t} \tag{2.60}$$

For the linear drag law (Equation 2.55), therefore, there are exact solutions.

Numerical solutions may also be obtained. We will demonstrate how using a slightly more refined method than before. Given velocity components u, w at time t, we obtain the velocity at a time $\delta t/2$ later by using the following difference equations:

$$u(t + \delta t/2) = u(t) + \delta u/2 \tag{2.61}$$
$$\delta u/2 = -k_1 u(t)\delta t/2 \tag{2.62}$$
$$w(t + \delta t/2) = w(t) + \delta w/2 \tag{2.63}$$
$$\delta w/2 = -[g + k_1 w(t)]\delta t/2 \tag{2.64}$$

Equations (2.62) and (2.64) are the difference forms of Equations (2.56) and (2.59) respectively, for a time increment of $\delta t/2$. We have calculated the velocities for the mid-point of the time increment, rather than for the whole

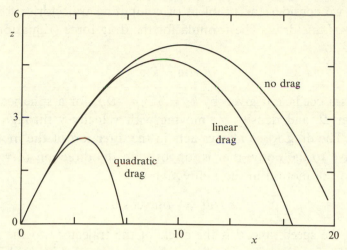

Fig. 2.13. Projectile trajectories computed for air using linear and quadratic drag, compared with that for no drag. The diameter was 1 mm, and $u_0 = w_0 = 10$ m s^{-1}.

increment, because we are going to use these velocities as typical of the whole time increment in order to calculate the distance travelled during that increment. If we used the velocities at the beginning of the time increment they would be too large, and if we used the velocities at the end they would be too small. The mid-point velocities may not be exactly right, but they will be close. (This method is sometimes called the "Feynman half-step method" because it is used by Feynman *et al.*, 1963, Chapter 9.)

Next, we use the mid-point velocities (call them u_m and w_m) to estimate the distance travelled in the whole time increment:

$$\delta x = u_m \delta t \qquad (2.65)$$

$$\delta z = w_m \delta t \qquad (2.66)$$

The process is then repeated for the next time increment.

For all time increments except the first, the mid-point velocity is determined from:

$$u_m(\text{new}) = u_m(\text{old}) + \delta u \qquad (2.67)$$

$$\delta u = -k_1 u_m(\text{old})\delta t \qquad (2.68)$$

and the corresponding formula for the z-component. These two equations are the same as Equations (2.61) and (2.62) except that they are calculated for a time difference of δt rather than $\delta t/2$.

Figure 2.13 compares results obtained using this method with those obtained analytically for no drag.

Finally, we consider the problem of computing trajectories for the more realistic *quadratic* drag. The formula for the drag force (Equation 2.50) can be reduced to

$$F_D/m = k_2 v^2 \qquad (2.69)$$

where k_2 is a coefficient given by $k_2 = 3C_D \rho/4D\rho_s$ for a spherical projectile of diameter D, and density ρ_s, moving with velocity **v** through a fluid of density ρ. The drag force always acts in the direction of the motion of the fluid relative to the body, that is, opposite to the direction of **v**. Therefore the equation of motion in the x-direction is

$$du/dt = -k_2 v^2 \cos \theta \qquad (2.70)$$

where v is the speed, and θ is the angle of the trajectory to the horizontal, and is itself a function of t. For linear drag, we solved this problem by using the identity

$$v \cos \theta = u$$

However, if we make this substitution in Equation (2.70), we still have a term in the equation (v or $\cos \theta$) that is a function of t. The result is that it is very difficult to obtain an analytical solution for equations of motion involving quadratic drag. Ballistics is a science that involves nontrivial mathematics: some idea of its complexity can be obtained by reading the summary in the book by Sutton (1957). Numerical integration, however, may be readily applied to the problem. In fact, one of the first applications of electronic computers was to solving problems in ballistics. The Mark I computer, developed by IBM and Harvard University, was used for this purpose in 1944. (It is worth reflecting that this machine, which was 15 m long and 2.4 m high, had much less computing power than a briefcase-size personal computer does today.)

The numerical method proceeds as follows:

(1) Determine the small change in horizontal velocity component u over a small time interval by using the difference form of Equation (2.70):

$$\delta u = -k_2 v^2 \cos \theta \, \delta t \qquad (2.71)$$

For the first calculation we set the time increment to be one-half the interval used in subsequent calculations, as we did before, and we use the initial values of v and θ. Then the new value of the horizontal velocity component is

$$u(\text{new}) = u(\text{old}) + \delta u \qquad (2.72)$$

Do the same for the vertical component, w.

(2) Determine the new value of the speed v from

$$v = \sqrt{(u^2 + w^2)} \tag{2.73}$$

where u and w are the new values, computed in step 1. Determine the new value of θ from

$$\tan \theta = w/u \tag{2.74}$$

(3) Use the new values of u and w to calculate the position at the end of the time increment, as before (Equations 2.65 and 2.66).

(4) Repeat steps 1–3 for the next time increment.

Typical results are shown in Figure 2.13. In this case, we cannot compare the numerical result with an exact result, because the analytical solution is not known. The precision of the result can be assessed, however, by running the program for several different sizes of time interval: as the time interval gets smaller the solution should converge on the correct result (subject to an accumulation of rounding errors in the computer). The assessment of numerical errors is an essential part of the numerical solution of differential equations – but not a topic that we can pursue further in this book (for more on this subject, see Press *et al.*, 1989).

2.14 Rotation using the vector product

Before ending our review of elementary mechanics we shall consider how to apply vector notation and Newton's laws to the problem of rotatory motion. Rotatory motion is obviously very important in the earth sciences: surface forces, such as drag, rarely act through the centre of mass, and so they tend to produce rotation rather than simple sliding. A pebble on the bottom of a stream may slide, but it may also roll along the bed of a stream. Saltating sand grains, observed by high-speed photography, can be seen to be spinning rapidly (hundreds of revolutions per second) – and the spinning motion can be shown to produce additional fluid lift forces. Lithospheric plates move at the earth's surface by slow rotation about axes passing through the centre of the earth. The earth itself is rotating about its axis. We generally ignore this rotation when we consider local motion at the earth's surface, but it produces important dynamic effects on all large-scale motions such as the movement of air masses or ocean currents.

Rotation about an axis can be defined by specifying the axis of rotation and the angular rate of rotation about that axis. In Figure 2.14, let the plate

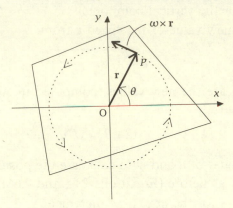

Fig. 2.14. Rotation of a plate about an axis at O.

be rotating about an axis, O, normal to the plane of the paper, at an angular rate of rotation ω. Consider a single point P on the plate: its position can be represented in circular coordinates as (r, θ). r is the length of the vector **r** pointing from the axis of rotation to P, and θ is the angle measured (in radians) from some reference axis. By definition, the rate of rotation is a vector $\boldsymbol{\omega}$ whose magnitude is given by

$$\omega = d\theta/dt \tag{2.75}$$

Assuming right-handed axes, the vector $\boldsymbol{\omega}$ points upwards, parallel to the axis of rotation. The point P, is moving in a circular orbit around O, with a velocity **u** in a direction normal to the vector **r** at a speed equal to ωr.

One example, of particular interest to the earth scientist, is the rotation of the earth. Its angular rate of rotation is $\Omega = 7.29 \times 10^{-5}$ radians per second. At a latitude ϕ, the distance from the earth's axis is $R\cos\phi$, where R is the radius of the earth. So at 49 degrees north, for example, the velocity at the earth's surface is almost 305 m s^{-1}.

A special vector notation has been devised to take account of rotation: the *vector* (or *cross-*) *product*. Using this notation we write

$$\mathbf{u} = \boldsymbol{\omega} \times \mathbf{r} \tag{2.76}$$

We discuss vector products thoroughly in the next section.

In Chapter 7, on strain, we will see that a vector may be used not only to represent a *rate* of rotation, but also the rotation itself, provided that the angle is very small. Rotations through large angles cannot be represented by a vector, because they do not conform to the algebra of vectors.

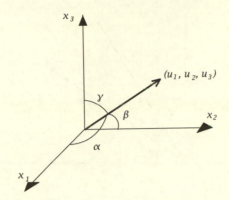

Fig. 2.15. Index notation for vector coordinates.

2.15 The vector product (cross-product)

The vector product **u** of two vectors **v** and **w** is given by

$$\mathbf{u} = \mathbf{v} \times \mathbf{w} \tag{2.77}$$

u points normally upwards from the plane formed by rotating anticlockwise from **v** to **w**. The direction of **u** may be remembered by the *right-hand rule*: **v** and **w** are represented by the first and second fingers of the right hand. The thumb then points in the direction of **u**. If **v** is perpendicular to **w** then **u**, **v**, **w** form a right-handed set of axes. It follows that **w** × **v** is a vector in the direction opposite to **u**:

$$\mathbf{v} \times \mathbf{w} = -\mathbf{w} \times \mathbf{v} \tag{2.78}$$

Thus the cross-product of a vector with itself is zero.

The *magnitude* of the vector product is given by $vw \sin \theta$, where θ is the angle between **v** and **w**. Thus an alternative way of writing the vector product is

$$\mathbf{v} \times \mathbf{w} = \mathbf{n}vw \sin \theta \tag{2.79}$$

where **n** is a unit vector whose direction is given by the right-hand rule.

At this point it is convenient to introduce a new way of representing vector coordinates, and a new notation (Figure 2.15). Instead of calling the axes x, y, and z, we will call them x_1, x_2, and x_3. So a vector **u** can be represented by the coordinates (u_1, u_2, u_3).

The vector product in Equation (2.77) can be given as the relation between the three components of **u** and the three components of **v** and **w** by the equations

$$u_1 = v_2 w_3 - v_3 w_2$$

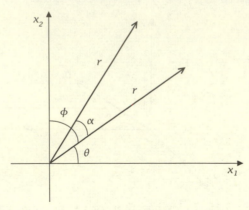

Fig. 2.16. Rotation of a vector about the x_3 axis by an angle α.

$$u_2 = -(v_1 w_3 - v_3 w_1) \qquad (2.80)$$

$$u_3 = v_1 w_2 - v_2 w_1$$

If you are familiar with determinants, you will recognize the right-hand side of each of these equations as being an order-two determinant (if you are not, see Appendix C). Using determinants, the whole set of equations may be written:

$$\mathbf{u} = \begin{vmatrix} v_2 & v_3 \\ w_2 & w_3 \end{vmatrix} \mathbf{i} + \begin{vmatrix} v_3 & v_1 \\ w_3 & w_1 \end{vmatrix} \mathbf{j} + \begin{vmatrix} v_1 & v_2 \\ w_1 & w_2 \end{vmatrix} \mathbf{k} \qquad (2.81)$$

or as the third-order determinant

$$\mathbf{u} = \mathbf{v} \times \mathbf{w} = \begin{vmatrix} \mathbf{i} & \mathbf{j} & \mathbf{k} \\ v_1 & v_2 & v_3 \\ w_1 & w_2 & w_3 \end{vmatrix} \qquad (2.82)$$

These, and many other algebraic properties of vectors are listed and proved in standard texts on vector algebra, and in reference works such as the *Handbook of Chemistry and Physics*, published by The Chemical Rubber Publishing Company (and updated frequently). A particularly useful compendium of mathematical information is the encyclopedia edited by Gellert *et al.* (1977).

2.16 Rotation of vectors about an axis

First we consider the simple case of a rotation in two dimensions. Figure 2.16 is similar to Figure 2.14 in that it shows rotation of a position vector about the x_3-axis, which is normal to the plane of the diagram (the $x_1 x_2$-plane).

The vector has length r, and makes an angle θ with the x_1-axis and ϕ with the x_2-axis. The cosines of these angles, ℓ_1 and ℓ_2, are called the *direction cosines* of the vector. The coordinates of the vector are

$$x_1 = r \cos \theta = r\ell_1 = r \sin \phi \tag{2.83}$$
$$x_2 = r \cos \phi = r\ell_2 = r \sin \theta \tag{2.84}$$

Now rotate the vector by an angle α. The new coordinates (indicated by primes) are

$$x_1' = r \cos(\theta + \alpha) \tag{2.85}$$
$$x_2' = r \cos(\phi - \alpha) \tag{2.86}$$

Using angle-sum relations from basic trigonometry, the first pair of equations above may be substituted into the second pair to give the relations between the new and old vector coordinates:

$$x_1' = x_1 \cos \alpha - x_2 \sin \alpha \tag{2.87}$$
$$x_2' = x_1 \sin \alpha + x_2 \cos \alpha \tag{2.88}$$

This pair of formulae is useful not only for plotting rotating vectors (for example, on a computer screen) but also for transforming one coordinate system to another. This is because the transformation of coordinate systems is itself simply a matter of rotating reference axes (which may be represented as vectors). Think of α as the angle between the new axis (say x_1') and the old axis (x_1). The cosine of this angle can be represented by ℓ_{11}. The angle between the new x_1-axis, and the old x_2-axis is then $\pi/2 - \alpha$. It can be represented by ℓ_{12}, where the first subscript refers to the new axis, and the second to the old one. Note that

$$\ell_{12} = \cos(\pi/2 - \alpha) = -\sin \alpha$$

Using this notation Equations (2.87) and (2.88) can be written as

$$x_1' = \ell_{11}x_1 + \ell_{12}x_2 \tag{2.89}$$
$$x_2' = \ell_{21}x_1 + \ell_{22}x_2 \tag{2.90}$$

Suppose that we want to rotate a vector by an angle θ relative to a given set of reference axes. It is clear that the same relative result can be obtained by rotating the axes relative to the vector by an equal but opposite angle $-\theta$. The new coordinates of the vector are then given by the equations listed above.

Generalizing this result to three dimensions, the new coordinates of a vector $\mathbf{u}'$ are related to the old coordinates of the same vector $\mathbf{u}$ by the following set of equations:

$$
\begin{aligned}
u_1' &= \ell_{11}u_1 + \ell_{12}u_2 + \ell_{13}u_3 \\
u_2' &= \ell_{21}u_1 + \ell_{22}u_2 + \ell_{23}u_3 \\
u_3' &= \ell_{31}u_1 + \ell_{32}u_2 + \ell_{33}u_3
\end{aligned}
\tag{2.91}
$$

We can represent this set of equations more compactly as

$$
u_i' = \sum_j \ell_{ij}u_j
\tag{2.92}
$$

where it is understood that both i and j vary from 1 to 3. We could make the notation even more compact if we were to specify that a repeated index (such as the subscript j in the equation shown) means that the expression must be summed over that subscript. Then we could simply leave out the summation sign (large Greek capital sigma) in the above equation. Both of these notations are, in fact, in common use: the use of a repeated index to indicate summation is called the *Einstein summation convention*. The repeated index is called a *dummy index*, because any symbol could be used. Dummy indices can appear only twice in a single term (otherwise their meaning is ambiguous). An index that appears only once in a given term is called a *free index*. The same free indices must occur on both sides of the equation. Of course, using a simple notation does not get around the fact that actually calculating the new coordinates is fairly complicated, but it does make the logic of the equation easier to follow. Index notation is widely used in the professional literature, and we will use it throughout this book.

From looking at the full equations we might conclude that we need to know nine direction cosines to determine the effect of a rotation. But in fact we only need three, because direction cosines must follow the following rules.

(1) The sums of squares of cosines related to a single vector must add up to unity:

$$
\begin{aligned}
\ell_{11}^2 + \ell_{12}^2 + \ell_{13}^2 &= 1 \\
\ell_{21}^2 + \ell_{22}^2 + \ell_{23}^2 &= 1 \\
\ell_{31}^2 + \ell_{32}^2 + \ell_{33}^2 &= 1
\end{aligned}
$$

(this means that three of the cosines depend on the rest).

(2) The coordinate axes are orthogonal, i.e., at right angles to each other.

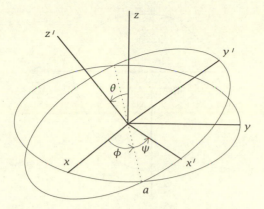

Fig. 2.17. One definition of the Euler angles.

Therefore they are related by three equations:

$$\ell_{11}\ell_{21} + \ell_{12}\ell_{22} + \ell_{13}\ell_{23} = 0$$
$$\ell_{11}\ell_{31} + \ell_{12}\ell_{32} + \ell_{13}\ell_{33} = 0$$
$$\ell_{21}\ell_{31} + \ell_{22}\ell_{32} + \ell_{23}\ell_{33} = 0$$

Note that we can represent all six of these equations compactly using index notation:

$$\ell_{ik}\ell_{jk} = \delta_{ij} \tag{2.93}$$

In this equation δ_{ij} is *Kronecker's delta* which has the property that

$$\delta_{ij} = 1 \quad \text{for } i = j$$
$$= 0 \quad \text{for } i \neq j$$

To become familar with the Einstein index notation, expand Equation (2.93) to yield the six equations given in the text.

Any desired rotation of a set of axes x, y, z to a set x', y', z' may be achieved as follows (see Figure 2.17). The set x, y, z is first rotated by an angle ϕ (phi) about the z-axis, then by an angle ψ (psi) about an intermediate axis a which lies in the xy-plane, and, finally, by an angle, θ (theta), about the z' axis. The three angles are generally called the *Euler angles*, but the same rotation can be carried out in different ways, and not all authors define the angles in the same way. One definition is shown in Figure 2.17. A good discussion (with very clear diagrams) is given by Goldstein (1980).

We might be tempted to think that ℓ_{ij} would be the same as ℓ_{ji}. But in fact this is not the case: the angle between the new first coordinate axis

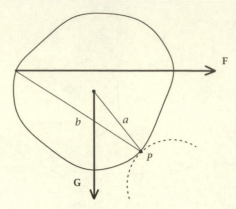

Fig. 2.18. A pebble rotated about a pivot P by a drag force **F**.

and the old second coordinate axis is not generally the same as the angle between the new second coordinate axis and the old first coordinate axis.

It was shown by Leonhard Euler in 1776 that rotations of rigid bodies can always be represented by rotation about a single appropriately chosen axis (this is generally not one of the coordinate axes). Euler's theorem applies to the movement of rigid lithospheric plates over the surface of the earth, and provides a convenient way to represent their motion (for a full discussion of the application to plate tectonics, see Le Pichon *et al.*, 1973).

Rotation about reference axes is a technique much used in computer graphics: the formulae and their use in computer programs are explained by Foley and Van Dam (1990) and Newman (1979). A good way to become familar with these techniques is to experiment with the computer program ROTAXES.

2.17 Moments and torque

Newton's laws are about the balance of forces on a point object or its acceleration under an unbalanced force, but they do not tell us about the balance of moments on an extended object or the relation between moment and rate of rotation. The basic principles for the equilibrium of rotating rigid bodies were discovered by James Bernoulli and Leonhard Euler, and first clearly enunciated by Euler in 1771, almost a hundred years after Newton's work (Truesdell, 1968, p. 172). We will illustrate them with an example.

Consider a rigid body which is being rotated around some pivot P by a fluid drag force **F** (Figure 2.18). The pivot is the axis (shown normal to the figure) about which rotation will take place when the force is large enough to overcome gravity: its position is determined by the geometry of the bed underlying the pebble. The gravity force **G** acts down through the centre of mass, and tends to rotate the pebble anticlockwise about the pivot, whereas the total drag force **F** acts some distance above the centre of mass (because of the velocity gradient in the flow above the bed). How strong does the drag force have to be to roll the pebble over the pivot?

The answer is that the pebble will start to rotate when the *torque* about the pivot produced by drag is larger than that produced by gravity. The *moment* produced by a force **F** acting at a point defined by a position vector **x** whose origin is the axis of pivot *P* is defined as

$$\mathbf{M} = \mathbf{x} \times \mathbf{F} \tag{2.94}$$

The torque (total moment) **T** produced by drag is the vector sum of all the moments produced by the surface drag forces acting on the pebble. We assume for simplicity that this can be represented as being produced by a single drag force, as shown in the figure.

The condition for equilibrium is that the sum of all the torques is zero, i.e. $\sum \mathbf{T} = 0$, which in this case is

$$(\mathbf{a} \times \mathbf{G}) + (\mathbf{b} \times \mathbf{F}) = 0 \tag{2.95}$$

We see, therefore, that the condition for static equilibrium of a rigid body is not only that the sum of all the forces acting on a body be equal to zero (i.e., there is no translation), but also that the sum of all the torques must be equal to zero (i.e., there is no rotation). In elementary mechanics, the complication of torques is generally avoided by considering only point masses, rather than extended rigid masses. We consider an application of moments to plate flexure in Section 8.10.

Torque here refers to the resultant of a system of moments of forces. Moment may refer to other vectors: for example, one can refer to the moment of velocity or momentum. One can apply the same idea to scalars, in which case the moment is a scalar. For example, the moment of a mass about a point is the product of the mass and the distance from the point (the term is used in a different, but analogous manner in statistics).

2.18 Angular momentum and moment of inertia

Bodies that rotate, like the earth, tend to continue to do so: it takes the application of a torque to increase or decrease the rate of rotation. This suggests that it is possible to extend the concept of momentum to apply specifically to rotating bodies. We define *angular momentum* (also called the "moment of momentum") **I** as follows:

$$\mathbf{I} = m\mathbf{x} \times (d\mathbf{x}/dt) = m\mathbf{x} \times \mathbf{v} \qquad (2.96)$$

where **x** is a position vector, giving the location of the mass m. The rate of change of angular momentum is:

$$dI/dt = m\mathbf{v} \times \mathbf{v} + m\mathbf{x} \times \mathbf{a}$$
$$= m\mathbf{x} \times \mathbf{a} \qquad (2.97)$$

Remember that the cross-product of a vector with itself is equal to zero.

The relation to the torque **T** can be seen by noting that

$$m\mathbf{x} \times \mathbf{a} = \mathbf{x} \times m\mathbf{a} = \mathbf{x} \times \mathbf{F} = \mathbf{T} \qquad (2.98)$$

So just as force is equal to the rate of change of (linear) momentum, so torque is equal to the rate of change of angular momentum.

The analogy with Newton's second law can be made even clearer by writing the relation of torque to angular velocity, **ω** or angular acceleration, **α**:

$$\mathbf{T} = M(d\boldsymbol{\omega}/dt) = M\boldsymbol{\alpha} \qquad (2.99)$$

where M is called the *moment of inertia*. It is defined as the torque necessary to produce unit angular acceleration. Note that it is a scalar (with units of mass times length squared) not a vector like the angular momentum. An application of these ideas is given when we consider the symmetry of stresses in Section 4.8.

The moment of inertia of regular geometric bodies can be derived using integral calculus. For example, the moment of inertia of a sphere, of mass m and radius r, composed of material of uniform density, is given by

$$M = (2/5)mr^2 \qquad (2.100)$$

For irregular bodies M can be calculated by numerical methods.

Though the earth is not a perfect sphere, and its internal density distribution is not known directly, its moment of inertia can be determined accurately from the precession (slow "wobble") of its axis of rotation. The theory is given by Stacey (1992, Chapter 3).

2.19 Centrifugal and tidal forces

It follows from Newton's second law that circular motion, which corresponds to a centripetal acceleration, implies the existence of a centripetal force, that is, a force directed toward the centre of the circular orbit. In the case of bodies at the surface of the earth, this centripetal force is provided by the force of gravity. Without this force, bodies would continue to travel in a straight line (relative to the fixed stars, which constitute an "absolute" or "inertial" reference frame for Newtonian mechanics), or, in other words, they would fly tangentially off the earth's surface into space.

Sometimes, however, it is convenient to treat a rotating point mass as being *at rest* in its own rotating frame, the real centripetal force on it being balanced by an imaginary *centrifugal force*.

Centrifugal force is variously called a "virtual", "fictive", or "inertial" force (see Section 2.2); such forces arise whenever we use an accelerating (non-inertial) frame, because Newton's second law only holds in inertial frames. (French, 1971, pp. 507–11 gives a good discussion of centrifugal force and illustrates the passions that this subject can arouse among some physicists.)

In the earth sciences, we generally prefer to use a rotating reference frame, fixed with respect to the surface of the earth, because it corresponds to our subjective feeling that we ourselves are at rest, rather than revolving in space at speeds of several hundred metres per second! When we consider a point mass *moving* in such a rotating frame, a further virtual force is brought into play, as will be now seen. The equation of motion is then written (per unit mass):

$$\text{Acceleration of mass in inertial frame} = \sum(\mathbf{F}/m)$$

We can re-express this as

$$\text{acceleration of mass in rotating frame} - \text{centrifugal acceleration} - \text{Coriolis acceleration} = \sum(\mathbf{F}/m)$$

Generally, however, it is not written this way: instead the acceleration of the mass relative to the rotating frame is placed on the left-hand side, and the other accelerations are moved to the right-hand side and called the *centrifugal force* (per unit mass) and the *Coriolis force* (per unit mass, named for the French physicist G.-G. Coriolis 1792–1843). The Coriolis force is derived in detail in the next section. It is very small: the term is generally used for the horizontal components, but there is also a vertical component which at its maximum is about four orders of magnitude less than gravity. The vertical component may safely be neglected, and so may the horizontal components for most applications, where there are much larger horizontal forces at work. But if there are no such forces, or if they are balanced by other forces, then the Coriolis force becomes important. This is generally the case for large-scale motion of the atmosphere and oceans, and is the basic reason why circular motions (cyclones, anticyclones, ocean gyres, etc.) are common geophysical phenomena.

An important case where centrifugal force is not balanced by gravity arises in the rotation of the earth and moon about their common centre of mass. Because the earth is so much larger, this is actually located within the earth, though not at its centre. The centrifugal forces are directed away from this earth–moon centre and thus have horizontal as well as vertical components, relative to the earth's surface. The vertical components are negligible (much smaller even than the Coriolis components, because the earth–moon system takes two weeks to rotate) but the horizontal components are the "tide-generating forces" that, in an ideal ocean covering all of the earth, cause flow towards the point on the earth directly below the moon and directly opposite it on the other side of the earth. Flow continues until it is balanced by a horizontal component of pressure produced by the sloping water surface. As the horizontal components are very small, balance is achieved by water "bulges" (tidal waves) that are only about one metre high. The earth's rotation carries these two bulges around the earth once each day. In theory, therefore, oceanic tides due to the moon should be semidiurnal, with an amplitude of about a metre (for a clear discussion of the tides, see Strahler, 1971).

Of course, there are other rotary motions that might also be considered. Not only are we rotating about the earth's centre, and about the earth–moon centre of mass, but also about the centre of mass of the solar system (which is, effectively, the centre of the sun). Rotation about the sun produces a solar component of tidal forces that is on average about 20 per cent of the strength of the lunar component. Luckily, other larger rotations (of the galaxy) can be neglected in terrestrial mechanics.

2.20 Coriolis forces

2.20.1 Rotation and the vector product

To understand the Coriolis force we first return to the use of the vector cross product to represent rotation about an axis (Figure 2.19(a)). We saw that, from the definitions of the angular rate of rotation and the cross-product, the tangential velocity $\mathbf{v}$ is given by:

$$\mathbf{v} = \boldsymbol{\omega} \times \mathbf{r} \qquad\qquad (2.101)$$

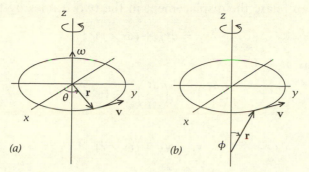

Fig. 2.19. Tangential velocity **v**, related to rotation of **r**: (a) with **r** in the *xy*-plane; (b) with **r** not in the *xy*-plane.

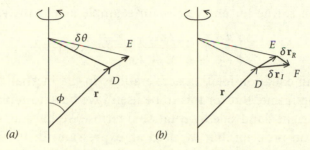

Fig. 2.20. Change in position of **r**: (*a*) due to rotation alone; (*b*) due to movement produced not only by rotation.

Since **r** and **v** are at right angles, the magnitudes are related by

$$v = (d\theta/dt)r = \omega r$$

In the more general case, shown in Figure 2.19(*b*), the same vector equation applies, but the magnitudes are now related by $v = \omega r \sin \phi$.

Now consider a small change in position of **r** with time, without change in magnitude (Figure 2.20(*a*)). Due to rotation in the *xy*-plane, **r** moves from D to E, and $DE = r \sin \phi \delta\theta$. But $\omega = \delta\theta/\delta t$, i.e., $\delta\theta = \omega\delta t$. So we can write:

$$\overrightarrow{DE} = (\boldsymbol{\omega} \times \mathbf{r})\delta t$$

Note that this change, which is due only to rotation, would *not be seen* by an observer in the rotating reference frame.

Now suppose that there is an additional displacement EF, which is not due to rotation alone. Together with the displacement DE, this carries the position from D to F. In the "absolute" (inertial) frame this is seen as $\delta\mathbf{r}_I$, but in the rotating frame it is seen only as a movement from E to F, i.e., as

$\delta \mathbf{r}_R$. So we can relate the displacement in the two frames by the equation:

$$\delta \mathbf{r}_I = \delta \mathbf{r}_R + (\boldsymbol{\omega} \times \mathbf{r})\delta t$$

In the limit as $\delta t \to 0$:

$$\left(\frac{d\mathbf{r}}{dt}\right)_I = \left(\frac{d\mathbf{r}}{dt}\right)_R + (\boldsymbol{\omega} \times \mathbf{r}) \qquad (2.102)$$

or

$$\mathbf{v}_I = \mathbf{v}_R + (\boldsymbol{\omega} \times \mathbf{r}) \qquad (2.103)$$

2.20.2 *Coriolis acceleration*

The derivation of Equation (2.102) assumed $\mathbf{r}$ was a position vector, but the same equation is true for any vector, for example a velocity, $\mathbf{v}_I$:

$$\left(\frac{d\mathbf{v}_I}{dt}\right)_I = \left(\frac{d\mathbf{v}_I}{dt}\right)_R + (\boldsymbol{\omega} \times \mathbf{v}_I) \qquad (2.104)$$

So we have an equation relating accelerations in the inertial frame to those in the rotating frame. But for this to be useful we have to replace *all* inertial terms on the right-hand side by rotational terms. We have an expression for $\mathbf{v}_I$ so that is no problem, but we need an expression for its time derivative in the rotating frame. To obtain this we differentiate Equation (2.103) in the rotating frame:

$$\left(\frac{d\mathbf{v}_I}{dt}\right)_R = \left(\frac{d\mathbf{v}_R}{dt}\right)_R + (\boldsymbol{\omega} \times \mathbf{v}_R) \qquad (2.105)$$

Now we use this equation and Equation (2.103) to eliminate the inertial terms from the right-hand side of Equation (2.104):

$$\left(\frac{d\mathbf{v}_I}{dt}\right)_I = \left[\left(\frac{d\mathbf{v}_R}{dt}\right)_R + (\boldsymbol{\omega} \times \mathbf{v}_R)\right] + [(\boldsymbol{\omega} \times \mathbf{v}_R) + \boldsymbol{\omega} \times (\boldsymbol{\omega} \times \mathbf{r})]$$

$$= \left(\frac{d\mathbf{v}_R}{dt}\right)_R + 2\boldsymbol{\omega} \times \mathbf{v}_R + \boldsymbol{\omega} \times (\boldsymbol{\omega} \times \mathbf{r}) \qquad (2.106)$$

The terms in this equation are all acceleration terms:

$$\begin{array}{ccccc} \text{acceleration} & & \text{acceleration} & & \\ \text{in inertial} & = & \text{in rotating} & - & \text{Coriolis} & - & \text{centrifugal} \\ \text{frame} & & \text{frame} & & \text{acceleration} & & \text{acceleration} \end{array}$$

Centrifugal acceleration. It can be seen that the third term in Equation (2.106) is minus the centrifugal acceleration since $\boldsymbol{\omega} \times \mathbf{r} = \mathbf{v}$, the tangential velocity, and $-\boldsymbol{\omega} \times \mathbf{v}$ is a vector of magnitude v^2/r directed away from the axis of rotation.

Coriolis acceleration. The second term in Equation (2.106), if we include a minus sign, is $-2\omega \times v_R$. This is a vector normal to v_R, the velocity vector in the rotating frame. Therefore, it represents an acceleration tending to deflect the motion, as seen in the rotating frame.

In general, at latitude Φ, where $\Phi = 90 - \phi$, we can resolve v_R into components u and v. It can then be shown that the Coriolis acceleration has the following components:

$$\text{east component}: \quad fv = 2\Omega \sin \Phi v \qquad (2.107)$$

$$\text{north component}: \quad -fu = -2\Omega \sin \Phi u \qquad (2.108)$$

where Ω is the magnitude of the angular rotation of the earth ($\Omega = 7.29 \times 10^{-5}$ rad s^{-1}). f is called the *Coriolis parameter*. It is a maximum at the poles and falls to zero at the equator. There is also a vertical component of the Coriolis acceleration, which can always be ignored because it is very small compared with gravity.

The horizontal components cannot be ignored for large-scale motions where there are no major overriding forces. They tend to deflect such motions to the right in the northern hemisphere (positive latitudes) and to the left in the southern hemisphere (negative latitudes). If a body is at rest relative to the surface of the earth $u = v = 0$, and there is no Coriolis acceleration.

Coriolis forces are discussed thoroughly at an elementary level by Stommel (1989) and at a more advanced level by Pedlosky (1987: our discussion is largely condensed from this work).

2.21 Review problems

(1) **Problem:** Suppose a lineation is found to have length L, azimuth A (east of north) and plunge d degrees.

(a) Derive the formulae for expressing this as a vector (x, y, z), where the x-axis points east, the y-axis points north and the z-axis points down.

(b) Is this system of axes right- or left-handed?

Answer: Figure 2.21 shows the problem. The lineation is shown by the bold arrow, and its projection onto the horizontal plane by the bold line. The length of the projection is $L \cos d$. The projection onto the x-axis is $(L \cos d) \sin A$, and the projection onto the y-axis is $(L \cos d) \cos A$. The

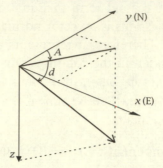

Fig. 2.21. Lineation (arrow) in two different reference systems.

projection of the lineation onto the z-axis is $L \sin d$. So the required coordinates are

$$x = (L \cos d) \sin A$$
$$y = (L \cos d) \cos A$$
$$z = L \sin d$$

If the x-axis is aligned with the first finger and the y-axis with the second finger, then the thumb, corresponding to the z-axis, points down – so this set of axes is left-handed. An alternate test is which way a clockwise motion from x to y would drive a (normal, right-handed) screw. The answer is upwards – so this set of axes must be left-handed.

The fact that a reference system often used by geologists is left-handed is unfortunate, as almost all other scientists use right-handed axes. However, it is easily changed to a right-handed system by making north the x-direction and east the y-direction.

(2) **Problem:** Geologists can reconstruct ancient current directions ("paleo-currents") by measuring sedimentary structures such as cross-bedding and current lineations. The dip direction of cross-beds indicates the direction of current flow, current lineations indicate one of two possible directions (e.g., towards either north or south). The cross-bedding dip directions (azimuths), measured on separate sets at a single locality, are as follows: 10, 250, 261, 265, 274, 281, 285, 290, 293, 297, 300, 301, 302, 303, 307, 311, 320, 326, 333, 352.

(a) Plot these data as a rose diagram.
(b) Determine the vector mean and strength.
(c) If the data were current lineations, what difference would that make to the method of processing the data?

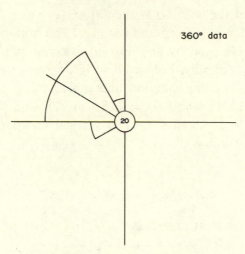

Fig. 2.22. Output from VECSTAT for data of Problem 2; there are a total of 20 observations.

First carry out the calculations using a calculator or spreadsheet, and showing the steps in the computation. Then verify the results using the program VECSTAT.

Answer: The vector statistics and rose diagrams are shown in the output from VECSTAT (Figure 2.22). Current lineations indicate only the line of movement. An azimuth of 45 degrees might indicate flow towards that direction, or away from it (towards $180 + 45 = 225$ degrees). Therefore the data are periodic over 180 degrees, not over 360 degrees. The normal techniques may be used to process the data, provided they are converted to a 0–180 degree scale, then doubled. After the mean of the transformed data has been found it should be restored to its true value by dividing by two.

(3) **Problem:** In a volcanic eruption from a cone which rises 300 m above the surrounding (flat) countryside, large bombs are observed to be thrown a maximum distance of 3000 m. Neglecting air resistance, and assuming that the initial angle of ejection was 45 degrees, calculate

(a) the initial speed of the bombs;

(b) the speed at which they hit the ground.

Answer: The height at time t is given by:

$$z = z_0 + w_0 t - g t^2 / 2$$

Using the top of the volcano as the datum ($z = z_0 = 0$), when the bomb hits the ground $z = -300$ m and $t = t_{max}$. The horizontal component of velocity is not changed by any applied force, so it remains constant at $u_0 = 3000/t_{max}$. Initially, because $\theta = 45$ degrees, $w_0 = u_0$. Substituting $w_0 = 3000/t_{max}$ in the equation above and using $g = 9.8$ m s^{-2}, we obtain $t_{max} = 25.95$ s and $w_0 = 115.6$ m s^{-1}, and therefore the speed is $\sqrt{2} \times 115.6 = 163.5$ m s^{-1}. The vertical component of velocity at $t = 25.95$ s is given by

$$w = -gt + w_0$$
$$= -254.3 + 115.6 = -138.7 \text{ m s}^{-1}$$

So the total velocity at impact is

$$|V| = \sqrt{(-138.7)^2 + (115.6)^2} = 180.6 \text{ m s}^{-1}$$

(4) **Problem:** How high does a projectile rise, when it is thrown vertically upward with an initial speed w_0? Neglect air resistance, and derive the result in two different ways: (a) using the equations of motion; (b) from energy considerations.

Answer:

(a) A derivation from the equation of motion was given in Section 2.7. After two integrations of the vertical equation of motion

$$dw/dt = -g$$

we obtain

$$z = -gt^2/2 + w_0 t$$

Using the fact that $w = 0$ at the top of trajectory, when $t = w_0/g$ we obtain the solution

$$h = w_0^2/2g$$

(b) If we use $z = 0$ as the level at which the potential energy is zero, then the work W required to raise a mass m to a height h, against a force $F = mg$ is:

$$W = \int_0^h mg\,dz = mgh$$

The kinetic energy, initially $mw_0^2/2$, decreases to zero at the top of the trajectory, where the speed is equal to zero. Because we have

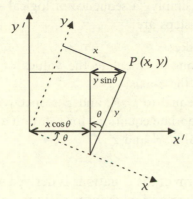

Fig. 2.23. Effect of anticlockwise rotation of axes on a vector P (or clockwise rotation of vector).

neglected friction, the gain in potential energy must equal the loss in kinetic energy:

$$mgh = mw_0^2/2$$

so $h = w_0^2/2g$ as before.

(5) **Problem:** Show that clockwise rotation of a vector about the z-axis by an angle θ transforms x, y, z into the new coordinates:

$$x' = x \cos \theta + y \sin \theta$$
$$y' = y \cos \theta - x \sin \theta$$
$$z' = z$$

Design an algorithm that will give the new coordinates of a point P, with old coordinates (x, y, z), after sequential rotation about the x-, y-, and z-axes through the three angles ψ, ϕ, θ (psi, phi, theta). We assume that the old coordinates and the angles are specified, and the new coordinates are to be determined.

Answer: Clockwise rotation of a vector is equivalent to anticlockwise rotation of the axes. Figure 2.23 shows a point P and two sets of axes: the broken lines are the original axes before rotation by an angle θ about the z-axis (which points upwards out of the plane of the paper). Note that the positive sense of rotation is anticlockwise (the same as the direction necessary to rotate the x- towards the y-axis). The derivation of x' in terms of x and y is clear from the diagram, and a similar construction will also make clear the derivation of y'.

An algorithm is simply "a sequence of logical steps necessary to solve the problem". The steps are:

(a) specify $x, y,$ and z;
(b) use the equations to calculate the values of $x', y',$ and z' after the rotation about the z-axis;
(c) set $x, y,$ and z equal to $x', y',$ and z' respectively, and repeat step (b) using the appropriate equations for rotation about the y- and x-axes;
(d) display the final x', y' and z';
(e) stop.

The appropriate form of the equations is derived simply by interchanging the z's and y's for the rotation about y, and the z's and x's for the rotation about x. For example, for the rotation about y the equations will be:

$$x' = x \cos \phi + z \sin \phi$$
$$z' = z \cos \phi - x \sin \phi$$

(6) **Problem:**

(a) Derive the equation for the vertical component of velocity in the ballistics problem, assuming linear drag, i.e., that the drag force per unit mass is $k_1 v$ (Equation 2.55);
(b) Derive an expression giving the (x, z) coordinates of a bomb at time t for this case.

Answer:

(a) The equation to solve is Equation (2.59):

$$dw/dt = -g - k_1 w$$

Separating the variables,

$$dw/(-g - k_1 w) = dt$$

Tables of integrals give

$$\int \frac{dx}{(a + bx)} = \frac{\ln(a + bx)}{b} + C$$

so:

$$\int \frac{dw}{(-g - k_1 w)} = -\frac{\ln(-g - k_1 w)}{k_1} = t + C$$

Solving for C using $w = w_0$ at $t = 0$, and rearranging gives

$$-\frac{\ln(-g - k_1 w) - \ln(-g - k_1 w_0)}{k_1} = t$$

or

$$\ln\left[\frac{(-g - k_1 w)}{(-g - k_1 w_0)}\right] = -k_1 t$$

and converting from a logarithm to an exponent gives

$$\frac{(-g - k_1 w)}{(-g - k_1 w_0)} = \exp(-k_1 t)$$

then rearranging yields the solution:

$$w = -(g/k_1) + [w_0 + (g/k_1)]\exp(-k_1 t)$$

(b) Because $u = dx/dt$ and $w = dz/dt$ we derive an equation for x and z by integrating Equation (2.58): $u = u_0 \exp(-k_1 t)$ and the equation for w derived in (a). Remember that $\int \exp(at)dt = \exp(at)/a$. Set $u = dx/dt$, then:

$$x = \int u\,dt$$

$$= \int u_0 \exp(-k_1 t)dt$$

$$= -\frac{u_0}{k_1}\exp(-k_1 t) + C$$

at $x = 0, t = 0$ so $\exp(-k_1 t) = 1$ and $C = u_0/k_1$. So

$$x = (u_0/k_1)[1 - \exp(-k_1 t)]$$

Integrating the equation for w is a little more messy, but follows the same lines and gives

$$z = \frac{(w_0 k_1 + g)}{k_1^2}[1 - \exp(-k_1 t)] - gt/k_1$$

(7) **Problem:** Use the program EJECTA to explore the effects of linear and quadratic drag on small particles. Roughly, linear drag is due to viscosity and quadratic drag to formation of a turbulent wake.

For example, run the program for a diameter $D = 0.1$ mm (0.0001 m) for an initial velocity V having horizontal and vertical components ($u = w$) of 5, 10, and 20 m s^{-1}. Note that suitable scales must be chosen for the graph, in order to show sufficient detail for the linear and quadratic drag cases (the no-drag case will be off scale for most of the trajectory).

After this experience with computer experimentation, consider the general question of how to determine whether linear or quadratic drag is more important in a particular case. What does "more important" mean, and how can you quantify the answer?

Answer: Several answers are possible:

(a) We may say that a particular form of drag is more important if the drag force is larger. However, the answer may change as the velocity changes during a single trajectory. The equations for the drag force per unit mass on a sphere in air (given in the program listing, in SI units) are

$$\text{linear}: \quad F_D = 1.3 \times 10^{-7}(V/D^2)$$
$$\text{quadratic}: F_D = 1.7 \times 10^{-4}(V^2/D)$$

If we divide the two we get:

$$\frac{\text{linear drag}}{\text{quadratic drag}} = \frac{7.6 \times 10^{-4}}{VD}$$

This shows that if V is large, D must be very small for linear drag to be important relative to quadratic drag.

(b) Another answer would be that linear drag is more important than quadratic drag if the total distance of transport is less with linear than with quadratic drag. If we set $D = 0.0001$ m and $V = 7.6$ m s^{-1}, then the equation given in (a) tells us that initially the two components of drag will be the same. But as the velocity decreases, the linear drag will have a proportionately larger effect, and so the total distance of transport will be less.

We cannot determine the condition for equal distance of tranport analytically, because we have no analytical expression for distance when there is quadratic drag. Experimenting numerically suggests that an initial velocity of 28 m s^{-1} (horizontal and vertical components of 20 m s^{-1}) comes close to giving equal distances of travel in the two cases. For velocities less than that, the effects of linear drag, when integrated over the whole trajectory, are more important.

2.22 Suggested reading

The following are good general texts on elementary mechanics (the Holton book is easier than the others):

Holton, G. J., 1973. *Introduction to Concepts and Theories in Physical Science.* Revised with new material by S. G. Brush. Reading MA, Addison-Wesley, 589 pp. (QC23.H758)

Feynman, R. P., R. B. Leighton and M. Sands, 1963. *The Feynman Lectures on Physics. Vol. 1: Mainly Mechanics, Radiation, and Heat.* Reading MA, Addison-Wesley (QC21.F43 v.1)

French, A. P., 1971. *Newtonian Mechanics.* New York, W. W. Norton and Co., 743 pp. (QC125.2.F74)

French, A. P. and M. G. Ebison, 1986. *Introduction to Classical Mechanics.* New York, Van Nostrand Reinhold, 310 pp. (A revised and shortened edition of French, 1971. QC125.2.F74)

Pippard, A. B., 1972. *Forces and Particles: An Outline of the Principles of Classical Physics.* New York, John Wiley and Sons, 321 pp. (QC21.2.P56)

Smith, P. and R. C. Smith, 1990. *Mechanics.* New York, John Wiley and Sons, second edition, 321 pp. (QA805.S673)

The standard advanced treatment, which nevertheless contains very clear expositions of some elementary topics, is

Goldstein, H., 1980. *Classical Mechanics.* Reading MA, Addison-Wesley, second edition, 672 pp. (QA805.G6)

A part of elementary mechanics is learning the mathematical techniques necessary to represent objects and their motions in space: the same techniques are useful for computer graphics. Good introductions are given by:

Foley, J. D. and A. Van Dam, 1990. *Computer Graphics: Principles and Practices.* Reading MA, Addison-Wesley, second edition, 1174 pp. (T385.C587)

Newman, W. M., 1979. *Principles of Interactive Computer Graphics.* New York, McGraw-Hill, second edition, 541 pp. (T385.N48)

Vectors are described in many texts, for example,

Gellert, W. *et al.*, eds., 1977. *The VNR Concise Encyclopedia of Mathematics.* New York, Van Nostrand Reinhold, 760 pp. (QA40.V18)

Hoffmann, B., 1966. *About Vectors.* New York, Dover, 134 pp. (QA161.H63)

A good introduction to numerical methods, with computer programs, is given by:

Press, W. H., B. P. Flannery, S. A. Teukolsky and W. T. Vettering, 1989. *Numerical Recipes in Pascal: The Art of Scientific Computing.* Cambridge University Press, 759 pp. (QA76.73.P2N87)

For the history of mechanics see:

Mach, E., 1960. *The Science of Mechanics.* Translated by T. J. McCormack, sixth edition, LaSalle IL, Open Court, 634 pp. (QA802. M1313)

Truesdell, C., 1968. *Essays in the History of Mechanics.* New York, Springer-Verlag, 384 pp. (See particularly Chapter II. QC122.T7)

Westfall, R. S., 1971. *Force in Newton's Physics: the Science of Dynamics in the Seventeenth Century.* New York, American Elsevier, 579 pp. (QA802.W48)

Westfall, R. S., 1977. *The Construction of Modern Science: Mechanisms and Mechanics.* Cambridge University Press, 171 pp. (QA 802.W47)

Westfall, R. S., 1980. *Never At Rest: A Biography of Isaac Newton.* Cambridge University Press, 908 pp. (QC16.N7W35)

For other references cited in the text see the list of references at the end of the book.

3

Dimensional analysis and the theory of models

3.1 Introduction

Dimensional analysis is a powerful method that can be applied to the solution of physical problems that cannot be solved by more direct, analytical methods. It is also an important guide to the design of experiments, and to the construction of scale models of natural phenomena and engineering designs.

The basis of dimensional analysis is very simple: the two sides of an equation must refer to the same type of physical entity, that is they must have the same dimensions. Masses can only be equal to masses, they cannot be equal to forces, and so on. Therefore in an equation such as Newton's second law, if there are forces on one side of the equation, and mass times acceleration on the other, and if the equation is true, force must have the same dimensions as mass times acceleration. If work is indeed equivalent to heat, they must both have the same dimensions.

It turns out, rather unexpectedly, that it is possible to make considerable simplifications to dynamical equations, based only on this simple idea. In some cases, if we know the main variables that are important in a given system, and what their dimensions are, we can write down the form of the equations that must relate those variables to one another, even without any further understanding of the physics involved. Of course, since there is no free lunch in physics any more than in the rest of life, the catch is that we must understand the system at least well enough to be able to list a minimum set of variables that are the important ones in that system.

The method dates back to Joseph Fourier (1768–1830) and Lord Rayleigh (1842–1919) and is explained in the rest of this chapter, and more fully in the works listed in the bibliography. There are still relatively few published examples of applications in the earth sciences, but the method deserves to be more widely used.

In this chapter, we also discuss the theory of scale models. If we want to investigate phenomena experimentally at a smaller (or larger) than natural scale, then dimensional analysis tells us the conditions necessary for dynamic similarity, that is the conditions that must be followed so that the proportions between the forces in the model are everywhere the same as the corresponding proportions in nature. Only if this is true will the dynamic phenomena be similar. The results are often counter-intuitive; for example, a small model of the continental crust would have to be made of very "weak" material, such as a soft mud. Far from being "unrealistic" such models are the only ones which can hope to reproduce realistically some of the large-scale phenomena of the crust, such as mountain ranges or sedimentary basins.

3.2 Dimensions and dimensional homogeneity

Certain quantities have dimensions. For example, it is not sufficient to say that a crystal is 10 long: we would immediately ask, 10 what? The answer could be 10 mm, or 10 inches, but some unit of length must be specified. The same applies to volumes, but in this case we recognize that a volume can be specified in terms of length units cubed, e.g., mm^3. In this way we find that most physical quantities can be described in terms of a small number of "fundamental" dimensions: length [L], mass [M] and time [T]. Alternatively, force [F] may be substituted for mass. In mechanics, these three "fundamental dimensions" are sufficient to describe all the quantities used. In thermodynamics temperature [Θ] is also needed, and in other branches of science, other dimensions may be required. The square bracket is used to signify that only the dimension is being specified, not the particular units or numerical value.

It is a fundamental principle of dimensional analysis that all physically meaningful equations must be dimensionally homogeneous. For example, an equation of the type:

$$A = B + C \tag{3.1}$$

can be valid only if A, B and C all have the same physical dimensions, i.e., if

$$[A] = [B] = [C] \tag{3.2}$$

We can use this principle to discover the dimensions of quantities from their definition. For example:

volume $\qquad [V] = [L^3]$

mass density $\quad [\rho] = [m/V] = [ML^{-3}]$

velocity $\qquad [u] = [dx/dt] = [LT^{-1}]$
(dx has units of length and dt has units of time)

acceleration $\quad [a] = [d^2x/dt^2] = [LT^{-2}]$
(dt^2 has units of $[T^2]$ but d^2x has units of $[L]$)

force $\qquad F = ma$ so $[F] = [ma] = [MLT^{-2}]$

stress $\qquad \sigma = F/A$ so $[\sigma] = [F/A] = [ML^{-1}T^{-2}]$
(stress and pressure can also be expressed as $[FL^{-2}]$)

Note that some quantities are dimensionless, i.e., pure numbers or ratios. For example, π is defined as the ratio between the circumference and the diameter of a circle, i.e., it is a length divided by a length and is therefore a pure number. We do not have to specify what units we are using when we specify pure numbers.

Strain, for example, the extension of a wire when it is loaded, is measured as the ratio between the increment of length δL and the original length, L. So strain is dimensionless.

Knowledge of dimensions is useful in converting from one system of units to another. For example, the cgs unit of force is the dyne, the SI unit is the newton.

$$[F] = [MLT^{-2}]$$
$$1 \text{ newton} = 1\text{kg m s}^{-2}$$
$$= (1000 \text{ g}) (100 \text{ cm}) \text{ s}^{-2}$$
$$= 10^5 \text{ g cm s}^{-2}$$
$$= 10^5 \text{dynes}$$

Some equations appear not to be dimensionally homogeneous. This happens particularly if they are empirical, or derived from statistical analysis. For example, the Chézy equation for river flow is:

$$U = C\sqrt{RS} \qquad (3.3)$$

where U is the average speed, R is the hydraulic mean depth (the average depth in a wide river), S is the slope, and C is a coefficient determined empirically. Analysis of the equation shows that C is not a dimensionless coefficient but must have dimensions. The dimensions of the other terms in the equation are: $[U] = [LT^{-1}]$ and $[R] = [L]$. S is dimensionless (slope is the tangent of the angle of slope, i.e., a length divided by a length). From the Chézy equation,

$$C = U/\sqrt{RS}$$
$$[C] = [LT^{-1}]/[L^{1/2}] = [L^{1/2}T^{-1}] \tag{3.4}$$

These are the dimensions of the square root of an acceleration, so we can obtain a dimensionless form of Chézy's C by dividing by an acceleration. We know that the acceleration due to gravity must be important in the flow of rivers, because it is the downslope component of gravity that causes the water to flow. So we might well suspect that a better form of the Chézy equation would be:

$$U = \left(\frac{C}{\sqrt{g}}\right)\sqrt{RSg} = C_0\sqrt{RSg} \tag{3.5}$$

where

$$C_0 = C/\sqrt{g} \tag{3.6}$$

This form is better in the sense that it might be expected to hold even if g varies, e.g., to be true for Mars as well as Earth. It is also far more convenient to use than the original Chézy equation because any consistent set of units may be used. As long as the same units of length and time are used for all variables in the equation, the numerical value of C_0 remains the same. In the original Chézy equation, if the units change from the English to the metric system, so does the value of C.

3.3 Dimensionless products and the pi theorem

Suppose there is a physical problem which we cannot solve analytically but for which we can write a complete list of physically important variables.

3.3.1 *Example 1: Flow of a viscous fluid down a plane surface*

Suppose we are interested in a sheet of lava flowing down the side of a volcano. To simplify the problem, assume that the side of the volcano is a plane with a (locally) uniform slope S. We are interested in the way that the velocity u varies with perpendicular height y above the base of the lava flow. The velocity may be expected to depend not only on y, but also on the downslope component of the lava's weight. For a unit volume of lava, this is equal to the density ρ times the component of gravity acting downslope gS, or γS, where γ is the specific or ("unit") weight of the lava. The depth d should also be a factor, because it determines the total component of weight acting on the bed. It has been observed that lava generally flows like a viscous fluid, so the viscosity μ is also important. The first step in the analysis is to list the variables and their dimensions, as shown in the following table.

Variables	Dimensions
velocity, u	$[LT^{-1}]$
height above plane, y	$[L]$
depth, d	$[L]$
downslope component of specific weight, (γS)	$[ML^{-2}T^{-2}]$
viscosity, μ	$[ML^{-1}T^{-1}]$

The dimensions of viscosity are obtained from the definition of viscosity as the coefficient in Newton's law of viscosity, which relates the stress tending to shear a fluid to the rate at which it shears:

$$\tau = \mu \frac{du}{dy} \tag{3.7}$$

where τ is the shear stress (Chapter 9 gives a fuller discussion of viscosity). Though we are here using mass as one of the fundamental dimensions, the units of viscosity can also be expressed as a pressure times time. As this is simpler, the SI unit is commonly given as one pascal second (Pa s).

We can state that there must be some kind of functional relationship between these variables. We write this formally:

$$f(u, y, d, (\gamma S), \mu) = 0 \tag{3.8}$$

It is possible to rearrange the variables to form combinations (products) that are dimensionless. An example is y/d (or d/y).

A set of such dimensionless products is said to be *complete* if each product in the set is independent of the others i.e., it cannot be expressed as a product of others in the set, and every other dimensionless product can be expressed as a product of the dimensionless products in the set. The *Buckingham pi theorem* states that if an equation is dimensionally homogeneous, it can be reduced to a relationship among a complete set of dimensionless products.

It is not obvious that this is true, but it can be proved, though it is a curiosity of the history of dimensional analysis that many published "proofs" of the pi theorem have subsequently been shown to be incomplete or invalid.

It is a consequence of the pi theorem that, if the original N variables involve n dimensions, it is generally possible to reduce the relationship to a function of $N - n$ dimensionless products.

Though the pi theorem guarantees a complete set of dimensionless products, it does not state that the set is unique. In general, there is more than one way to reduce a list of dimensional variables to a complete set of dimensionless products, so different choices are possible. In order to make an appropriate choice of products, it is best to proceed in a systematic way.

(1) List all the variables and their dimensions (as in the table).

(2) Select three variables (the *repeating variables*) which have independent dimensions, i.e., no one variable can be expressed as power products of the other two. One safe way to do this is to add a new dimension with each new basic variable: e.g., in the list above we could choose:

$$d([L]), \ u([LT^{-1}]), \ \mu([ML^{-1}T^{-1}])$$

However, another rule is that, if at all possible, the basic variables should not be those in whose behaviour we are most interested. In our example, we might be most interested in the way in which the velocity u, varies with perpendicular distance y from the base of the flow. If this were the case, we should not choose u as one of the repeating variables. Instead, we could select (γS). Although (γS) and μ both contain all three of M, L, and T, we can easily see that we cannot combine either one with d in any combination which will give the dimensions of the other.

For example, $(\gamma S)/\mu$ has dimensions $[ML^{-2}T^{-2}]/[ML^{-1}T^{-1}] = [L^{-1}T^{-1}]$ and no combination of (γS) and μ can be made which will give the dimensions of d, i.e., $[L]$. So d, (γS) and μ have independent dimensions.

(3) Use these three repeating variables ($n = 3$ in this example, but if only two dimensions were involved, only two repeating variables would be needed) in combination with each of the remaining variables to make them dimensionless.

For example, combine μ, (γS) and d with u to make a dimensionless product. If we arbitrarily set the exponent of u to be 1, the product would have the form:

$$\mu^a(\gamma S)^b d^c u$$

so the dimensions would be:

$$[ML^{-1}T^{-1}]^a[ML^{-2}T^{-2}]^b[L]^c[LT^{-1}]$$

For the products to be dimensionless the exponent of each of [M], [L], and [T] must be zero, so that

$$
\begin{aligned}
a + b &= 0 && \text{from the exponents of [M]} \\
-a - 2b + c + 1 &= 0 && \text{from the exponents of [L]} \\
-a - 2b - 1 &= 0 && \text{from the exponents of [T]}
\end{aligned}
$$

Add the equations for [M] and for [T] (*a* cancels out):

$$
\begin{aligned}
-b - 1 &= 0, \quad \text{so } b = -1 \\
a &= 1 \\
c &= -2
\end{aligned}
$$

So, the product required is

$$\mu u/(\gamma S)d^2$$

We might choose to call this "the dimensionless velocity".

Similarly, the other product is found to be y/d ("the dimensionless height above the plane").

So instead of the original functional relationship between five variables having dimensions,

$$f(u, y, d, (\gamma S), \mu) = 0$$

we can write instead the following functional relationship between two dimensionless variables:

$$f\left(\frac{\mu u}{(\gamma S)d^2}, \frac{y}{d}\right) = 0 \tag{3.9}$$

It is arbitrary whether we chose y/d or d/y, though we prefer y/d, because y is the variable of interest, which we have "scaled" by using the depth, d. Also it is common, but by no means universal, practice to avoid constructing products using fractional exponents of variables.

Dimensional analysis does not reveal the exact nature of the functional relationship between the two dimensionless variables: this must be determined by experiment. It does greatly reduce the number of experiments required. Suppose we wish to determine the dependence of u on $y, (\gamma S), \mu$, and d (and suppose we substitute some more easily handled viscous fluid for red-hot lava!), then we might measure u at 10 elevations above the bed for a given combination of $(\gamma S), \mu$, and d. Then we should really do the same for a series

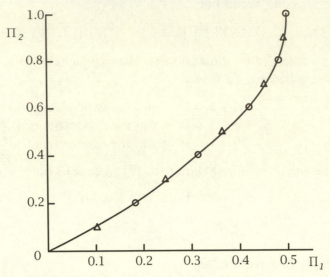

Fig. 3.1. Result of experiments on the flow of a viscous fluid down an inclined plane.

of different values of d – and also for different values of μ and (γS). If we investigated 10 values of each variable we would need 10^3 experiments, each with 10 observations – a total of 10^4 observations. If we trust the results of our dimensional analysis, however, we need only observe two dimensionless variables.

Figure 3.1 shows the (hypothetical) results of experiments to determine the functional relationship in the example analyzed above. For example, we might have investigated fluids of two different viscosities (e.g., two grades of motor oil, indicated by circles and triangles) and varied the depth and slope between two runs, making observations of velocity at five different elevations above the bed in each of the two runs (a total of only 10 measurements).

The functional relationship observed is a parabola. Setting

$$\Pi_1 = \frac{\mu u}{(\gamma S)d^2}$$
$$\Pi_2 = \frac{y}{d}$$

the curve has the equation:

$$\Pi_1 = \Pi_2 - 0.5\Pi_2^2 \tag{3.10}$$

The significant observation, which verifies the validity of the dimensional analysis, is that all the observation points fall along a single curve, even though several of the original five variables were not held constant.

Note that, to verify the dimensional analysis, we must vary most of the primary variables in the experiments, but we do not need to vary them all systematically by pairs while holding the others constant. Of course, the form of the function determined by experiment may only be true for the range of conditions that have been investigated.

In this example, we would find that the velocity distribution would not be parabolic if the viscosity were to become too low, or the depth too large, or the velocity too high: this is because the nature of the flow would change from viscous to turbulent. A perceptive experimentalist would be able to see that there was more than one "regime" of flow, separated by fairly abrupt transitions. We will see later that the condition for the change from viscous to turbulent flow, though it cannot be defined uniquely by any one of the original variables, can be defined by a single dimensionless variable. This variable is called the *Reynolds number*, and will be discussed again later, when we consider fluid resistance.

If the degree of scatter of experimental points about a single curve is more than expected from errors of observation, it probably indicates that a significant variable has been left out of the dimensional analysis, so that there has been uncontrolled variation from experiment to experiment. For example, in the analysis given above, we did not include the fluid density as a separate variable (it is included only in combination with g and S in γS). If the flow becomes turbulent, the functional relationship between Π_1 and Π_2 changes, so that velocities measured under turbulent flow conditions would not lie along the same trend. This would give rise to scatter. The scatter would be reduced by adding fluid density to the list of variables, which would produce a third dimensionless product including density. The physical reason why density must be included in turbulent flows is that such flows are characterized by large accelerations. Different forces are required to produce similar accelerations in fluids of different density, so density must be included. It can be neglected only for motions ("creeping flows") in which the accelerations produce negligibly small forces, i.e., where the inertia forces are negligible (Section 2.2).

Alternatively, we can make the problem simpler if we are not interested in all the variables. For example, if we are not interested in the velocity distribution (i.e., the variation of u with y) but only in the relation of the *average velocity U*, to d, (γS) and μ, we can simplify our original equation to the following

$$f(U, d, (\gamma S), \mu) = 0$$

and dimensional analysis gives

$$f\left(\frac{U\mu}{(\gamma S)d^2}\right) = 0$$

This equation states that the dimensionless product in the brackets does not vary for the problem at hand. So the equation can be written as

$$\frac{U\mu}{(\gamma S)d^2} = C$$

or

$$U = C\frac{(\gamma S)d^2}{\mu} \tag{3.11}$$

where C is a constant.

In this case, the problem has been practically solved, just by dimensional analysis. All that is left is to verify the result and determine the constant by experiment (the constant turns out to be 1/3 for viscous flow).

3.3.2 Example 2: The ballistic problem

We return to this problem to compare the results already obtained in Chapter 2 with those that can be obtained using only dimensional analysis.

We will try to obtain an expression for the horizontal distance travelled by a projectile, shot upwards at an angle θ to the horizontal, at a speed, v. The variables and their dimensions are given in the table.

Variables	Dimensions
horizontal distance, x	[L]
speed, v	$[LT^{-1}]$
gravitational acceleration, g	$[LT^{-2}]$
angle, θ	

As θ is dimensionless it must be one of the dimensionless variables. The other three variables involve only two dimensions, and can therefore be reduced to a single dimensionless product, which is easily found from inspection, so we have

$$f(v^2/xg, \theta) = 0$$

This is as far as one can go by dimensional analysis. Using the theory given in Chapter 2 (Equation 2.28), we could immediately derive the result that

$$v^2/xg = 1/\sin 2\theta$$

for the case where the projectile hits the ground at elevation $z = 0$ (e.g., Smith and Smith, 1990, p. 91).

Thus the actual function f is $v^2/xg - 1/\sin 2\theta$. If we were to take account of air resistance, we would have to include among the variables the density and viscosity of the air, because these produce a drag force on the projectile (see next example). We would then have three dimensionless products that needed experimental investigation.

3.3.3 Example 3: Fluid resistance (drag)

Suppose we want to study the resistance (drag) acting on a projectile. We are concerned, therefore, with a drag force F_D. It depends on the relative velocity between the fluid and the body, and will be a function of the size of the body, and the properties of the fluid. It depends on viscosity μ because that determines the frictional drag on the surface of the projectile, and on the density ρ because that determines the pressure at the stagnation point (see Chapter 2). The variables and their dimensions are given in the table.

Variables	Dimensions
drag force, F_D	$[MLT^{-2}]$
fluid velocity, u	$[LT^{-1}]$
size of solid body, D	$[L]$
density of fluid, ρ	$[ML^{-3}]$
viscosity of fluid, μ	$[ML^{-1}T^{-1}]$

We have

$$f(F_D, u, D, \rho, \mu) = 0$$

Which variables are we interested in? In this case, the answer is rather arbitrary: one possible answer would certainly be the drag force and its relation to viscosity.

If this is our choice then we designate the other three variables as the repeating variables. Do they have independent dimensions? Obviously yes, because only ρ has $[M]$, only u has $[T]$ and so it is not possible to construct the dimension $[L]$ of D from any combination of ρ and u.

A step-by-step elimination of dimensions soon gives the correct choice of dimensionless variables. For example, starting with the first variable we are interested in,

$$[F_D] = [MLT^{-2}]$$

- we divide by ρ to eliminate [M]:

$$[F_D/\rho] = [MLT^{-2}/ML^{-3}] = [L^4T^{-2}]$$

- then we divide by u^2 to eliminate [T]:

$$[F_D/\rho u^2] = [L^4T^{-2}/L^2T^{-2}] = [L^2]$$

So the required dimensionless product must be $F_D/(\rho u^2 D^2)$.

A similar process soon yields the second product and the result

$$f\left(\frac{F_D}{\rho u^2 D^2}, \frac{\mu}{\rho u D}\right) = 0 \qquad (3.12)$$

For historical reasons, this result is generally written:

$$f(C_D, Re) = 0 \qquad (3.13)$$

where

$$C_D = \frac{F_D}{(\rho u^2/2)A} \qquad (3.14)$$

$$Re = \frac{\rho u D}{\mu} \qquad (3.15)$$

and A is the cross-sectional area of the projectile (proportional to D^2). The product $\rho u D/\mu$ is called the *Reynolds number Re*, after the Irish engineer, Osborne Reynolds (1842–1912), who first showed that the transition from laminar to turbulent flow generally takes place at a critical value of the Reynolds number (see Section 11.2). There are now a large number of named dimensionless products – for an extensive list see Catchpole and Fulford (1973).

In Chapter 2 (Section 2.11) we derived the form of the drag coefficient C_D from simple energy considerations, and indicated that experiments show that it is constant over a wide range of flow conditions. The same result could have been obtained by dimensional analysis, if we had ignored the fluid viscosity. With this assumption, there are only four variables with dimensions, and only one dimensionless variable C_D is required to represent them, so it must be a constant. If we include viscosity in our list, the drag coefficient cannot be a true constant, but must be a function of a dimensionless number that includes the viscosity (the Reynolds number). This is confirmed by experiment, which gives the specific form of the function shown in Figure 3.2.

There is a certain range of Reynolds numbers (10^4 to 10^5) for which the drag coefficient of many shapes is roughly constant. Observation of the

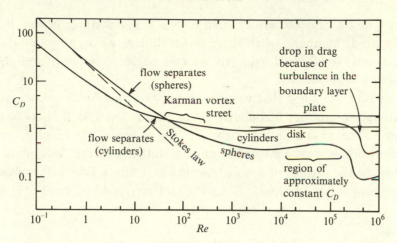

Fig. 3.2. Drag coefficient versus Reynolds number, for several different shapes of a body moving through a fluid.

experiments shows that in this range, there is a well-developed turbulent wake. In contrast, at very low Reynolds numbers there is no turbulent wake, the flow is laminar and there is a definite relationship between drag coefficient and Reynolds number. In this range, for spheres, we have Stokes' law:

$$C_D = 6/Re \tag{3.16}$$

This law, named for G. G. Stokes (1819–1903), who gave the first theoretical derivation, can also be written as

$$F_D = 3\pi\mu Du \tag{3.17}$$

or, in dimensionless terms,

$$\frac{F_D}{\mu Du} = 3\pi$$

Notice that one of our original variables, fluid density, does not appear in this product. Stokes' law could easily have been derived from dimensional analysis by neglecting density.

Although we cannot know the exact form of the function relating drag coefficient to Reynolds number, except by experiment, we could have made some progress in deducing its shape just from dimensional analysis. The Reynolds number can be interpreted as a ratio between two types of force: the inertia forces (due to accelerations), which are proportional to the density, and the forces due to friction, which are proportional to the viscosity. It would be a reasonable guess that at low Reynolds numbers, for laminar flows, we could neglect density (and so we could deduce Stokes' law). At high Reynolds numbers, when there is a turbulent wake, we guess that drag is dominated by inertia forces, so we can neglect viscosity, and the drag coefficient is a constant.

In actuality, the guess that we can neglect viscosity for Reynolds numbers high enough to produce turbulence is not strictly true. At Re between about 10^5 and 10^6 there is a sudden drop in drag (Figure 3.2) which is an effect dependent on viscosity (development of turbulence in the "boundary layer" very close to the surface of a body moving in a fluid). For further discussion of this example, see Section 11.2, and Middleton and Southard (1984).

3.3.4 Example 4: Impact craters

What determines the size of meteorite craters? We might be tempted to list a large number of factors, but by using some knowledge of the phenomena themselves and their basic physics, the list can be drastically reduced. For example, it is known that large meteorites strike planets at very high speeds, so that the mechanism of crater formation is essentially one of penetration, producing intense compression of both the meteorite itself and the "target" (the rock at the impact site), followed by extremely rapid decompression, that is by an intense explosion. So meteorite craters are much like craters excavated by large, near-surface explosions (produced by TNT or nuclear devices). It can be concluded that the most important property of the meteorite is its kinetic energy $E = mu^2/2$, and that the angle of approach, etc., are of minor importance. The size of the crater can be measured by its diameter D, and depends on how the impact energy is dissipated. It might be used up mainly in throwing rock out of the crater, against gravity, in which case D depends on the specific weight of the rock, γ, and the volume thrown out. If we assume that all craters of the same size have the same shape, then the volume is simply a function of the size. In this case, the three variables E, D and γ can form only one dimensionless product, so:

$$\frac{E}{\gamma D^4} = \text{constant} \qquad (3.18)$$

Since the specific weight of rock does not vary much on a given planet, the diameter should be directly proportional to the fourth root of the kinetic energy, and craters should be larger on smaller planets, where g is smaller (we assume that the gravitational attraction of the planet does not significantly affect the approach velocities of the faster meteorites.)

Things are not that simple, however, because a good part of the energy is used not to throw rock out of the crater, but to creat seismic (shock) waves: their energy depends on the density of the rock, and the speed of the waves c (Chapter 8). If all the energy were used to create shock waves, therefore, the significant variables would be E, ρ, c and D. This list includes variables with all three of the fundamental dimensions [M], [L], and [T], and so it can be reduced to a single dimensionless product:

$$\frac{E}{\rho c^2 D^3} = \text{constant} \qquad (3.19)$$

The analysis can be considerably elaborated (Holsapple and Schmidt, 1987) but it is not surprising that studies of experimental craters show that the diameter varies as a power of the energy that lies between 1/4 and 1/3.

A final word of caution on the use of dimensionless ratios for representing experimental results. Though they have many advantages, there is one serious disadvantage. Spurious results may be obtained if statistical techniques and tests (e.g., linear regression, commonly used for fitting a straight line to data plots) are applied to ratios rather than to the original data. The reason is best seen from an example. Suppose we have three variables x, y, and z, all with the same dimensions, and suppose further that x and y do not vary much (and are uncorrelated) but z varies a great deal, in a way that is not related to x and y. If we plot x/z versus y/z, then the plot will show a fairly linear line relationship, even if there is, in fact, no real relationship between any of the three variables. The reason is, of course, that we are essentially plotting C_1/z versus C_2/z, where the C's are nearly constant – in other words we are plotting the reciprocal of z against itself. The result is a straight line, with a slope of one. A more complete discussion of this pitfall is given by Chayes (1971), Yalin and Kamphuis (1971) and Waythomas and Williams (1988).

3.4 Scale models

An important use of dimensional analysis is to design scale models. In a true scale model not only is the geometry of the original reproduced, but all the forces involved remain in the same relative proportions to each other. In other words, the model is not only *geometrically similar* to the original but also *dynamically similar*. The ratio between a characteristic length in the model and in the original is called the *scale factor (sf)* for length. However, we may also want to choose a scale factor for (say) velocity that is different from that of length. The design problem for the model is then to set all the other scale factors so that the model is dynamically similar to the original.

3.4.1 Glacier model

Consider the simple case where we can reduce the original variables to a single dimensionless product, as in Example 1 above:

$$\frac{U\mu}{(\gamma S)d^2} = \frac{1}{3} \qquad (3.20)$$

Because the equation is dimensionless, it is true whatever the scale, i.e., whatever value d takes. We can write

$$\left(\frac{U\mu}{(\gamma S)d^2}\right)_{\text{original}} = \left(\frac{U\mu}{(\gamma S)d^2}\right)_{\text{model}} = \frac{1}{3} \qquad (3.21)$$

Suppose we want to make a model of a glacier, and we hypothesize that a glacier is essentially a viscous fluid flowing slowly down a plane. Taking values appropriate for a moderate size glacier,

$$d \approx 100 \text{ m}$$
$$(\gamma S) \approx 10^2 \text{ kg m}^{-2}\text{s}^{-2}$$
$$U \approx 1 \text{ m day}^{-1} \approx 10^{-5} \text{ m s}^{-1}$$

we can calculate the viscosity of glacier ice:

$$\mu \approx \frac{1}{3}\left(\frac{10^2 \times 10^4}{10^{-5}}\right) = \frac{1}{3} \times 10^{11} \text{ Pa s}$$

As water has a viscosity of about 10^{-3} Pa s, this is a very high viscosity.

In the model we would like d reduced to 0.1 m (i.e. reduced by a scale factor of 10^{-3}). We also want to speed up the flow, so that it can be observed over a short time. So we set U to be 10^{-3} m s^{-1} (it has thus been increased by a scale factor of 100). What would the viscosity have to be in the model?

It is not possible to change (γS) much, for two reasons: most viscous liquids have specific gravities close to that of ice, and changing the slope S changes the geometry of the model – thus violating geometrical similarity. If (γS) is held constant, Equation (3.21) reduces to:

$$\left(\frac{U\mu}{d^2}\right)_{\text{glacier}} = \left(\frac{U\mu}{d^2}\right)_{\text{model}} = \frac{1}{3} \qquad (3.22)$$

Rearranging, and using subscripts g and m to refer to "glacier" and "model" respectively:

$$\frac{\mu_m}{\mu_g} = \frac{U_g}{U_m}\left(\frac{d_m}{d_g}\right)^2 = \frac{(sf \text{ for } d)^2}{(sf \text{ for } U)} = 10^{-8}$$

So the viscosity in the model should be 10^{-8} times its value in the glacier, or about $(1/3) \times 10^3$ Pa s, which is about the viscosity of a very thick oil (see Appendix B).

3.4.2 Tectonic models

Hubbert (1945) gave another example: suppose we are interested in building a model of a block of the earth's crust about 1000 km (10^6 m) across. The model is to be 0.5 m across. What sort of material should we use?

The major variables controlling the behaviour of the crust are assumed to be the scale (designated by some characteristic length l), the strength of the crust σ_s, gravity g, and density ρ. These can be grouped into a single dimensionless product:

$$\frac{\sigma_s}{\rho g l} = \text{constant} \tag{3.23}$$

Gravity cannot easily be varied from its value at the earth's surface. The density ρ of natural materials varies by only a small amount. So eliminating these two variables, we can write

$$\left(\frac{\sigma_s}{l}\right)_{\text{earth}} = \left(\frac{\sigma_s}{l}\right)_{\text{model}} \tag{3.24}$$

The crushing strength of granite is about 2×10^8 kg m^{-1} s^{-2}. The scale factor for length is $0.5/10^6$, so the corresponding strength of the model material should be about 100 kg m^{-1} s^{-2}.

A pressure large enough to exceed this strength would be produced by a very thin model. The model thickness h that produces this pressure is given by $\rho g h = 1000$ kg m^{-1} s^{-2}. Choosing appropriate values ($\rho = 2000$ kg m^{-3}, $g \approx 10$ m s^{-2}) we calculate $h = 0.005$ m (0.5 cm). In other words a column of the required material would not be able to support its own weight if it were more than 0.5 cm thick. This is obviously a very weak material: perhaps a rather sloppy mixture of clay and water. This makes clear why such weak materials are commonly used in experiments on the formation of large geological structures.

Using such materials does raise experimental difficulties. One problem is that we are interested not only in the strength of the material but also in its other properties, such as viscosity. More generally, we want the way in which the model materials respond to relatively low stresses (at low temperatures), to be similar to the way in which rocks respond to much higher stresses (at higher temperatures). In other words, we are interested in *rheological similarity* (see Chapter 10). Modern experimental work on large structures has focussed on the use of large centrifuges, which generate centrifugal accelerations of as much as 10 000 *g*. A centrifuge with a diameter of 1 m permits experimentation on a model several centimetres across (Dixon and Summers, 1985). A search has been made for model materials, such as plasticine and silicone putty, whose properties are suitable for use in such experiments (Weijermars and Schmeling, 1986).

One further point may be made. We generally assume an equivalence between mass and force, because of Newton's second law. If motion is very slow, however, as it is in large-scale tectonic processes, it may be that inertia forces are negligible. Nevertheless, mass (and density) may still be important because of gravity, and there are still frictional forces, etc., to consider. In this case [M] and [F] may be treated as independent dimensions, i.e., there are four fundamental dimensions, not three. In dimensional analysis, therefore, we may hope to reduce the number of original variables by four in this case (this is the approach taken by Hubbert). Alternatively, we may consider that mass, as such, is not really important to the physical problem: it is only mass times the acceleration due to gravity, i.e., weight, that is important. This is what we did in a previous example, and as it reduces the number of variables by one, while keeping the number of fundamental dimensions as three, it leads to the same result.

Other applications to earth-science problems and models are listed in the bibliography.

3.5 More on modeling

The word "model" has been used in science with a large range of different meanings. One hears of laboratory, or scale models, of physical models, of mathematical models, of numerical models, and of conceptual models. Many books discuss the "art" of modeling (e.g., Harte, 1985). What do the different kinds of model have in common? Some writers maintain that scientific theories are themselves "models" of reality. If this is so, how is the "art of modeling" different from the old-fashioned scientific method?

A scale model is an attempt to reproduce some phenomenon at a different physical scale – but in reality it is generally less than that, because very few modelers are interested in reproducing *all* the aspects of the original phenomenon. Some things are ignored, and some are simplified. In constructing a model of a glacier by using a viscous fluid, we chose to ignore known properties of glacier ice that are different from those of viscous fluids: ice is a polycrystalline solid that can fracture and slip over a solid surface, and a true fluid cannot behave that way. Though we may be forced to build our model from materials that are simpler than the original materials, we must also be aware that some model materials may have properties that are not present (or not so important) in the original. For example, in choosing a soft mud to model continents, we paid attention only to the property of strength: but muds and continents have other properties such as elasticity, cohesion (tensile strength) and viscosity, that may not be properly scaled in a small model, and that may produce some effects not seen at larger scales and fail to produce others that are observed.

Not all physical models are scale models; some are built to the same physical dimensions as the prototype. Examples include wind tunnels that are large enough to test full-scale airplanes and flumes that are the size of small rivers. The model still provides the advantages of simplicity and explicit control of the independent variables, but without the difficulties and simplifications that are often involved with scale models. The primary drawback is the relatively large expense.

The fact that a model is constructed at all generally means that the original is incompletely understood – and therein lies the "art" of modeling. Success depends on the "physical intuition" or experience of the modeler. He or she must know which parts of nature may be isolated for study, and which aspects of the original must be included in the model and which may safely be ignored. In dimensional analysis, as we have seen, success depends largely on identifying the main physical variables, and excluding those that are less important.

What has been written of scale models might equally well be written of the other types of model: all depend on some degree of simplification of reality. A mathematical model of a glacier produced using the theory of viscous fluids is generally more highly simplified than a physical model, because the equations can be derived only for simple boundary conditions.

Some of the complexity of the physical model may be unrepresentative of reality and, therefore, undesirable. The relative uses of physical and numerical models are changing as numerical techniques and computers become ever more powerful. Numerical models of large bays and estuaries, and drainage systems, are generally found to be less expensive and more flexible than scale models. Field observations are needed to verify that the model provides reasonable results, whatever type of model is used.

In a mathematical or numerical model, the equations *alone* do not constitute the model. A differential equation, for example,

$$dz/dt = -gt + w_0$$

(see Section 2.7) does not constitute a model of the real world, until it is linked to the world conceptually, that is by an idea. In this example, the model is the idea that the vertical flight of a volcanic bomb can be represented as the motion of a point mass in a constant gravitational field. Using this idea, we can easily derive the equation, and it is the idea that gives physical meaning to the symbols used, and makes the equation more than just a special example of a simple class of differential equations.

We can distinguish not only between models and equations, but also between models and theories. Models simplify reality, but they generally elaborate theories, using concepts which are extraneous to the theory itself. Models are therefore not theories, but physical or mental constructs, from which theories may be deduced or which may be used to evaluate theories. Even if a successful theory is deduced from a model, however, the model may still be quite misleading. For example, Navier originally deduced the Navier–Stokes equations for a viscous fluid (which will be derived in Chapter 9) from a mechanical model of a fluid that is now known to be erroneous. Our present model of a viscous fluid is quite different from Navier's, but his equations of motion remain. Generally, we prefer models that are consistent with, or derived from, more general models and theories. Glacier flow models should be consistent with what is known about mechanics and the behaviour of polycrystalline ice: they should not be simple "ad hoc" models that have been constructed only from macroscopic observations of a few glaciers.

One test of a successful model is how well it reproduces the essential phenomena that are being modeled: that is, how good are the predictions yielded by the theories embedded in the model? We are particularly impressed if theories can be deduced from a model, and verified by observation, which were not anticipated at the time that the model was originally constructed. It is also important that theories deduced from models should yield predictions which are precise and can be verified or falsified. Some models, for example, those developed from Newtonian mechanics for the movement of the planets, yield predictions that are very precise and have been accurately verified by repeated observations. Other models, even if they lead to mathematical equations, give predictions of limited precision and accuracy (e.g., general circulation models of the atmosphere). Yet other models (e.g., "facies models") are almost completely qualitative, and thus very hard to falsify. In such models, art may be more important than science.

One characteristic of successful models is that they deal with aspects of nature which can be isolated from the whole. Much of nature is not like this, but consists of many different parts, each subject to different physical laws, but each connected to the others by complex "feedback" relationships. In such cases, we are concerned more with the whole *system* than with particular models. Landscape is an example: models might be developed to explain many particular features (soil, slopes, river channels, etc.) yet we would still not be able to explain how the whole landscape develops. Conversely, it may be possible to explain some general properties of landscapes, even if all the component models in the system are not fully understood (e.g., Huggett, 1985).

Even inadequate models are generally better than none at all, since they may be refined into models which are better founded conceptually, and which yield more quantitative and thus more rigorously tested predictions. Models, like scientific hypotheses, generally express our present level of understanding of the world. If they are inadequate it is probably either because our understanding is limited, or because we are attempting the impossible (perhaps some phenomena, such as the weather, simply cannot be accurately predicted).

A few models are interesting in themselves, even though it is known that they do not correspond closely to reality. This is because they exhibit complex behaviour, which may be similar in general (though not in particulars) to phenomena observed in nature. A good example is the Lorenz model of thermal convection, which will be discussed in Chapter 12. This was originally proposed as a simple mathematical/numerical model of the atmosphere. Though it has many deficiencies in this respect, study of this model has played an important part in the development of the modern science of nonlinear dynamics.

3.6 Uses of dimensional analysis

The following list of the uses of dimensional analysis also summarizes much of this chapter.

(1) All physically correct equations must be dimensionally balanced. This can be of considerable use when checking the results of computations (including those made to solve assignments).

(2) Dimensional analysis is used to reduce the number of variables to be investigated in any physical problem by the number of independent dimensions in the problem. The governing equations for many physical problems often do not readily yield analytical or numerical solutions. Sometimes, it is difficult even to formulate the correct governing equations. There are also many occasions when the general form of the equations is known, but the problem is so complex that some empirical observations are needed to determine particular coefficients in the equations. In all these cases, laboratory or field observations are needed and

the expense of this exercise depends directly on the number of variables to be investigated. A careful dimensional analysis serves to reduce this list. In addition, dimensional analysis provides a useful means of organizing and controlling experiments, so that one is careful to minimize the variation in potentially important variables while investigating the effects of changing other variables.

(3) Dimensional analysis provides scaling laws for model experiments. Without scaling laws developed from dimensional analysis, we could not demonstrate that scale models provide both geometric and dynamic similarity.

(4) Dimensional analysis serves to organize our thinking about a physical problem. It not only reduces the number of variables investigated experimentally, but also gives useful insight into a physical problem. It may lead almost to the final solution of some problems, with measurement needed only to confirm the analysis and provide a numerical coefficient. Such thinking is particularly useful where physical models are difficult or impossible, or where neither time or money are available for experiments.

(5) Results presented in dimensionless form are directly generalizable and may be used with any consistent set of units. Although it is not true that any results presented in dimensionless form are generally valid, confusion and mistakes incurred in applying physical relationships derived from one location to another are much reduced when dimensionally consistent formulae are used and the related constants are dimensionless.

(6) The governing equations of any physical problem can be nondimensionalized, that is, expressed in a form in which all of the terms in the equation are dimensionless. Doing this gives the scaling relations, and helps us understand the conditions under which each of the terms is likely to be important. When the governing equations of a physical problem are nondimensionalized, one result is that dimensionless coefficients are found as coefficients of some of the terms.

For example, when the governing equations for a viscous fluid (the Navier–Stokes equations, Section 9.4) are nondimensionalized, one of these coefficients is $1/Re$, which is found as a coefficient multiplying the terms representing viscous forces in the fluid. This makes clear the nature of the Reynolds number as the ratio of inertia and viscous forces. It explicitly demonstrates the reduced importance of viscous forces at large values of Re: where Re is very large, the numerical values of the viscous terms are orders of magnitude smaller than other terms in the equation and can be neglected. Such reasoning must be applied with discrimination: in a large flow, we may have to consider the Reynolds number not just for a single, large, length scale, but also for small length scales typical of some important *parts* of the flow (e.g., in the layer close to a solid boundary). Such scaling arguments are frequently made and play a central role in attempts to simplify the governing equations so that they may be solved analytically or numerically.

3.7 Review problems

(1) **Problem:** It is observed that some strong currents carry high concentrations of grains, particularly close to the bed. Suppose that when a concentrated dispersion of grains is sheared, there is a force per unit area P, the "dispersive pressure", which tends to move the grains in the direction z, normal to the plane of shearing (the xy-plane). P is produced by collisions between grains, and should be related to the mass of individual grains, that is to their grain size D, and density ρ, and to the rate of shear du/dz.

(a) Use dimensional analysis to determine the nature of the function $P = f(D, \rho, du/dz)$.

(b) Grains forming part of a single lamina (in the xy-plane) should have been deposited under conditions of uniform P and du/dz. If grains of quartz ($\rho = 2650$ kg m^{-3}) and garnet ($\rho = 4200$ kg m^{-3}) are both present, and the quartz grains are 1 mm in diameter, what will be the size of the garnet grains?

(c) What would the diameter of the garnet grains have been if the lamina had been deposited by settling (rather than from a mass of shearing grains)? Assume that the grains in the lamina all sank through the fluid at the same constant *settling velocity*, and that the drag coefficient is a constant.

Answer: The dimensions of the variables are as follows:

$$\begin{array}{lll}
P & [ML^{-1}T^{-2}] & \text{(force per unit area)} \\
D & [L] & \\
\rho & [ML^{-3}] & \\
du/dz & [T^{-1}] & \text{(velocity divided by length)}
\end{array}$$

There are four variables, with three dimensions, so we should be able to reduce them to a single dimensionless group. First eliminate [M]:

$$[P/\rho] = [L^2T^{-2}]$$

To eliminate $[L^2]$ divide by D^2, and to eliminate $[T^{-2}]$ divide by $(du/dz)^2$. Then the dimensionless product is

$$\Pi = \frac{P}{\rho D^2 (du/dz)^2}$$

It follows that

$$P = (\text{constant})\rho D^2 (du/dz)^2$$

If we have used all the important variables, this constant should be the same for all fluids and grains. If P and du/dz are constant in a given lamina then

$$(\rho D^2)_{\text{garnet}} = (\rho D^2)_{\text{quartz}}$$

or

$$D_{\text{garnet}} = \sqrt{2650(0.001)^2/4200} \text{ m}$$
$$= 0.794 \text{ mm}$$

The settling velocity (also called the "fall velocity") is obtained by equating the submerged weight of a grain and the drag force acting on it. The submerged weight of the grain is proportional to the density, minus the density of water, and the cube of the diameter, so the ratio of weights for quartz and garnet is

$$\frac{\text{wt}_{\text{quartz}}}{\text{wt}_{\text{garnet}}} = \frac{(\rho_q - \rho_w)D_q^3}{(\rho_g - \rho_w)D_g^3} = \frac{1650}{3200 D_g^3}$$

The drag force is proportional to the diameter and the velocity, both squared: but we assume the two settling velocities are the same, so the ratio of the drag forces is simply

$$(F_D)_q/(F_D)_g = D_q^2/D_g^2 = 1/D_g^2$$

The drag-force ratio and the weight ratio must be the same to give the same settling velocity, so

$$\frac{1650}{3200 D_g^3} = \frac{1}{D_g^2}$$

and

$$D_g = 1650/3200 = 0.52 \text{ mm}$$

(2) **Problem:** In the text, we derived an equation for the flow of a viscous fluid down a slope. We assumed that density was only important when combined with gravity to produce the downslope component of specific weight (γS).

(a) Using dimensional analysis, derive an expression for the way in which the average velocity U depends on the depth d, if the flow is turbulent and density must now be considered independently of specific weight.

(b) Simplify this for the case where viscosity may be neglected, and compare your result with the Chézy equation.

Answer:

(a) The important variables and their dimensions are

$$U \ [LT^{-1}] \qquad d \ [L] \qquad (\gamma S) \ [ML^{-2}T^{-2}]$$
$$\mu \ [ML^{-1}T^{-1}] \qquad \rho \ [ML^{-3}]$$

We are interested in U and d, so we want to choose the other three variables as repeating quantities. First, we verify that they have independent dimensions: $(\gamma S)/\mu$ has dimensions $[L^{-1}T^{-1}]$, therefore it is clear that no combination of γS and μ can give the dimensions of density. So these three variables do have independent dimensions. Now make a dimensionless product involving U:

$$\Pi_1 = \mu^a (\gamma S)^b \rho^c U$$

a, b, c are exponents to be determined, and the exponent of U is arbitrarily set to unity. Writing the dimensions gives

$$[\Pi_1] = [0] = [ML^{-1}T^{-1}]^a [ML^{-2}T^{-2}]^b [ML^{-3}]^c [LT^{-1}]$$

$$a + b + c = 0 \quad \text{from the exponents of [M]}$$
$$-a - 2b - 3c + 1 = 0 \quad \text{from the exponents of [L]}$$
$$-a - 2b - 1 = 0 \quad \text{from the exponents of [T]}$$

From the last two equations $c = 2/3$; from the first and last $b = -1/3$, so $a = -1/3$. Multiply through by 3 to remove fractions. Then the required product is

$$\Pi_1 = \mu^{-1} (\gamma S)^{-1} \rho^2 U^3$$

Similarly, the dimensionless product involving d is

$$\Pi_2 = \mu^{-2} (\gamma S) \rho d^3$$

So, the final result is

$$f\left(\frac{\rho^2 U^3}{\mu(\gamma S)}, \frac{(\gamma S)\rho d^3}{\mu^2}\right) = 0$$

(b) If μ can be neglected we need to eliminate it from these two dimensionless products. Since we are reducing the number of variables to four, we should be able to use a single dimensionless product, which will be equal to a constant. The easiest way to do this is to divide Π_1^2 by Π_2:

$$\Pi = \frac{\Pi_1^2}{\Pi_2} = \frac{\rho^3 U^6}{(\gamma S)^3 d^3}$$

We could have arrived at the same result by combining the original four variables. Dividing the exponents by three and solving for U gives:

$$U = (\text{constant})(gdS)^{1/2}$$

This compares with the Chézy equation:

$$U = C(RS)^{1/2}$$

where $R = d$ for a two-dimensional flow down a plane. It is, in fact, the *dimensionless form* of the Chézy equation discussed in the text.

(3) **Problem:** In making hydraulic models (for example, small-scale models of rivers or ships) it is generally considered to be very important to observe Froude number scaling. The *Froude number* is given by

$$Fr = U/\sqrt{gd}$$

where d is a characteristic length, such as the depth, or the size of the ship. The Froude number, named for the English engineer William Froude (1810–79: the name is pronounced "Frood") is important because it is the ratio between a characteristic velocity, such as the speed of a ship, and the speed of any (long) surface waves that may result from a free-surface flow, such as the waves seen in the wake of a ship (Subsection 9.4.2).

(a) What does Froude scaling imply about the velocity, if the scale of the model is 1/100 that of the original?

(b) Suppose we want a model that is scaled by both Froude and Reynolds number, where the Reynolds number is

$$Re = \rho U d/\mu$$

The Reynolds number is important because it is proportional to the ratio between inertia and viscous forces.

If g is held constant, what does this imply about the density ρ, and viscosity μ, of the model fluid, compared with the original fluid (water)? If the length scale factor is 1/100, is it possible to find a liquid with these properties?

Answer:

(a) Froude scaling means that

$$\left(\frac{U}{\sqrt{gd}}\right)_{\text{model}} = \left(\frac{U}{\sqrt{gd}}\right)_{\text{original}}$$

If g cannot change, this implies that

$$(U_m/U_0)^2 = d_m/d_0 \qquad\qquad (3.25)$$

So if the length ratio is 1/100, the velocity ratio must be 1/10.

(b) Reynolds scaling means that

$$(\rho U d/\mu)_{\text{model}} = (\rho U d/\mu)_{\text{original}}$$

or

$$\frac{d_m}{d_0} \frac{U_m}{U_0} = \frac{\mu_m}{\mu_0} \frac{\rho_0}{\rho_m}$$

Substituting from Equation (3.25) gives

$$\left(\frac{d_m}{d_0}\right)^{3/2} = \frac{\mu_m}{\mu_0} \frac{\rho_0}{\rho_m}$$

For a length scale of 1/100, this gives

$$(1/100)^{3/2} = \frac{\mu_m}{\mu_0} \frac{\rho_0}{\rho_m}$$

or

$$(\mu_m/\mu_0) = 10^{-3}(\rho_m/\rho_0)$$

Almost all common liquids have densities that are close to that of water (within a factor of two). So one is looking for an experimental liquid with a viscosity some 1000 times less than that of water. Such liquids do not exist. For example, gasoline (a low-viscosity liquid) has a viscosity only about one third that of water.

The consequence is that it is impossible in practice to achieve Froude and Reynolds number scaling simultaneously. So, open-channel flow models, in which both are potentially important, never show true dynamic similarity to the originals. Luckily, most forces are not a strong function of the Reynolds number, once it is well above a critical value (i.e., above about 10^5). This value is easily attained in scale models, provided they are not made too small.

(4) **Problem:** Rock salt has a density considerably less than most other rocks: it is also capable of flowing under pressure in a manner similar to that of ice. Layers of salt deeply buried in sedimentary basins tend to flow and rise through the surrounding sedimentary rocks as dome-like masses called "diapirs". We can use the height L of the diapir above its "parent" salt layer to indicate the length scale. The other important variables are the buoyancy of the diapir, per unit volume, $\gamma' = \delta\rho g$, where ρ is the density of the salt and $\rho + \delta\rho$ is the density of the surrounding rocks, the viscosity of the salt μ, and the speed of rise v (or characteristic time $t = L/v$). In salt domes, L is about 1 km, μ about 10^{15} kg m^{-1} s^{-1}, and the density of salt is about 2100 kg m^{-3} and of other sedimentary rocks about 2400 kg m^{-3}.

(a) If only the above variables are important, what can we deduce about salt domes? Evaluate the constant by making use of the geological observation that we need about one million years (3×10^{13} s) for salt domes to form.

(b) If we make a model only 0.1 m in size, and place it in a centrifuge where we can achieve accelerations of 100g, and we want to see the model salt dome grow in 10^4 s (about three hours) what properties should the model material have?

(c) Why would it be unrealistic to try to build a model of a salt dome without using a centrifuge?

(d) What do you think is the most important variable omitted from the analysis given above?

Answer:

(a) Simple dimensional analaysis leads to

$$\Pi = \gamma' L^2 / \mu v = \text{constant}$$

We can evaluate the constant by entering estimated values for real salt domes:

$$\frac{300 \times 10 \times 10^6}{10^{15} \times 1000/(3 \times 10^{13})} = \text{constant} \approx 10^5$$

(b) In the model $L = 0.1$ m, and $v = 0.1/10^4 = 10^{-5}$ m s^{-1}. The model law is

$$\left(\frac{\gamma' L^2}{\mu v}\right)_{\text{model}} = \left(\frac{\gamma' L^2}{\mu v}\right)_{\text{original}}$$

Rearranging, and using the values of L and v set in the problem, gives

$$\frac{(\gamma'/\mu)_m}{(\gamma'/\mu)_0} = \frac{(L^2/v)_0}{(L^2/v)_m} = 3 \times 10^{13}$$

$\delta\rho$ cannot be changed much from the original to the model, so assume they are the same. Then

$$\gamma'_m/\gamma'_0 = g_m/g_0 = 100$$

and

$$\mu_m/\mu_0 = 100/(3 \times 10^{13}) \approx 3 \times 10^{-12}$$

Using $\mu_0 = 10^{15}$ Pa s gives $\mu_m = 3300$ Pa s (kg m^{-1}s^{-1}). This is a reasonable viscosity for a model material – it could be achieved by using a mixture of oil and putty and would make it possible to cut up the model after the experiment to observe the results.

(c) If we try to run the experiment without increasing g, however, then the ratio of viscosities changes from 3×10^{11} to 3×10^{13}, and the required viscosity becomes 33 Pa s, which is a pretty low viscosity (about that of honey) – it would make it difficult to cut up the model. The problem can be resolved by enclosing the experiment in a glass box, using transparent materials and taking stereo pictures to study the evolving shapes of the interfaces. It is not impossible to use low-viscosity fluids and small density differences, but it is easier to use materials that have higher viscosities. Note that we cannot really improve the situation much by increasing the density contrast in the model (instead of g), since common liquids do not vary much in density (and therefore in density contrast).

(d) The most important variable omitted is probably the strength (or perhaps viscosity) of the rocks surrounding the rising diapir, since this determines the resistance to the motion of salt through them.

3.8 Suggested reading

3.8.1 General works on dimensional analysis and physical models

Baker, W. E., P. S. Westine and F. T. Dodge, 1991. *Similarity Methods in Engineering Dynamics: Theory and Practice of Scale Modeling.* New York, Elsevier, revised edition, 384 pp. (Many fully discussed examples, including many related to the space program. Also included are discussions of impact phenomena and soil mechanics. Fairly advanced. TA177.B35)

Bridgman, P. W., 1921. *Dimensional Analysis.* Cambridge MA, Harvard University Press, reprinted by Dover. (Good discussion of fundamental principles. QC39.B85)

Kline, S. J., 1965. *Similitude and Approximation Theory.* New York, McGraw-Hill, 229 pp. (Advanced, sophisticated treatment. QA401.K65)

Langhaar, H. L., 1951. *Dimensional Analysis and Theory of Models.* New York, John Wiley and Sons, 166 pp. (Good practical text with two chapters on applications to stress and strain and to fluid mechanics. QC39.L35)

Macagno, E. O., 1971. Historico-critical review of dimensional analysis. Jour. Franklin Institute, v.292, pp. 391–402. (The history traced back to Fourier and Rayleigh.)

Sedov, L. I., 1959. *Similarity and Dimensional Methods in Mechanics.* New York, Academic Press, 363 pp. (Advanced, many examples from fluid mechanics. QC39.S44)

Yalin, M. S., 1971. *Theory of Hydraulic Models.* London, Macmillan, 266 pp. (Application of scaling theory to hydraulics, including sediment transport. TC163.Y35)

Yalin, M. S., and J. W. Kamphuis, 1971. Theory of dimensions and spurious correlation. Jour. Hydraulic Research, v.9, pp. 249–65.

3.8.2 Works on other types of model

Books on physical models are listed above. There are many books on mathematical models, including some that combine instruction in the art of modeling with an introduction to differential equations and numerical analysis.

Aris, R., 1978. *Mathematical Modeling Techniques.* San Francisco, Pitman Advanced Publishing Program, 191 pp. (Philosophical rather than practical. QA401.A68)

Harte, J., 1985. *Consider a Spherical Cow: a Course in Environmental Problem Solving.* Los Altos CA, William Kaufman, 283 pp. (Practical, examples include both physical and chemical models. QH541.15.M34H37)

Geographers, who span the boundary between the natural and social sciences, have an extensive literature dealing with the use of many different kinds of models. Some recent books are:

Huggett, R. J., 1985. *Earth Surface Systems.* New York, Springer-Verlag, 270 pp. (QE33.2.M3H75)

Thorn, C. E., 1988. *An Introduction to Theoretical Geomorphology.* Boston, Unwin Hyman, 247 pp. (GB401.5.T54)

Geologists have written much less on the theory of models in the earth sciences, but the following are exceptions:

Greenwood, H. J., 1989. On models and modeling. Canadian Mineralogist, v.27, pp. 1–14.

Von Engelhardt, W. and J. Zimmermann, 1988. *Theory of Earth Science.* Translated by Lenore Fischer. Cambridge University Press, 381 pp. (QE33.E5313)

Walker, R. G., 1990. Facies modeling and sequence stratigraphy. Jour. Sedimentary Petrology, v.60, pp. 777–86.

Almost all books on the philosophy of science discuss models.

Hesse, M. B., 1963. *Models and Analogies in Science.* University of Notre Dame Press, 184 pp. (An original and sophisticated discussion. Q175.H56)

O'Hear, A., 1989. *An Introduction to the Philosophy of Science.* Oxford, Clarendon Press, 239 pp. (A recent nontechnical introduction. Q175.O454)

3.8.3 Applications of dimensional analysis to the earth sciences

Church, M. and D. M. Mark, 1980. On size and scale in geomorphology. Progress in Physical Geography, v.4, pp. 342–91.

Dixon, J. M. and J. M. Summers, 1985. Recent developments in centrifuge modelling of tectonic processes: equipment, model construction techniques and rheology of model materials. Jour. Structural Geol., v.7, pp. 83–102.

Hubbert, M. K., 1937. Theory of scale models as applied to the study of geologic structures. Geol. Soc. America Bull., v.48, pp. 1459–520.

Melosh, H. J., 1989. *Impact Cratering: A Geologic Process.* Oxford University Press, 245 pp. (QB603.C7M45)

Ramberg, H., 1981. *Gravity, Deformation and the Earth's Crust.* New York, Academic Press, second edition, 452 pp. (Chapters 1 and 6 discuss scale models of tectonic phenomena. Much of the book describes the use of centrifuges to increase *g* in tectonic models. QE601.R25)

Southard, J. B., L. A. Boguchwal and R. D. Romea, 1980. Test of scale modelling of sediment transport in steady unidirectional flow. Earth Surface Processes, v.5, pp. 17–23.

Strahler, A. N., 1958. Dimensional analysis applied to fluvially eroded landforms. Geol. Soc. America Bull., v.69, pp. 279–300.

Weijermars, R. and H. Schmeling, 1986. Scaling of Newtonian and non-Newtonian fluid dynamics without inertia for quantitative modelling of rock flow due to gravity (including the concept of rheological similarity). Phys. Earth Planetary Interiors, v.43, pp. 316–30.

Willis, J. C. and N. L. Coleman, 1969. Unification of data on sediment transport in flumes by similarity principles. Water Resources Research, v.5, pp. 1330–6.

4

Stress

4.1 Introduction

In Section 2.2 we saw that it is useful to distinguish two classes of force: *surface forces* and *body forces*. Body forces act on every element of mass, so they are generally expressed as a force per unit mass, i.e., an acceleration. For example, the force of gravity per unit mass is the acceleration due to gravity *g*. Surface forces act along or across real or imaginary surfaces and are generally expressed as forces per unit area, or *stresses*. It is convenient to resolve a force acting at a surface into components acting normal to the surface and one or two components acting parallel with the surface. Dividing by the surface area yields the corresponding components of stress: a *normal stress*, or *pressure* acting normal to the surface, and one or two *shear stresses* acting parallel to the surface.

We begin our discussion of stresses by considering two examples that can readily be studied in laboratory experiments: (*a*) shear stress produced by friction between two solid bodies, such as masses of rock or soil; and (*b*) stress systems produced by compression of a solid body (Figure 4.1). It might be thought that the first example involves only shear stress, and the second only pressure, but both types of stress are involved in both examples.

Friction is a force that resists the sliding of two solid bodies past each other along some pre-existing surface, such as a fault, joint, or slide surface. To produce friction, the surfaces must be pressed together. In an experiment on friction, therefore, a normal stress is first applied across the slide surface, and then a shear stress is applied until sliding begins.

In a compression test, the sample is generally prepared as a cylinder with flat ends. The curved surface of the cylinder is compressed by a uniform "confining pressure" *P* and then a piston or "ram" is used to apply a higher "load pressure" *L* to the two flat ends of the sample. Thus all the applied

100

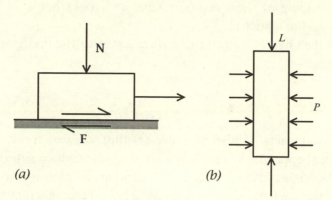

Fig. 4.1. Experiments on surface forces: (*a*) Friction at the base of a sliding block; (*b*) triaxial compression of a rock sample.

forces are normal forces: the result is first, a shortening of the sample along the axis of the cylinder and some expansion in the directions normal to the axis, and then failure of the sample by shearing along planes that are generally oblique to the axis. Application of unequal normal forces therefore produces shear forces along inclined surfaces within the sample, and when these shear forces exceed the strength of the material it breaks, yielding blocks which begin to slide along these surfaces.

Clearly, there is a close connection between friction along fractures that already exist in, say, a rock, and formation of fractures in the first place. The shear strength of a rock can be regarded as due to its "internal friction". By analogy, the concept of internal friction may also be extended to materials that deform plastically, and to true fluids. The resistance to flow produced by shearing is measured by *viscosity*, and it may therefore be regarded as a measure of internal friction. The analogy is, however, far from perfect, because solids respond to shear forces differently from fluids. Except for small elastic deformations, solids do not shear until some critical shear strength has been exceeded, but true fluids deform continuously in response to a shear stress, no matter how small. As we will see, most rocks under large confining pressures and most soils show a mechanical behaviour intermediate between solids and fluids.

After considering the experimental evidence on friction and strength, we develop the concepts and notation necessary to describe stress in the earth. We then describe techniques to measure strength in the earth. We apply these techniques by showing that there are regional stress patterns related to the movement of lithospheric plates. On a smaller scale we show how an understanding of the stresses in soils and rocks helps us to understand landslides and avalanches.

We end the chapter by returning once again to the mathematical description of stress.

4.2 Friction

Friction always acts parallel to a pre-existing surface, and it always acts in the direction opposite to the forces tending to produce slip (shear) on that surface. As one solid body slides past another, friction is not the only force acting on the surface. There must also be a force normal to the surface, which presses the two bodies together. This second surface force is called the normal force, or the load acting normal to the surface. Friction may be measured as a force per unit area, in which case it is called a frictional shear stress, and the normal force may also be measured per unit area, in which case it is called a normal stress, or pressure.

4.2.1 Amontons' law

It has long been known that the ratio between the frictional force F and the normal force N is roughly constant:

$$F/N = f = \tan \phi \qquad (4.1)$$

f is the *coefficient of friction*, often expressed as the tangent of an angle ϕ called the *angle of sliding friction*. Equation (4.1) should be called *Amontons' law of friction* after Guillaume Amontons (1663–1705), the French scientist who first established the relationship in 1699. Often it is called *Coulomb's law*. C. A. Coulomb (1736–1806) was a French physicist who made important contributions to the theory of electricity. He also worked on strength and friction. He used Amontons' law and extended it to include cohesion (see Section 4.12).

For a block to start sliding down an inclined plane (see Figure 4.2), the downslope component of gravity $W \sin \alpha$, must equal or exceed the *static* frictional force F:

$$F = W \sin \alpha \qquad (4.2)$$

and from the law of friction,

$$F/N = \frac{W \sin \alpha}{W \cos \alpha} = f$$

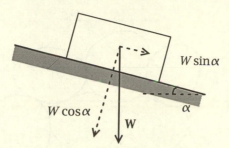

Fig. 4.2. Block sliding down an inclined plane.

It follows that

$$f = \tan \alpha$$

which explains why the coefficient of friction is frequently replaced by an angle of friction.

The static friction, which must be overcome to start movement, is different from the *dynamic* friction, which is exerted on a block that is already sliding. The static friction is generally larger, so a block which has started to slide cannot be brought to rest by friction until the slope flattens out.

4.2.2 Rugosities, dilatation and dispersive pressure

One of the older theories for the origin of friction was that it is due to the interlocking of the "rugosities" or individual roughness elements making up a rough surface. This is certainly not the full explanation of friction, for example, between smooth metal or glass surfaces, but it is part of the explanation of the internal friction of granular materials, such as sand. Consider the forces involved when one layer of nearly spherical grains slides past another (Figure 4.3). If we assume that there is no friction at the small tangential contacts between the grains themselves, we can resolve the forces producing and opposing sliding into their components parallel to the plane of movement *AB*, which is tangent to the contact between grains:

$$F \cos \alpha = N \sin \alpha$$

or

$$F/N = \tan \alpha$$

This geometrical analysis therefore predicts a law of friction which has the same form as the observed law. The values of α predicted by the theory are a function of the grain packing, and therefore also of the porosity

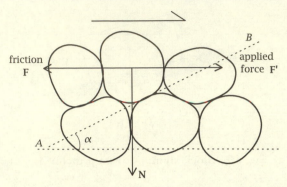

Fig. 4.3. Sliding between two layers of grains. The shearing force **F′** is applied to the upper layer of grains and opposed by a frictional force **F**. There is also a normal force **N** which might simply be the weight of the upper layer of grains (or might include the weight of higher layers of grains not shown in the figure). The broken line *AB* is tangential to the central two grains and shows the plane along which grains will slide past each other.

(Allen, 1985): they are generally much less than those observed in real sands. Geometrical analysis explains only part of the internal friction: it considers the effect of large roughness elements (the grains) but ignores the effect of friction on the surfaces of contact between the grains.

Shearing of a mass of sand grains implies movement not only parallel but also normal to the plane of shear. If the shearing extends throughout the mass, this implies a general expansion in volume, called *dilatation*. Such a dilatation can be demonstrated in a simple experiment first described by Osborne Reynolds (1842–1912) and shown in Figure 4.4. The rubber bag is filled with packed sand saturated with water. When the bag is squeezed, the water level in the tube goes down (not up, as one might expect) because deformation involves dilatation, and therefore produces an increase in pore space. Only if a sand is very loosely packed can shearing take place without dilatation.

Dilatation is an important phenomenon in the shearing of rocks. It is probably the reason why the ground "swells" slightly before shearing forces are relieved by the faulting that causes earthquakes. Dilatation not only produces the ground swelling through bulk expansion, but also causes a drop in groundwater levels, and allows trapped radiogenetic gases to be released. All these phenomena have been used (with limited success) to predict earthquakes.

Though the analysis given above refers to shearing solid masses of sand, there are reasons to suppose that similar forces also apply to shearing a

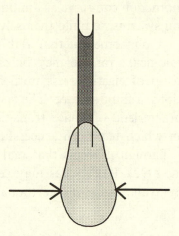

Fig. 4.4. Reynolds' experiment on dilatation.

dispersion of solid grains provided the grains are relatively close together, even if they are not always in contact. Bagnold (1956) performed experiments which indicated that forces normal to the plane of shearing could exist in dispersions with as little as nine per cent solids by volume. He called these normal forces, per unit area, the *dispersive pressure*. If one reflects on the matter, it is not surprising that such forces exist in a dispersion of solid grains, but it *is* surprising that they do *not* exist in the shearing of ordinary fluids. For example, the internal friction (viscosity) of a gas is not much affected by the pressure.

4.2.3 *Angle of repose*

The geometrical analysis of Figure 4.3 may also be applied to the sloping surface of a pile of grains. The angle of repose is the maximum angle of slope of such a pile. More precisely, one can recognize:

the angle of initial yield. A pile cannot have its surface built up beyond this angle;

the angle of residual shear. Avalanching down a surface ceases at this angle.

The first angle is generally similar to that measured by shearing sand in a box, that is, to the angle of internal shearing resistance. For experimental results see Statham (1977) and Allen (1985).

Though the angle of repose has been studied in sedimentology and soil mechanics for many years, sand avalanching has recently aroused renewed interest among theoretical and experimental physicists. Bak *et al.* (1988) have suggested that this phenomenon can serve as a model for a general tendency, observed in many natural systems, to build themselves up repeatedly towards a "critical" condition – a phenomenon that Bak has called *self-organized criticality*. Another geological example may be earthquakes, which result from the gradual build-up of elastic stress until friction is overcome and there is a sudden slip along a fault surface. The sudden transition from one state to another is characteristic of phase transitions (e.g., the melting of a solid), a phenomenon which has been successfully analyzed by physical theory in recent years. Though it seems that real sandpiles do not fit the Bak model well (Mehta, 1992), Bak's ideas may yet be used to throw light on exactly how and why avalanching and earthquakes take place.

4.2.4 Theory of friction

If a geometrical analysis of roughness is not an adequate basis for a theory of friction, what is? Although friction is of great economic importance and has been extensively studied, no comprehensive theory has yet emerged. But the following is another partial theory worth noting.

Even apparently "smooth" surfaces are not smooth when examined closely at high magnifications. Consequently, when such surfaces are "in contact" there is actual contact between only a small fraction (perhaps less than 0.01%) of the surfaces.

The actual area of contact depends mainly on the magnitude of the normal force N. This is because the points of contact bear a very large normal force. Suppose that the actual points of contact have a total area A, and a certain strength per unit area S. It is reasonable to suppose that A increases as N increases: for simplicity assume A is directly proportional to N.

Slip will take place when F exceeds the total strength of all the contacts, that is, when:

$$F = AS$$

and as A is proportional to N, i.e., $A = kN$:

$$F/N = kS$$

kS is itself a constant for a given material.

A detailed review of modern theories of friction is given by Scholz (1990).

4.3 Rock falls and avalanches

We consider now another geological phenomenon in which friction plays an important part: rock avalanches. Many avalanches begin when a slab of

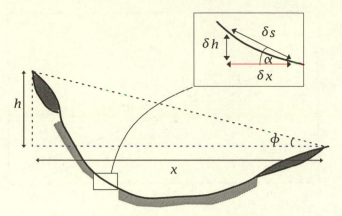

Fig. 4.5. Movement of a rock avalanche.

rock (or snow) breaks loose on a mountain slope and starts to slide down the slope; the frictional forces that previously resisted movement have been overcome. The reason why a particular avalanche begins may be rather obscure. Commonly, though, water pressure builds up as rain or snow-melt infiltrates a potential failure plane. As we will see in the next chapter, an increase in water pressure decreases the effective normal stress and therefore reduces the strength. Another possibility is that the region is shaken by an earthquake. Ground motion produced by the earthquake may temporarily double or triple the component of gravity acting down the slide plane.

Whatever the triggering event, suppose that a mass m, begins to slide down a slope angle α at a speed v, in the direction s of the local slope (Figure 4.5). Consider the energy of the moving debris. For a small distance δs moved by the debris, we can write the energy balance as

$$\text{KE gain} = \text{PE loss} - \text{energy lost to friction}$$
$$\delta(mv^2/2) = mg \, \sin\alpha \, \delta s - fmg \cos\alpha \, \delta s \qquad (4.3)$$

Setting $\sin\alpha \, \delta s = \delta h$ and $\cos\alpha \, \delta s = \delta x$:

$$\delta(v^2/2g) = \delta h - f\delta x \qquad (4.4)$$

Integrating over the total distance of travel, from $x = 0$ to $x = x_{\text{max}}$, and assuming $v = 0$ at $x = 0$, gives

$$0 = h - fx_{\text{max}}$$

or

$$f = h/x_{\text{max}} = \tan\phi \qquad (4.5)$$

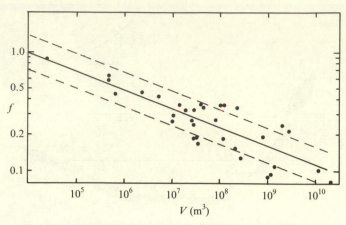

Fig. 4.6. Decrease in coefficient of friction f with volume V for large rock avalanches (after Scheidegger, 1975).

In other words, the coefficient of friction for the whole avalanche may be expressed as the drop in elevation divided by the horizontal distance of travel. One advantage of considering the energy balance, rather than the equations of motion, is that we do not need to be concerned with the detailed history of the slide, for example, the way in which it accelerates at the source and decelerates at the terminus. Instead, we obtain a result for friction averaged over the whole path of the slide.

Typical values for small avalanches are 0.6 to 0.8, corresponding to angles of friction of 30–40 degrees. These are close to the normal angles of dynamic friction for sliding rocks. But for avalanches where the volume of the slide exceeds 10^5 m^3 the angle of friction progressively decreases (Figure 4.6). In other words, these large rock avalanches are much more "mobile" than one would expect. Various theories have been proposed to explain this. Until recently, one of the most popular was that such large avalanches travel on a "cushion" of air trapped beneath the moving mass of debris, which acts as a lubricating layer. It was argued that because large avalanches travel at very high speeds (hundreds of metres per second) such a cushion of air could not escape during the very short period of movement.

High mobility seems to be a general rule for large avalanches, and it is hard to believe that a cushion of air was trapped under all of them. Furthermore, large mobile rock slides have also been identified on Mars and the Moon, where air was either absent or much thinner than on Earth. Thus many scientists now doubt the validity of the air cushion hypothesis, and think that the high mobility must be due to some other friction-reducing factor. What this factor is remains a mystery. It cannot be a simple reduction in friction due to an increase in pore space, because Bagnold (1956) showed that the coefficient of friction in grain flows remained just as high as it is in loosely packed sand, even for porosities as high as 90% (and see the review of subsequent research by Campbell, 1990).

4.4 Strength

By the *strength* of a material we mean the stress necessary to produce significant permanent deformation. To be more precise, we need to define some condition of deformation as *failure*: then the stress producing failure is the strength. This is not always easy to do, however, because there is a wide variation in the style and rate of deformation in natural materials. In order to illustrate how some natural materials deform, we will first describe one example of the relationship between stress and deformation as measured in the laboratory.

4.4.1 Measuring stress–strain behaviour

A simplified, controlled simulation of natural conditions at depth within the earth is obtained using a *triaxial compression test*. The design of the apparatus depends on the pressure range (large for rock mechanics, relatively small for soil mechanics) and on the porosity of the sample. If the material has pore spaces, even in a very small proportion (e.g., 0.1%), then its mechanical behaviour depends strongly not only on the pressure applied to its surface, but also on the fluid pressure developed in the pores. The pore pressure is generally carefully controlled, especially in triaxial tests of soils and porous rocks. We return to the effects of pore pressure in the next chapter.

A typical triaxial testing apparatus is shown schematically in Figure 4.7. The specimen A is prepared as a cylinder with flat ends, and enclosed in a flexible jacket (rubber at low pressures, metal at high pressures). A chosen confining pressure is applied by increasing the pressure of the fluid B surrounding the jacket. The pressure on the ends of the specimen, is then increased in increments by means of a piston (ram), driven by a hydraulic jack. The confining pressure σ_2 ($\equiv \sigma_3$ in Figure 4.8) acts equally in all horizontal directions, and the pressure applied in the vertical direction σ_1 is calculated from the measured pressure, after subtracting resistance due to friction in the apparatus. If the specimen is porous, the pore pressure may rise (or fall) as the specimen is compressed. With a soil specimen, porous disks at each end of the specimen allow the entry or escape of pore fluids and

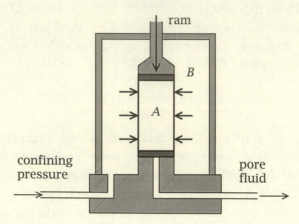

Fig. 4.7. Triaxial testing apparatus (schematic).

the monitoring of pore pressure. The strength of the material is generally determined in many tests covering the range of confining pressures relevant to the particular field application.

The deformation is measured by the *strain ε*, that is, the proportional change in length of the specimen, and is generally plotted against the *differential stress* $\sigma_1 - \sigma_2$. Figure 4.8(a) shows typical results of a series of tests run on marble at different confining pressures.

The way that strain varies with stress in natural materials has been found to be influenced by the materials, the pore pressure (if pores are present), the confining pressure, the rate of strain (i.e., the rate at which the specimen is compressed by the piston), and the temperature. With soil specimens, a distinction is made between tests in which the pore pressure is allowed to equilibrate to some constant pressure (called *drained tests* because it is necessary to allow pore fluids to drain into or out of the specimen), and those in which the specimen is sealed in, so that fluid cannot enter or leave it (*undrained tests*). In undrained tests, the pore pressure may undergo large changes as specimens deform during the test. Drained tests are generally easier to interpret, but more difficult to carry out, as it may be necessary to use very low strain rates to allow time for the pore fluids to move into or out of the specimen.

In the typical test of a non-porous rock material shown in Figure 4.8(b) several different types of response can be observed:

• At low confining pressures (e.g., (i) in the figure), the rock compresses elastically until it fails. In the elastic regime, strain is directly proportional to differential stress. Failure takes place at relatively small strains, generally

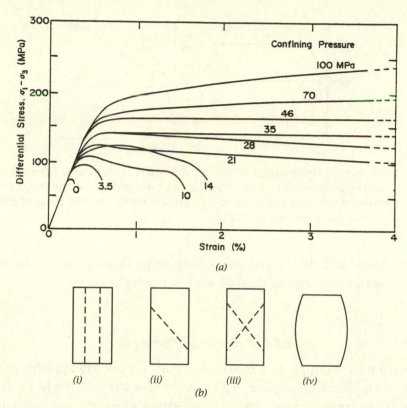

Fig. 4.8. Experimental results for triaxial compression of Wombeyan marble (after Paterson, 1958). (*a*) Stress–strain curves for various confining pressures. (*b*) Appearance of specimen after failure. The confining pressures were (*i*) 0 MPa, (*ii*) 3.5 MPa, (*iii*) 35 MPa, (*iv*) 100 MPa.

by breaking along fractures roughly parallel to the axis of the test cylinder (i.e., by "axial splitting").

• At higher confining pressure (e.g., (*ii*) and (*iii*) in the figure), a similar response is observed, except that the rock becomes stronger (higher stresses and strains are reached before failure) and fails along one inclined shear surface (*ii*), or along two "conjugate" surfaces (*iii*). Deformation continues, mainly by sliding along these surfaces, at lower stresses than were necessary to cause failure.

• At very high confining pressure (e.g., (*iv*) in the figure), there is a smooth transition from elastic strain to *ductile* deformation. In the ductile regime, deformation continues and (for a given strain rate) is generally either proportional to the differential stress (line sloping upwards on the stress–strain plot), which is called *work-hardening*, or independent of it (horizontal line). Some materials, particularly soils, show an inverse relation between

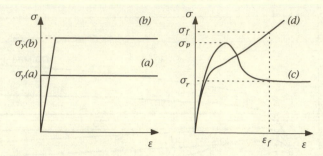

Fig. 4.9. Examples of stress–strain behaviour for different materials: (*a*) ideal plastic; (*b*) ideal elasto-plastic; (*c*) material showing work-softening; (*d*) material showing work hardening. σ is stress, ε is strain; ε_f is a critical strain derived from engineering criteria, not from the stress–strain curve.

deformation and differential stress (line slope downward on the stress–strain plot), a phenomenon called *work-softening*.

4.4.2 Definitions of strength

We see from the example of a triaxial test that it is not always easy to define exactly when failure takes place. To illustrate how strength might be defined, it is useful to consider examples of stress–strain diagrams for four kinds of ideal material (Figure 4.9).

(1) In the first material (Figure 4.9(*a*)), the definition of both failure and strength is unambiguous. No deformation takes place so long as the stress is smaller than the yield stress $\sigma_y(a)$. When the stress reaches $\sigma_y(a)$, the material deforms instantly and continuously as long as the stress is applied. This is what is meant by failure, so we define $\sigma_y(a)$ to be the strength. Real materials do not show this type of behaviour, but we can use it to define a *plastic* material. It is usual to distinguish between plastics for which the *rate* of deformation depends on the stress (i.e., all real plastics), and *ideal plastics* for which there is no such dependence.

(2) The second material (Figure 4.9(*b*)), shows behaviour closer to that of some real materials. Stresses smaller than $\sigma_y(b)$ produce small elastic deformations. These deformations are entirely recoverable, and are generally proportional to the applied stress. They do not constitute failure. In the example, there is an abrupt yield stress at which the material deforms continuously, so the strength can again be clearly

defined. The behaviour shown can be used to define an *ideal elasto-plastic*. Some soils and rocks show similar behaviour.

(3) The third material (Figure 4.9(*c*)) shows a type of behavior typical of many real materials. At low stresses there is elastic deformation, which merges at higher stresses into nonelastic deformation. There is a distinct peak (at σ_p) in the stress–strain plot at small values of strain. At larger values of strain, deformation continues at a lower stress (σ_r). These two stresses may be defined as the *peak* and *residual strengths*. This is an example of work-softening, similar to that evident at lower confining pressures, shown in Figure 4.8(*a*). A dense, compacted sand or a rock under moderate confining pressure might show such behavior.

(4) The fourth material (Figure 4.9(*d*)) also shows a transition from elastic to nonelastic deformation. In this case, however, the material always becomes stronger with further strain. This is work-hardening. A sand that is loose initially can be work-hardened: as the material strains, its volume decreases and its strength increases. A clear point of transition from elastic to nonelastic behaviour, if it exists, might be used to define the strength, even though the material can exhibit even greater resistance at larger strains.

A critical stress might be used for practical purposes. For example, in designing the foundation for a structure, any strain beyond a certain limit might damage the structure. So the strength might be defined as the stress σ_f that produces the critical value of strain ε_f (Figure 4.9(*d*)). We have seen in Figure 4.8(*a*) that some rocks can show work-softening or work-hardening, depending on the conditions of deformation.

In discussing these stress–strain examples, we have implicitly assumed that they represent the results of triaxial compression tests. But this is not necessarily the case. Other types of test apparatus can be devised that measure the response of materials to tensile, rather than compressive stress, or to shear stress. Corresponding to each type of test, it is possible to determine different values of stress at which failure takes place. Thus a single material may be characterized by several different types of strength.

Some values for unconfined strengths of typical rocks are given in Appendix B. In general, the tensile strengths of sediments and rocks are much smaller than their compressive strength. We see evidence of this in the common occurrence of joints and fractures near the earth's surface, where tensile stresses commonly prevail (Section 4.15).

4.4.3 Strength of soil

The stress–strain behaviour of most soils is related to the rearrangement of soil particles during shear and to the degree of prior consolidation of the soil. Heavily consolidated soils dilate and weaken during shear and show a peak strength at small strains. Underconsolidated soils compress and become stronger, so either no peak strength is observed or it is approached asymptotically.

For most soils, the volume change goes to zero at large values of strain, and the specimens show a value of strength that is nearly independent of its initial state of consolidation. This condition is called the "critical state", and it forms the basis of "critical state soil mechanics" (Roscoe *et al.*, 1958; Wood, 1990). This attempts to provide a unified approach to drained and undrained failure by analyzing the combination of soil volume and effective stress at which constant-volume yielding takes place. A sand that displays work-hardening behaviour over a range of strain values can approach a limiting stress at large strain along a curved stress–strain path that is approximated by the ideal elasto-plastic model.

The strength exhibited by natural materials may also depend on the rate at which the material is deformed. Many materials have a time-dependent component to their stress–strain behaviour, that is sometimes called *viscous*. In general, this means that the measured strength will decrease with decreasing strain rate. The reason for this is that the material can be thought of as slowly flowing in response to the applied stress: if the stress is applied more slowly, the material has more time to flow, or strain. Another example is ice, which deforms continuously at very slow strain rates (so it has essentially zero strength), but has a finite strength and eventually fractures at high strain rates (Chapter 10).

For further discussion of soil properties see Holtz and Kovacs (1981) or Lambe and Whitman (1969).

4.4.4 Terminology of deformation

We are now in a position to discuss the question of the terminology for different types of deformation and failure. Unfortunately there is no general agreement on the meaning of many of the terms used – in fact, the two authors of this book do not fully agree on the best terminology!

All natural materials show some *elastic* deformation. This type of strain is fully reversible, once the imposed stress is removed. It is generally small, and directly proportional to the stress. The *elastic limit* is defined as the stress at

Table 4.1. Nomenclature for deformation regimes (after Rutter, 1986)

Deformation mechanism	Localization	
cataclasis	cataclastic faulting	cataclastic flow
crystal-plastic	plastic shear zone	homogeneous plastic flow

which some of the strain becomes nonreversible. Near this limit, the relation between stress and strain generally also becomes nonlinear. At higher strains, all the strain may become irreversible (plastic behaviour), but there may be a regime where the behaviour is partly elastic and partly plastic. When failure takes place, it may be by *fracture* along relatively well-defined surfaces, or by shearing along groups of fractures in a more-or-less well-defined zone, or by a more-or-less continuous deformation that is frequently called *plastic flow*. It is possible to distinguish two major regimes of deformation (or types of deforming material): if failure takes place by fracture, the material is *brittle*, and if it takes place by flow it is *ductile*.

Terms such as "plastic" and "viscous" have the disadvantage that they mean different things to different people. Plastic deformation can mean simply permanent deformation without rupture (this is the way it is used in the phrase "plastic flow"). But metallurgists commonly use the term to refer to a particular mechanism of deformation in the solid state (by crystal gliding or recrystallization). And the term "plastic" can be applied, as it is in Figure 4.9 and Chapter 10, to any deformation which is characterized by a yield stress, followed by flow. The flow of real gases, liquids, and amorphous plastics is characterized by viscosity, measured as the ratio between the stress and the strain rate. Similar stress–strain-rate relations are observed in crystalline solids, where the mechanism of deformation is certainly quite different. Such deformation is nevertheless described by many workers as "viscous".

Rutter (1986) has suggested that, in rock mechanics and structural geology, the type of deformation beyond the elastic regime should be described using two separate criteria: (i) the deformation mechanism; (ii) its localization. With decreasing rate of strain, or increasing confining pressure and temperature, the deformation mechanism changes from *cataclastic*, i.e., due to breakage, to *crystal-plastic*, i.e., due to recrystallization and creep. Also, with decreasing total strain, or increasing confining pressure and temperature, the localization of deformation changes from being confined to well-defined

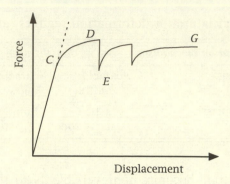

Fig. 4.10. Typical results from an experiment on rock friction. See explanation in text.

faults, or shear zones, to being unconfined. Unconfined deformation may be due to pervasive cataclasis, but more usually it is due to a form of plastic flow (Table 4.1).

Modern research on fracture tries to describe not only the conditions under which failure takes place, but also the mechanisms of failure, using the theory of cracks and solid-state creep. We do not have the space to review this work; for an introduction, see Atkinson (1987).

4.5 Experimental results for rock friction

Triaxial tests may also be used to study friction along fracture surfaces. For best results, however, the apparatus must be changed to allow for larger slippage along the surfaces than are permitted by normal triaxial presses. We will not discuss the experimental details (see Scholz, 1990) but simply present a typical result as a force–displacement (stress–strain) diagram (Figure 4.10) similar to those presented in the previous section. The rock deforms elastically to C, then there is some nonelastic deformation to *maximum friction* at D, followed by a sudden slip along fractures. The slip along fractures takes place instead of further elastic strain, so the strain at E (after slip) is not much larger than at D (before slip), but the stress is lower. The ratio of shear to normal force at C defines the *initial friction*. Continued deformation may produce several such slips, after which more or less continuous deformation is produced by application of a constant force (a constant *sliding friction*, G on the figure).

Many experimental results do not quite fit Amontons' law, but rather an equation of the type

$$F = aN + b \tag{4.6}$$

where *a* and *b* are experimental coefficients.

Byerlee (1978) reviewed the results obtained and found a large range of friction coefficients measured at low pressures (*N* divided by the area of contact). The friction coefficients depend on the rock type and on the roughness of the surface (cf. Scholz, 1990, pp. 56–73). For data at intermediate pressures (50–1000 bar, or 5–100 MPa) there is no strong dependence of initial or maximum friction on rock type. For maximum friction, points are clustered around the line

$$F = 0.85N \qquad (4.7)$$

For high pressures (2–20 kbar), the data for maximum friction are almost independent of rock type and are well fitted by the straight line

$$F = 0.5 + 0.6N \qquad (4.8)$$

This equation is now generally known as *Byerlee's law*. Friction coefficients of 0.85 and 0.6 correspond to friction angles of 40 and 31 degrees respectively.

At high confining pressures, there may be little distinction between strength (the stress required for failure) and the stress required to overcome friction on pre-existing failure surfaces having the same orientation as the surfaces that would be formed by new failures.

4.6 Definition of stress

After this general review of experiments on the response of soils and rocks to surface forces, we must now develop the theory necessary to analyze such data. The experimental results describe strains, or deformation, produced by changes in stress, or forces per unit area. We defer the consideration of strain to Chapter 7, but turn now to the proper definition of stress.

Consider a force acting in the direction normal to a plane. Let the magnitude of the force acting on area δA be δF. Then the magnitude of the average normal stress acting on the area is defined as

$$\bar{\sigma} = \delta F / \delta A$$

We can thus define the *normal stress magnitude at a point* as the limit of $\bar{\sigma}$ as δA goes to zero:

$$\sigma = \lim_{\delta A \to 0} \left(\frac{\delta F}{\delta A} \right) = \left(\frac{dF}{dA} \right) \qquad (4.9)$$

Of course, the normal stress has a direction as well as a magnitude, i.e., for a given surface it can be represented as a *stress vector* (also called a *traction*).

Stress vectors need not be parallel to or normal to a given plane, but they can of course be resolved into components parallel or normal to the plane (or any other plane). They can be added vectorially, as long as they all refer to a single plane. Thus stress can be treated as a vector quantity once we specify the plane on which it acts.

We would like the specification of the *state of stress* at a point to yield the stress vectors that would act on *any* plane passing through that point. These stress vectors will depend not only on the direction and magnitude of the force, but also on the orientation of the plane on which it is exerted. Thus, though a particular stress component acting on a particular plane may be described as a vector, the state of stress at a point is a more complicated entity, called a *stress tensor*. A stress tensor represents the stresses acting on three reference planes passing through a point. As we will show later, this is enough to define the stress acting on any other plane passing through that point. The stress tensor cannot be represented graphically, as force is, by a single arrow but may be represented as a *stress ellipsoid*. The surface of the stress ellipsoid is the locus of the points traced out by stress vectors acting on all possible planes passing through the point at the centre of the ellipsoid. The three principal axes of the ellipsoid have a special significance and are called the *principal stresses*. On planes normal to the principal axes, the principal stresses consist only of normal stresses, and there are no shear stresses (Section 4.9).

Forces applied to extended media (rocks, fluids) are transmitted as stresses. Even a force applied to a single point soon spreads out as a stress transmitted through the whole mass. Body forces such as gravity also produce stresses within extended masses. In this chapter we describe the mathematical notation and algebra needed to represent stress compactly.

For an extended, but elementary, discussion of the concept of stress in the earth, refer to Means (1976, Chapter 6). Other useful texts are listed at the end of this chapter.

4.7 Notation and sign convention

One problem encountered by those who take an interest in stress is the variety of notation used to represent the different types of stresses and their components.

Most authors use σ as a symbol for stress. Some distinguish the magnitude σ from the stress vector $\boldsymbol{\sigma}$. Many authors use σ only for the normal stresses, and use τ for shear stresses. We will generally follow this convention, though we will also use σ as the general symbol for all stresses (both normal and shear).

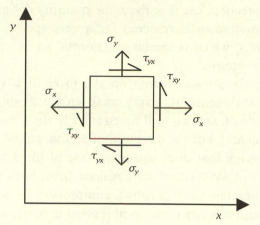

Fig. 4.11. Notation and sign convention for stresses (in two dimensions).

For complete representation of the state of stress at a point we need to consider three stress vectors, each acting on one of three mutually perpendicular planes. It is convenient to resolve each stress vector into a component of normal stress and two components of shear stress, so we need a notation for nine components of stress.

We can define these planes as normal to the three Cartesian axes x, y, and z (or axes 1, 2, and 3) i.e., the yz-, xz-, and xy-planes respectively. Then the stresses *normal* to these planes can be designated respectively as

$$\sigma_x, \sigma_y, \sigma_z$$

or

$$\sigma_1, \sigma_2, \sigma_3$$

For example, σ_x means "the stress component acting parallel to the x-axis, or normal to the yz-plane".

The positive direction can be defined either as the positive axis direction, or the reverse. The latter might be thought to be a peculiar choice, but it is often used in geology and rock mechanics because compressive normal stresses are then positive, not negative. The choice is, of course, purely a matter of convention.

The two stress components *parallel* to a given plane (say the xy-plane) can be defined in the following way (see Figure 4.11). The symbol τ_{xy} (or σ_{12}) means "the shear stress that acts on the plane normal to the x-axis (or x_1-axis) in the y-direction (or x_2-direction)".

As a natural extension to this notation, σ_{11} may be used for the normal stress in the x_1-direction, i.e., it is the same quantity designated as σ_1 above. Generally, the notation σ_1 is reserved for a very special normal stress, the first principal stress, which is the largest normal stress and acts on a plane that has no shear stresses.

If we consider the stresses acting on the faces of a very small cube of material (or, in two dimensions, a very small square, as shown in Figure 4.11) then the magnitude of, say, σ_{11} will be essentially the same on opposite faces of the cube (or square), but the direction of action will be opposite. We shall use a sign convention that does *not* define one of these as positive, and the other as negative: it says, either we consider them both positive (as drawn in Figure 4.11; then tension is positive, compression is negative) or as both negative (then tension is negative, compression is positive).

One way to avoid ambiguity is as follows: define positive faces to be those having an outward pole pointing in the positive direction. Then define components of stress as *positive* if they act in the positive direction on a positive face or in the negative direction on a negative face. All components shown on Figure 4.11 are then positive. Note that this convention has the consequence that compressional stresses (pressures) are *negative*.

The SI unit of stress is the pascal (Pa), which is the stress produced by a force of one newton acting normal or parallel to one square metre of surface. As the newton is not a very large force, and a square metre is quite a large area, a pascal is a very small stress. Atmospheric pressure (one atmosphere, about equal to one bar) is a normal stress of about 100 kilopascals (kPa). Pressures deep within the earth were once commonly expressed in bars (10^6 dynes cm^{-2}) but in the SI system should be expressed in MPa. One megapascal, equal to 10 bars, is the pressure at a depth of about 100 m in water, or about 30–40 m in rock.

4.8 Symmetry of stress components

Consider the equilibrium of stresses acting on a small square element in two dimensions, such as the one shown in Figure 4.12.

Let the side of the square be of length δx ($= \delta y$.) If this is small enough (infinitesimally small), then the stresses acting on one side of the square must be equal to those acting on the other side of the square (in the opposite direction), as shown in the figure. The normal forces will obviously cancel out, so there is no net normal force. But it is not quite so obvious what happens with shear stresses, because the shear forces act in different directions on opposite sides of the square and therefore tend to rotate the square.

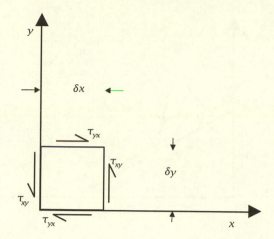

Fig. 4.12. Shear stresses acting on a small square.

If we assume that the square at equilibrium is not rotating, then it is clear that the shear stresses τ_{xy} acting on one pair of faces, and tending to rotate the square in the clockwise direction, must be balanced by the stresses τ_{yx} acting on the other pair of faces, which tend to rotate the square in the anticlockwise direction.

But there is an even more fundamental reason why the two shear stress components must be equal at a point. First calculate the clockwise and anticlockwise moments acting about the origin on a small stationary cube of side δx:

$$\text{clockwise:} \quad \gamma_{yx}(\delta x)^2 \qquad \text{anticlockwise:} \quad \gamma_{xy}(\delta x)^2$$

If these are not equal, the square will begin to rotate. The rate at which it rotates depends on the moment of inertia of the small cube. This is proportional to the density of the cube and, in all, to the fifth power of the side δx. So, as the side becomes smaller and smaller, the moment of the stresses will get smaller, in proportion to the square of the side, but the moment of inertia, which determines the rate of rotation, gets smaller much faster, proportional to the fifth power of the side. In the limit, the cube would tend to spin at an infinite rate, which is not possible. Therefore the components of shear stress acting on opposite sides of the cube (or square) must become equal as the side of the cube becomes infinitesimally small. The symmetry of the shear stress is sometimes called *Cauchy's second law* after A. L. Cauchy (1789–1857), the French mathematician who first gave a clear statement of the properties of stress (Truesdell, 1968). For Cauchy's first law see Section 9.4.

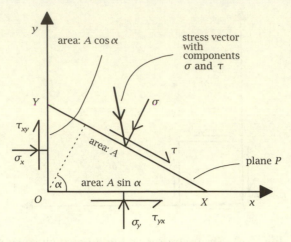

Fig. 4.13. Stresses acting on the surfaces bounding a small prism.

It is convenient to represent the stresses acting at a point by a square array:

- in two dimensions we have

$$\begin{array}{cc} \sigma_{11} & \tau_{12} \\ \tau_{21} & \sigma_{22} \end{array} \quad \text{or} \quad \begin{array}{cc} \sigma_{11} & \sigma_{12} \\ \sigma_{21} & \sigma_{22} \end{array}$$

- in three dimensions we have

$$\begin{array}{ccc} \sigma_{11} & \tau_{12} & \tau_{13} \\ \tau_{21} & \sigma_{22} & \tau_{23} \\ \tau_{31} & \tau_{32} & \sigma_{33} \end{array} \quad \text{or} \quad \begin{array}{ccc} \sigma_{11} & \sigma_{12} & \sigma_{13} \\ \sigma_{21} & \sigma_{22} & \sigma_{23} \\ \sigma_{31} & \sigma_{32} & \sigma_{33} \end{array}$$

These arrays are symmetrical, since we have seen that pairs of shear stresses must be equal:

$$\tau_{ij} = \tau_{ji}$$

Arrays of stress components are generally referred to as *stress matrices* or *stress tensors*. We will return later in this chapter to the mathematical techniques (matrix and tensor algebra) that are useful in manipulating these arrays. First we will develop some useful relationships between components, in two dimensions, and develop the concept of the stress ellipse.

4.9 Equilibrium of a small prism

To make practical use of the stress matrix, we need to be able to find the stress components in directions different from the reference directions.

Suppose that we know the stress components acting on two perpendicular surfaces. We want to find an expression for the stress components acting on some other surface P, whose normal makes an angle α with the x-axis. We assume unit length in the z-direction, so the three surfaces (represented in section by the lines OX, OY, and XY; see Figure 4.13) define a small prism. If the area of the surface P (the XY face) is A, the area of the OX face is $A \sin \alpha$ or mA, and the area of the OY face is $A \cos \alpha$, or ℓA, where ℓ and m are the direction cosines of the face P ($\ell = \cos \alpha$, $m = \cos \beta = \sin \alpha$).

Next, we need to make clear the sign convention to be used for stresses on a plane that is not normal to any coordinate direction. In Figure 4.13 we can see that the normal stresses shown acting on all faces tend to compress the prism. In this section we will use the convention that these compressive stresses are positive. On the two faces normal to the x- and y-axes, therefore, the two shear stresses shown are positive, because they act in the positive x- or y-direction on negative faces. But how about the shear stress acting on the plane P? We first consider a positive rotation of reference axes, so that the rotated x'-axis is normal to P. Then P would be a positive face, so for the shear stress to be positive (under the compression–positive convention) it must act in the negative rotated y'-direction. This direction is down to the right on the diagram. Note that some authors use a different sign convention (for details, see Malvern, 1969).

Now we obtain expressions for the two stress components on P by balancing the forces that act on the small prism (Figure 4.13). The known stresses act on the faces normal to the x- and y-axes, and the unknown stresses on the surface P.

What are the conditions for balance of the forces acting on this prism? We have already seen by considering moments that, for infinitesimal volumes, one condition is that pairs of shear stresses must be equal. Now we consider the balance of forces tending to push in a line, or translate, the prism. That is, in order for the prism to be at equilibrium, the sum of all the forces tending to translate the prism must be zero. We consider the force balance for components normal to and along the surface P. We calculate force as stress times area. From Figure 4.13 we can see that the shear force acting on OX is $\tau_{yx} A \sin \alpha$, and its component normal to the plane P is $\tau_{yx} A \sin \alpha \cos \alpha$. Writing the complete balance of components, we obtain

- normal to P:

$$A\sigma = \tau_{yx} A \sin \alpha \cos \alpha + \sigma_y A \sin \alpha \sin \alpha$$
$$+ \tau_{xy} A \cos \alpha \sin \alpha + \sigma_x A \cos \alpha \cos \alpha$$

Using $\tau_{yx} = \tau_{xy}$, and the identity $2 \sin \alpha \cos \alpha = \sin 2\alpha$,

$$\sigma = \sigma_x \cos^2 \alpha + 2\tau_{xy} \sin \alpha \cos \alpha + \sigma_y \sin^2 \alpha$$
$$= \sigma_x \cos^2 \alpha + \tau_{xy} \sin 2\alpha + \sigma_y \sin^2 \alpha \tag{4.10}$$

• parallel with P (lower right $+$):

$$A\tau = \tau_{yx} A \sin \alpha \sin \alpha - \sigma_y A \sin \alpha \cos \alpha$$
$$+ \sigma_x A \cos \alpha \sin \alpha - \tau_{xy} A \cos \alpha \cos \alpha$$

$$\tau = (\sigma_x - \sigma_y) \sin \alpha \cos \alpha - \tau_{xy}(\cos^2 \alpha - \sin^2 \alpha) \tag{4.11}$$

We now examine the condition that there exists some plane for which $\tau = 0$. Setting the expression for τ (Equation 4.11) equal to zero and dividing by $(\sigma_x - \sigma_y)$ gives

$$\tau_{xy}/(\sigma_x - \sigma_y) = \sin \alpha \cos \alpha/(\cos^2 \alpha - \sin^2 \alpha)$$
$$= \sin 2\alpha/2 \cos 2\alpha$$
$$= \tan 2\alpha/2 \tag{4.12}$$

This condition can be satisfied, so such planes exist. Also, if $\alpha = \alpha_o$ satisfies the condition, then so will $\alpha = \alpha_o + 90°$ because $\tan 2\alpha = \tan(2\alpha + 180°)$.

Differentiating the expression for σ, and applying Equation (4.11):

$$d\sigma/d\alpha = 2\sigma_x \cos \alpha(-\sin \alpha) + 2\tau_{xy} \cos \alpha \cos \alpha$$
$$+ 2\tau_{xy} \sin \alpha(-\sin \alpha) + 2\sigma_y \sin \alpha \cos \alpha$$
$$= 2(\sigma_y - \sigma_x) \sin \alpha \cos \alpha + 2\tau_{xy}(\cos^2 \alpha - \sin^2 \alpha)$$
$$= -2\tau \tag{4.13}$$

So a maximum or minimum of σ is obtained when $\tau = 0$.

We have therefore shown that there are two planes for which the shear stress is equal to zero: the planes are at right angles to each other, and for one of them the normal stress is a maximum (σ_1), and for the other it is a minimum (σ_2). These two stresses are called the *principal stresses*.

4.10 The stress ellipsoid

We now carry out a balance of force components in the same manner as in the previous section, but we choose the x- and y-axes to be in the direction of the principal stresses (so that there are no shear stresses acting on planes OX and OY; Figure 4.14).

Then, if we further resolve the stress vector acting on P into components

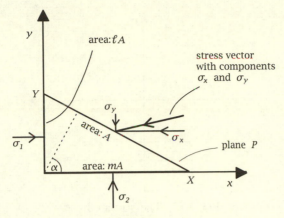

Fig. 4.14. Stress components in the directions of the principal stresses.

acting in the x-direction (σ_x) and y-direction (σ_y), we can show very easily that

$$\sigma_1 \ell A = \sigma_x A, \qquad \ell = \sigma_x/\sigma_1 \tag{4.14}$$
$$\sigma_2 m A = \sigma_y A, \qquad m = \sigma_y/\sigma_2 \tag{4.15}$$

and we know that the direction cosines ℓ, m have the property that

$$\ell^2 + m^2 = 1$$

so it follows that

$$\left(\frac{\sigma_x}{\sigma_1}\right)^2 + \left(\frac{\sigma_y}{\sigma_2}\right)^2 = 1 \tag{4.16}$$

This is the equation of an ellipse, with semimajor and semiminor axes of length σ_1 and σ_2 (Figure 4.15). The ellipse defines the locus of all points traced out by the stress vector σ, for all possible orientations of the plane on which it is acting. The coordinates of the stress vector (σ_x, σ_y) are related to the direction cosines of the plane by Equations (4.14) and (4.15). It can be shown (e.g., Jaeger, 1971) that in three dimensions the locus is defined by an ellipsoid, whose equation is similar to the equation of the ellipse given above. The stress ellipsoid is one way to provide a complete description of the state of stress at a point.

We can average the three principal stresses to obtain the *mean stress*:

$$\bar{\sigma} = (\sigma_1 + \sigma_2 + \sigma_3)/3 \tag{4.17}$$

If all three principal stresses are equal (and therefore equal to the mean stress) the stress field is said to be *hydrostatic*, and the mean stress is then

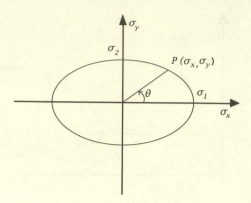

Fig. 4.15. The stress ellipse.

equal to the hydrostatic pressure. Pressures in the mantle or deep crust result from the weight of rocks, rather than of water, and are generally described as *lithostatic* rather than hydrostatic.

The *deviatoric stress* is defined as the difference between the actual normal stresses and the mean stress. Shear stresses are unaffected by variations in the mean stress, because they depend on the difference between normal stresses, not on their absolute values (Equation 4.11). For deformation of fluids, it is generally assumed that we can substitute the deviatoric stress matrix for the true stress matrix, and consider mean stress (pressure) separately. At great depths, in the lower part of the continental crust, and within the asthenosphere, rocks becomes so plastic that relatively small deviatoric stresses can produce "flow". In this case, deviatoric stresses are soon dissipated and the stress field may be very close to lithostatic. The important concept of pressure in fluids and rocks is discussed further in the next chapter.

As explained, if we know the magnitude and orientation of the principal stresses at a point, then we can very easily determine the stresses on any plane passing through that point. Extending this to a continuum, we can represent the variation of stress in two or three dimensions by plotting lines which are everywhere tangent to the principal stresses. Such lines are called *stress trajectories*, and are a very useful way of graphically representing the regional variations of stress in two dimensions. We return to this topic in Section 4.14.

4.11 The Mohr circle of stress

The general two-dimensional force balance considered in Section 4.9 can be applied to the special case considered in Section 4.10, for which the x-

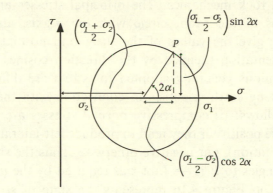

Fig. 4.16. The Mohr circle of stress. The principal stresses are σ_1 and σ_2. The point P represents the state of stress on the plane P shown in Figure 4.14.

and y-axes are parallel to the direction of the principal stresses. Then the shear stress on the planes normal to these axes would be zero, and the two equations for the normal and shear stress on the plane P (Equations 4.10 and 4.11) reduce to

$$\sigma = \sigma_1 \cos^2 \alpha + \sigma_2 \sin^2 \alpha \tag{4.18}$$

$$\tau = (\sigma_1 - \sigma_2) \sin \alpha \cos \alpha \tag{4.19}$$

Using

$$\cos 2\alpha = 2 \cos^2 \alpha - 1 = 1 - 2 \sin^2 \alpha$$

to rearrange Equation (4.18) and

$$\sin 2\alpha = 2 \sin \alpha \cos \alpha$$

to rearrange Equation (4.19) we obtain

$$\sigma = \left(\frac{\sigma_1 + \sigma_2}{2}\right) + \left(\frac{\sigma_1 - \sigma_2}{2}\right) \cos 2\alpha \tag{4.20}$$

$$\tau = \left(\frac{\sigma_1 - \sigma_2}{2}\right) \sin 2\alpha \tag{4.21}$$

If we use σ and τ as abscissa and ordinate, we find that these two equations define a circle, called the *Mohr circle of stress*, shown in Figure 4.16.

This representation of stress was first proposed in 1882 by the German engineer Christian Otto Mohr (1835–1918): it has proved to be a convenient graphical method for determination of two-dimensional stress fields (and also of strains; see Chapter 7). It is widely used in structural geology, and in soil and rock mechanics. The principal stresses are represented by the intersections of the Mohr circle with the σ-axis, and the points on the Mohr circle give the values of the shear and normal stress on planes of arbitrary orientation (defined by the direction cosines ℓ, m). Note that the sign conventions generally adopted for stress are different from those commonly used by earth scientists in connection with the Mohr diagram, which are as follows: (i) compressive normal stresses are positive; and (ii) shear stresses are positive if they tend to produce left-lateral displacement (or anticlockwise rotation), but negative otherwise. Thus the shear stress on the plane at right angles to P is $-\tau$ (not τ as required by the usual convention).

The point P on Figure 4.16 represents the state of stress on the plane P in Figure 4.14. The orientation of the plane P is given by the angle α between the σ_1-axis and the normal to P (or equivalently, the angle between the σ_2-axis and the trace (or projection) of the plane P in the xy-plane). Point P on the circle is found by drawing a radius at an angle of 2α from the positive σ_1-axis. The coordinates of point P in Figure 4.16 may be seen to be identical to the values of σ and τ in Equations (4.20) and (4.21) and, therefore, to Equations (4.10) and (4.11). Given σ_1, σ_2, and α, the state of stress for any plane can be determined graphically on Figure 4.16.

With the advent of powerful pocket calculators and computers, the Mohr circle is no longer commonly used for the graphical determination of stress; it is, nevertheless, a most useful device for demonstrating the state of stress at a point, particularly when used together with the strength criterion discussed in the next section.

4.12 Failure: the Navier–Coulomb criterion

It might seem logical that a material would be most likely to fail along planes for which the absolute values of the shear stresses are a maximum. On a Mohr diagram this condition would be as shown on Figure 4.17; there are two such planes: on one the shear stress is a positive maximum, on the other it is a negative minimum (note that if the normal to the plane is rotated in the clockwise, i.e., negative, direction from the σ_1-direction, the stress is negative, not positive). The two planes where the shear stress reaches a maximum absolute value make angles of 45 degrees with the directions of principal stresses.

This first guess does not take into account the increase in strength produced by increased normal stress. Experiments show that failure takes place on planes whose normals make an angle (α) somewhat larger than 45 degrees with the σ_1-direction. This means that the fractures form at angles somewhat

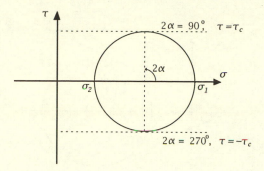

Fig. 4.17. Mohr diagram, showing planes of maximum shear stress at $2\alpha = 90$ degrees and 270 degrees.

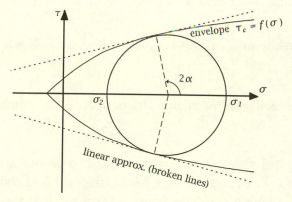

Fig. 4.18. Condition for fracture, as determined experimentally.

less than 45 degrees on either side of the σ_1-direction. Figure 4.18 shows this condition on a Mohr diagram. To determine the condition experimentally we use a triaxial testing apparatus. First we set the confining pressure σ_2, then increase the piston pressure σ_1 until failure takes place. In theory we should be able to measure α and determine the point on the Mohr circle that corresponds to failure, but in practice, the orientation cannot be measured accurately. Instead, we repeat the experiment for several different values of the confining pressure. A curve drawn tangent to all the different Mohr circles representing failure conditions defines an *envelope of failure*. All possible Mohr circles that produce failure are tangent to this envelope ($\tau_c = f(\sigma)$), and failure takes place when the stress circle expands so that it is just tangent to the envelope. Note that, in general, the shear stress needed to produce failure increases as the confining pressure increases.

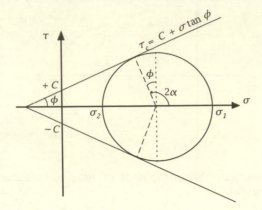

Fig. 4.19. Navier–Coulomb criterion for failure, represented on a Mohr diagram. The tangents to the Mohr circle are represented by $\tau_c = C + \sigma \tan \phi$.

One formula for the function $f(\sigma)$ is the *Navier–Coulomb* criterion,

$$\tau_c = \tau_o + \mu\sigma \qquad (4.22)$$

where μ is the coefficient of internal friction. It is also commonly written as

$$\tau_c = C + \sigma \tan \phi \qquad (4.23)$$

where C is called the *apparent cohesion* and ϕ is the apparent angle of internal friction. The corresponding Mohr diagram is shown in Figure 4.19. For this linear strength envelope, fracture takes place at the critical angle α_c, where

$$\alpha_c = (90 + \phi)/2 \qquad (4.24)$$

We add the adjective "apparent" to our definitions of C and ϕ because most natural materials do not have linear strength envelopes as given by Equation (4.23), but have concave-downward envelopes as suggested by Figure 4.18. Equation (4.23) is, however, often used as a linear approximation of a curved strength envelope over a limited range of principal stresses. On Figure 4.18, it is evident that a straight line would do a reasonably good job representing the strength envelope for large mean stresses. If this line is extrapolated to the τ-axis, however, it is clear that the apparent cohesion C (the intercept for the straight line) is much larger than the actual strength at zero normal stress.

Estimated values of μ for isotropic, compact, coherent rocks are in the range 0.7–1.7 (corresponding to ϕ angles of 35 to 60 degrees), with igneous rocks near the upper end of the range. The cohesion C measures the resistance to failure that exists even when the normal pressure is zero or negative (tensile). It is estimated to be about twice the tensile strength for most rocks. Tensile strength is rarely measured directly (though see Appendix B): theoretically it should be about 1/8 of the compressive strength but it is generally less.

Values of μ for loose sands and gravels vary over a range similar to that for intact rocks, depending on the degree of compaction and the range of grain sizes, both of which tend to increase the internal friction.

Clean, uncemented sands and gravels have essentially no cohesive strength. Soils with cohesive strength tend to contain silt and clay. Many have also been previously subjected to consolidation pressures larger than those acting at the present. Such "overconsolidated" soils (see next chapter) tend to have strongly curved strength envelopes so that the values of apparent C are much larger than the actual cohesive strength at zero normal stress.

Cohesive strengths for soils are several orders of magnitude smaller than those for intact rock (measured in kPa rather than MPa). The coefficients of friction for cohesive soils are generally smaller than for rocks or cohesionless soils, usually falling in a range 0.36–0.84 (20 to 40 degrees; see Appendix B).

Two important factors influence these representative values of shear strength. The first is that strength depends strongly on the consolidation history of the material. The second is that the measured frictional strength depends not only on the total stresses applied to a material, but also on the pressure of any fluid within the pores. These factors are discussed more fully in the next chapter.

4.13 Friction and faulting in the earth's crust

Earlier in this chapter we reviewed the results of experiments on rock friction, that is, the relation between shear stress and normal stress for sliding on existing planes in rocks. Friction might be expected to provide a "lower bound" for rock fracture, since we expect that it will take a higher stress to produce a new fracture in a rock, than to cause sliding on an old fracture. We saw that for normal pressures greater than 200 MPa (equivalent to 2 kilobars), experimental data suggested that $C = 50$ MPa (0.5 kilobars) and $\mu = 0.6$, for a wide range of rock types (Byerlee's law).

Some simple calculations can help us visualize what these numbers mean. The weight of a column of rock (density 2800 kg m^{-3}) is about $28h$ kPa, where h is the height of the column in metres. So each kilometre of burial in the crust corresponds, very roughly, to an increase in vertical rock pressure of about 28 MPa. 200 MPa corresponds to a depth of about 6–8 km.

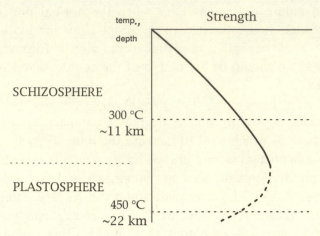

Fig. 4.20. Changes with depth in the strength of the lithosphere (after Scholz, 1990).

The distribution of earthquakes suggests that most faulting takes place at depths in the range from the earth's surface to about 15 km, and studies of lithospheric flexure (Chapter 7) indicate that the elastic part is about 20–50 km thick. Below depths of about 20 km, the lithosphere can deform slowly in a ductile manner ("flow") in response to relatively low shear stresses, rather than "break". So we are mainly interested in the strength of rocks, and in friction along faults, at depths down to about 20 km, corresponding roughly to vertical pressures up to about 600 MPa. Pressures on the nearly vertical surfaces typical of most faults might be considerably less. At 10 km, Byerlee's law suggests that shear stresses of the order of 100 MPa (1 kbar) are necessary to cause sliding along unlubricated faults. As we will see in the next chapter, much lower stresses can cause slip along faults lubricated by high fluid pressures.

Below depths of about 20 km, earthquakes are much less common, and deep focus earthquakes (which extend down to about 600 km) are found only in regions where relatively cold lithosphere is sinking into the ductile mantle along subduction zones. The geological and geophysical evidence indicate that the continental lithosphere may be divided into an upper region (called the *schizosphere* by Scholz, 1990) which is relatively elastic and brittle, and a lower region (called the *plastosphere* by Scholz) which is relatively ductile or "plastic", in the sense that it tends to flow, rather than fracture (Figure 4.20). Experiments have shown that ductile deformation is favoured both by higher confining pressures (though as we have seen, these tend to increase friction on fractures), and by higher temperatures. Both factors tend to make rocks more ductile at greater depths, particularly where temperatures increase

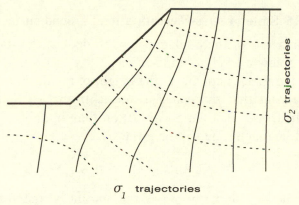

Fig. 4.21. Stress trajectories in rock close to a steep slope.

above about 300 °C, which in continental crust generally means at depths of about 15 km. There are, however, many complicating factors: the ductility of rocks depends on their mineralogy and water content, and on the rate of deformation (high shear rates produce more brittle behaviour).

4.14 Stress fields and trajectories

As we have seen, the state of stress at a point is fully described if we know the magnitude and orientation of the principal stresses at that point, i.e., if we know the size, shape and orientation of the stress ellipsoid.

A line whose tangent at every point is in the direction of a principal stress is called a *stress trajectory*. The way that the state of stress varies through space (the *stress field*) can be represented by a network of stress trajectories. As the principal stresses are always at right angles to each other, the stress trajectories form an orthogonal network.

With a little experience it is possible to sketch an approximation to the stress trajectories, by making use of the following rules-of-thumb:

(1) near a level surface, the trajectory of the maximum principal stress should be almost normal to the ground surface;
(2) stress trajectories for maximum and minimum principal stresses are normal to each other;
(3) at depth, the stress field tends to become hydrostatic, and the maximum stress trajectory is vertical.

A typical example is shown in Figure 4.21. Further discussion of the state of stress below the surface of the earth is given in Chapter 8.

4.15 State of stress beneath a level ground surface

We will now make use of these relationships to deduce the state of stress at shallow depths below a level ground surface.

Assume that the stresses are due to the weight of the near surface materials (e.g., sand, clay) and that the materials are isotropic, so that they do not tend to transmit stresses more in one direction than in another. Then σ_1 will be vertical, and its magnitude will be given by

$$\sigma_1 = \rho g z = \gamma z = p \tag{4.25}$$

where ρ is the density and $\gamma = \rho g$ is the specific weight of the material (weight per unit volume).

The other two principal stresses will be equal to each other and to some proportion k of the maximum principal stress:

$$\sigma_2 = \sigma_3 = k\sigma_1 = kp \tag{4.26}$$

k is called the *coefficient of pressure*.

Now consider an imaginary vertical wall within the ground. If we move the wall laterally it will tend to compress the ground on one side (increase σ_2) and relax it (decrease σ_2) on the other. At first the earth will respond elastically to the changes in σ_2 but soon failure will take place, and there will be permanent (ductile) deformation. Deformation will continue until a condition of equilibrium is re-established.

The state of equilibrium of such an earth mass is called a *Rankine state* after the British engineer W. J. M. Rankine (1820–72). In the case where the weight of the earth produces the ductile deformation that re-establishes equilibrium, the Rankine state is said to be the *active state*. On the other side of the retaining wall, where the earth mass is compressed by the action of the wall, the weight of the earth plays only a passive role, so the Rankine state is said to be passive. The active state is perhaps the state of most interest because in nature downslope movements will tend to relax confining pressures in regions of positive relief, so that an approximation to the active Rankine state of stress may be established.

The nature of the stresses in the active Rankine state are such that the earth is just about to fail. The Navier–Coulomb criterion defines this condition as

$$\tau_c = C + \sigma \tan \phi$$

The critical value of τ is achieved, for a given σ_1, as the value of σ_2 is gradually relaxed (cf. Figure 4.22). At this point, it is possible to show from

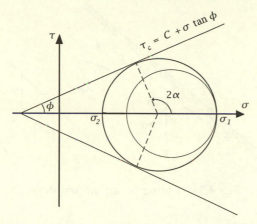

Fig. 4.22. Failure resulting from relaxation of σ_2.

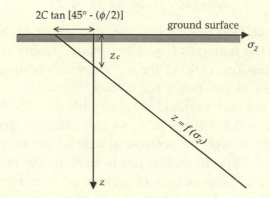

Fig. 4.23. Variation of σ_2 with depth in the active Rankine state.

the geometry of the Mohr diagram that

$$\sigma_2 \tan^2 \alpha = \sigma_1 - 2C \tan \alpha \qquad (4.27)$$

and because $\alpha = (90° + \phi)/2$, this is equivalent to

$$\sigma_2 = \sigma_1 \tan^2 [45° - (\phi/2)] - 2C \tan [45° - (\phi/2)] \qquad (4.28)$$

For the full derivations see Terzaghi (1943, pp. 29–41). This relationship is plotted in Figure 4.23. Equation (4.28) and Figure 4.23 show that σ_2 is negative (i.e., there is a state of tension) near the ground surface for materials that have any cohesive strength. At the surface the maximum tensile stress is $\sigma_2 = 2C \tan[45° - (\phi/2)]$. The zone of tension extends down to a critical depth, z_c which depends on the cohesion, the angle of internal friction, and

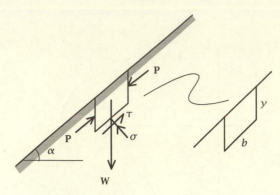

Fig. 4.24. Sliding on an infinite slope.

the specific weight of the material:

$$z_c = (2C/\gamma)\tan\left[45° + (\phi/2)\right] \tag{4.29}$$

As we might expect, the equation indicates that a state of tension is not possible for cohesionless materials (e.g., loose sand). Weakly cohesive materials commonly show tension cracks at the surface, which become closed by earth pressure only a few metres below the surface.

For a strong rock, for example granite, with $\phi = 60$ degrees, $C = 8 \times 10^6\,\mathrm{N\,m^{-2}}$, and $\gamma = 2.7 \times 10^4\,\mathrm{N\,m^{-3}}$, we can calculate that $z_c = 2.2 \times 10^3$ m. So, theoretically it might be possible to have a state of tension within the crust to a depth of 2 km. In reality, this is unlikely: the cohesion and angle of internal friction of large masses of granite are both likely to be reduced below the values given, due to the presence of fractures, joints, and so on. But more importantly, as we shall see later, actual measurements indicate the presence of large horizontal compressive stresses at depth within the crust.

4.16 Sliding on slopes

In the preceding sections we have been considering the state of stress below the ground surface. Now we wish to consider the conditions under which stress will produce failure – in other words, the stress required to exceed the strength and produce sliding or slumping.

4.16.1 Sliding on an infinite slope

First we consider sliding of a homogeneous material on a long slope, inclined at a constant angle α (Figure 4.24). If the material is uniform in weight, the shear stress increases uniformly with depth. Of course, real soils and rocks are rarely homogeneous, but the assumption is roughly true for poorly bedded clay or sand. If the strength is also uniform, sliding takes place along a plane surface. Both shear stress and strength increase with depth, and the failure plane is located where the stress just equals the strength.

Consider an element of the slope of unit width perpendicular to the figure, of breadth b down the slope, and of depth y. The forces acting on this element are its weight W, and the forces N, T due to the normal and shear stresses σ, τ acting on its lower surface. For an infinitely long slope the pressures acting on the sides P cancel out and need not be considered further. Considering the equilibrium normal and parallel to the slope gives

$$N = W \cos \alpha = [\gamma y (b \cos \alpha)] \cos \alpha$$

$$T = W \sin \alpha = [\gamma y (b \cos \alpha)] \sin \alpha$$

Sliding is resisted by the strength of the material, which we can represent by using the Navier–Coulomb criterion. If we express this in terms of force rather than stress, it becomes

$$S = \tau b = Cb + N \tan \phi$$

where S is the strength (expressed as a force) and ϕ is the angle of friction. The slope will fail when $T = S$ so that

$$\gamma y \sin \alpha \cos \alpha = C + \gamma y \cos^2 \alpha \tan \phi$$

or

$$\tan \alpha = (C / \gamma y \cos^2 \alpha) + \tan \phi \tag{4.30}$$

This equation gives the critical, or limiting condition for a landslide. Given the soil unit weight γ and the strength parameters C and ϕ, Equation (4.30) can be used to determine the slope angle α for which failure will occur at a given depth y, or the depth y at which failure will occur for a given slope α. The latter problem will often not have a solution: existing slopes are evidently stable at all depths for the present values of C, ϕ, γ, and α.

Inspection of Equation (4.30) suggests that a slope may become unstable in several ways:

- α increases (e.g., when a river undercuts and steepens a slope, which increases T while decreasing N and, therefore, S);
- either C or ϕ decreases (e.g., from chemical weathering of the near-surface materials);
- γ increases (e.g., by loading with a structure or landslide debris from upslope).

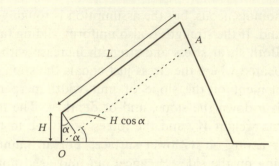

Fig. 4.25. Sliding of a rock slab.

It turns out that the most common cause of landslides is not any of these. Rather, most landslides are probably caused by an increase in pore water pressure from rain or snow melt, which decreases the fraction of N that is effective in producing the frictional component of strength. We consider this problem further in the next chapter.

Two special cases of Equation (4.30) are the following.

(1) If $C = 0$, then $\alpha = \phi$. For cohesionless material, the slope can stand no steeper than the angle of internal friction (a result we mentioned earlier). This is a simple and striking result: the largest possible slope angle in a slope of cohesionless material is exactly equal to the rate at which frictional strength increases with normal stress, as can be found in laboratory experiments such as those discussed in Section 4.4.

(2) If $\phi = 0$ then $\sin \alpha \cos \alpha = C/\gamma y$ or

$$(\sin 2\alpha)/2 = C/\gamma y \qquad (4.31)$$

For example, wet clays may have very low angles of internal friction. Over a short time period, the apparent friction angle may actually be zero (i.e., there is no increase in strength with N). We explain how this can happen when we consider the distinction between the total stress and the "effective" stress in the next chapter.

4.16.2 *Sliding of a rock slab (after Cruden, 1976)*

Next we consider the case where a slab of rock slides down a slope because the friction along a weak plane, e.g., a bed or joint plane, is exceeded by the downslope component of the gravity force. This differs from the previous case, because now we assume that the relative weakness of bedding or joint surfaces determines the direction in which sliding can take place.

Figure 4.25 shows, schematically, a situation often closely approximated in the Canadian Rocky Mountains. The side of a mountain is a dip slope, of length, L with both dip and slope towards a valley at an angle of α degrees. At the base of the slope, there is an undercut of height H (produced, for example, by the lateral erosion of a river flowing in the valley).

For L much larger than H, the volume of the slab at risk of sliding is approximately $LH \cos \alpha$, and the gravity force acting on it, per unit length, is $\gamma H \cos \alpha$. The problem is identical to that of sliding on an infinite slope, with H replacing y.

Dangerous rock slides will be the ones for which H can become quite large before sliding takes place. If H is small when sliding takes place the volume of the slide will be small, and so the effects are not likely to be severe. Substituting H for y in Equation (4.30) and solving for H gives

$$H = C/[\gamma \cos \alpha (\sin \alpha - \cos \alpha \tan \phi)] \qquad (4.32)$$

Cruden (1976) points out that, if $\phi = 30$ degrees, then a slide taking place on an angle $\alpha = 31$ degrees will have 30 times the volume of one taking place on an angle $\alpha = 60$ degrees. So catastrophic slides will tend to occur on bedding planes dipping almost at the angle of friction.

Cruden has shown that most of the known historic and prehistoric large slides in the Rocky Mountains fit this model. The most destructive historic slide is that which destroyed the town of Frank, Alberta in 1903. Most early reports (and most textbooks) show the slide surface as a joint, but a reinvestigation by Cruden and Krahn (1973) shows that it was actually a bedding plane.

4.16.3 Sliding on finite slopes: rotational slumps

Consider the case of landsliding on a slope that is not very long and has no weaker planes likely to form slide surfaces. The stress distribution beneath a finite slope is more complex than that below an infinite slope. Nevertheless, we can sketch in the stress trajectories, as we did in Figure 4.21, or we can determine them from elasticity theory. From our study of the Mohr circle we know that the angle between the trajectories and the failure surface will be $45° + (\phi/2)$, so it is possible to determine this failure surface as the probable slide surface (Figure 4.26).

From Figure 4.26 we can see that the failure surface must be a curved, concave-up, surface. The failure surface must break the ground surface close to the base of the slope, otherwise movement on the failure surface would have to raise a large mass of material beyond the base of slope. The actual

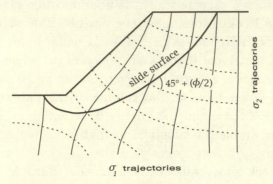

Fig. 4.26. Sliding on a finite slope. The slide surface is assumed to make an angle of $45° + \phi/2$ with the σ_2 trajectories.

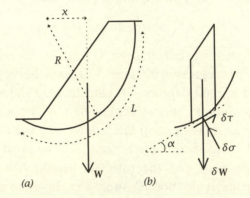

Fig. 4.27. Rotational slide on a cylindrical surface: (*a*) general section of slope. In (*b*) the approximate base length of the vertical slice is ℓ_i; its radius of curvature is R.

failure surface can be approximated by a circular (cylindrical) surface as shown in Figure 4.27. By equating moments about the axis of the cylinder, we find that the equilibrium condition for a circular slide on the verge of failure is

$$\bar{\tau}LR = Wx$$

where $\bar{\tau}$ is the average shear stress required for failure along the slide surface. For a slide to take place, the force causing failure, given by the right-hand side, must just equal the resistance to failure, given by the left-hand side. For a stable slope, the resistance must exceed the disturbing force. The ratio between the two forces (left-hand side/right-hand side) is called the *factor of safety F*.

For a purely cohesive material, the strength does not vary with position along the failure surface and the equation above can be easily solved. If we substitute the cohesive strength C for $\bar{\tau}$, then the factor of safely is given by

$$F = CLR/Wx$$

Typically, F is calculated for a large number of potential failure circles to determine the smallest value for a particular slope. This corresponds to the most likely failure surface. Most materials have a frictional component to their strength. Because the normal stress σ varies along the slip surface, so does the strength. A simple moment balance can no longer be performed. In these cases, the slope is divided into a series of thin slices and the moment determined for each slice (a process analogous to calculating the area under a curve by dividing the area into many thin slices and summing the area of individual slices). For this case, overall moment equilibrium is calculated from the sum of the moments for all the individual slices. Equating the disturbing and resisting moments for the slope just at the point of failure (limiting equilibrium) gives (Figure 4.27(*b*))

$$\sum \delta W R_i \sin \alpha = \sum \delta \tau \ell_i R_i$$

where R_i and ℓ_i are the radius and base length for individual slices. In this analysis, we have neglected any contribution to the moment balance from forces acting on the sides of the slices. If, as before, we replace τ with an expression for the material strength (in this case the Navier–Coulomb criterion $\tau = C + \sigma \tan \phi$), the ratio of the resisting to the disturbing (sums of) moments is again the factor of safety:

$$F = \frac{\sum (C\ell_i + \delta W \cos \alpha \tan \phi)}{\sum \delta W \sin \alpha}$$

where we have substituted $\delta W \cos \alpha$ for $\sigma \ell_i$.

Neglecting the inter-slice forces can underestimate F by as much as 60 per cent. A number of more complicated methods have been developed that provide a more complete treatment of the static equilibrium of the slope and more accurate estimates of F. These methods are all iterative and require more calculations than can be conveniently carried out by hand, although they may easily be performed by modern personal computers (Nash, 1987). The computer program SLICES gives examples of two simple algorithms. Improving the analysis requires information about the distribution of stresses, the pore pressures, and the material strength beneath a slope. Iverson and Reid (1992) give a sophisticated theoretical analysis that shows how some of these factors affect stability. Often, knowledge of these factors is so poor that it is difficult to justify the use of advanced methods.

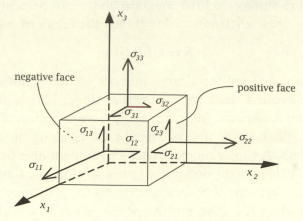

Fig. 4.28. Sign convention for faces and stresses in three dimensions. All stresses shown are positive.

4.17 Tensor components of stress

Now that we have had the opportunity to become familar with some applications of stress components, we will discuss the mathematical techniques that are required to describe the state of stress in three dimensions and which are employed to derive the principal stresses from an array of components measured in directions that do not coincide with the principal stress directions.

4.17.1 Conventions

We will choose right-handed reference axes. Remember that, in Section 4.7, we defined positive faces to be those having a pole pointing in the positive direction. Components of stress will be defined to be positive if they act in the positive direction on a positive face, or in the negative direction on a negative face (Figure 4.28). It is a consequence of this definition that compressional stresses are negative. This convention differs from that adopted by some geologists and geophysicists, but it is the usual convention in other disciplines. Remember that the first subscript gives the pole (normal) to the plane on which the stress acts, and the second gives the direction of the stress component.

4.17.2 Components of stress

We have seen that we can resolve the stress vector acting on any arbitrary plane into three stress components (one normal to the plane, two within the plane and parallel to two reference directions). So, we can write down the stress vectors acting on each of the orthogonal planes normal to the x-, y- and z-directions:

$$\boldsymbol{\sigma}_x = \sigma_{xx}\mathbf{i} + \sigma_{xy}\mathbf{j} + \sigma_{xz}\mathbf{k}$$
$$\boldsymbol{\sigma}_y = \sigma_{yx}\mathbf{i} + \sigma_{yy}\mathbf{j} + \sigma_{yz}\mathbf{k} \tag{4.33}$$
$$\boldsymbol{\sigma}_z = \sigma_{zx}\mathbf{i} + \sigma_{zy}\mathbf{j} + \sigma_{zz}\mathbf{k}$$

We commonly refer to the coefficients as the components of stress (see Section 4.8), and write them as an array of numbers:

$$\sigma_{ij} = \begin{bmatrix} \sigma_{xx} & \sigma_{xy} & \sigma_{xz} \\ \sigma_{yx} & \sigma_{yy} & \sigma_{yz} \\ \sigma_{zx} & \sigma_{zy} & \sigma_{zz} \end{bmatrix} \quad \text{or} \quad \begin{bmatrix} \sigma_{11} & \sigma_{12} & \sigma_{13} \\ \sigma_{21} & \sigma_{22} & \sigma_{23} \\ \sigma_{31} & \sigma_{32} & \sigma_{33} \end{bmatrix} \tag{4.34}$$

We will show that this array has the mathematical properties (yet to be defined) of a *matrix*, or *second-order tensor*, and that it is sufficient to define the state of stress at a point.

Each row of the stress tensor consists of the stresses acting at one reference plane, and each column consists of the stresses acting in one reference direction. The stress vector acting on a plane normal to one of the axes is a *row vector*, a special type of matrix, with three columns but only one row:

$$\sigma_i = [\sigma_{i1}, \sigma_{i2}, \sigma_{i3}] \tag{4.35}$$

Similarly, the stress components in a given direction, say the j-direction, form a *column vector* σ_j:

$$\sigma_j = \begin{bmatrix} \sigma_{1j} \\ \sigma_{2j} \\ \sigma_{3j} \end{bmatrix} \tag{4.36}$$

(In this case, the subscript notation used does not indicate whether the vector is a row or column vector.) We may regard a matrix as a generalization of the concept of a vector.

4.17.3 Components in two dimensions

Earlier in this chapter, we resolved the components of stress acting on a plane surface. We now repeat the process to demonstrate how the components of different stress vectors are related to each other. Let us write the components

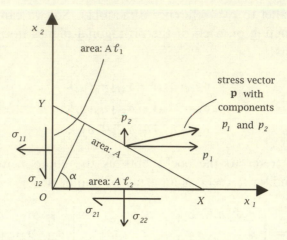

Fig. 4.29. Components of a stress vector (in two dimensions).

of a stress vector **p** acting on a surface of area A as shown in Figure 4.29. Note that in this diagram we use the numerical index notation, and ℓ_1 and ℓ_2 for direction cosines. This is just the same type of figure as Figures 4.13 and 4.14 except for a change in the notation and/or the direction of the reference axes. We resolve the stress components in the direction of the reference axes, but we do not assume that the reference axes coincide with the principal stresses, as we did in Figure 4.14.

As before, we convert stresses into forces by multiplying them by the area over which they act. We then equate forces; for equilibrium in the x_1-direction we have

$$p_1 A = \sigma_{11}\ell_1 A + \sigma_{21}\ell_2 A$$

or

$$p_1 = \sigma_{11}\ell_1 + \sigma_{21}\ell_2 \tag{4.37}$$

In the x_2-direction we have

$$p_2 A = \sigma_{12}\ell_1 A + \sigma_{22}\ell_2 A$$

or

$$p_2 = \sigma_{12}\ell_1 + \sigma_{22}\ell_2 \tag{4.38}$$

We know that $\sigma_{21} = \sigma_{12}$, so we could also write this system of equations as

$$p_1 = \sigma_{11}\ell_1 + \sigma_{12}\ell_2$$
$$p_2 = \sigma_{21}\ell_1 + \sigma_{22}\ell_2 \tag{4.39}$$

This equation is the two-dimensional form of *Cauchy's formula*. It is one of the fundamental equations of continuum mechanics, because it enables us to determine the stresses in the coordinate directions (however they are chosen) for any plane from the stress matrix in that coordinate system.

This form of Cauchy's formula has the advantage that it can easily be converted into matrix or tensor notation. In fact, such notation has been developed specifically for the purpose of handling sets of linear equations, of which Equation (4.39) is a simple example.

4.17.4 Matrix notation

The following are three different ways of writing Equations (4.39) in notation of varying degrees of compactness, using the definition of *matrix multiplication*:

$$\begin{bmatrix} p_1 \\ p_2 \end{bmatrix} = \begin{bmatrix} \sigma_{11} & \sigma_{12} \\ \sigma_{21} & \sigma_{22} \end{bmatrix} \begin{bmatrix} \ell_1 \\ \ell_2 \end{bmatrix} \tag{4.40}$$

or

$$[p_i] = [\sigma_{ij}][\ell_j] \qquad i,j = 1,2 \tag{4.41}$$

or

$$\mathbf{p} = \Sigma \mathbf{l} \tag{4.42}$$

where

$$\mathbf{p} = \begin{bmatrix} p_1 \\ p_2 \end{bmatrix}, \quad \mathbf{l} = \begin{bmatrix} \ell_1 \\ \ell_2 \end{bmatrix}, \quad \Sigma = \begin{bmatrix} \sigma_{11} & \sigma_{12} \\ \sigma_{21} & \sigma_{22} \end{bmatrix}$$

The square brackets indicate that the contents are a matrix (some authors use parentheses instead of square brackets). We use the normal convention that bold-face capital letters refer to the matrix and bold-face lower-case letters refer to a row or column vector (a matrix with only one row or column) and the subscripted lower-case letters refer to a typical element of the matrix or vector. In this notation, a matrix is not simply an array of numbers, but one that can be manipulated according to a set of rules (matrix algebra). The set of rules has been designed so that it is useful for the manipulation of linear equations. In general,

$$\mathbf{C} = \mathbf{AB}$$

is defined to have elements c_{ij} given by

$$c_{ij} = \sum_{k=1}^{N} a_{ik} b_{kj}$$
$$= a_{i1}b_{1j} + a_{i2}b_{2j} + \cdots + a_{iN}b_{Nj} \tag{4.43}$$

Note that matrix multiplication is defined only if the number of columns of the first matrix, **A**, is equal to the number of rows of the second matrix, **B**. Note also that the rules for matrix multiplication differ substantially from those of ordinary algebraic multiplication, and that the order of multiplication makes a difference. Division by a matrix is not defined, but an analogous operation can be carried out by multiplying by the inverse of a matrix (see below). Many of the other rules are, however, relatively straightforward. In the following paragraphs, we introduce a few of these rules, and some of the notation. For a more complete treatment of matrix algebra see a text such as that by Potter and Goldberg (1987).

Matrices with the same number of rows as columns are said to be square. If we make a new matrix by interchanging the rows and columns of a square matrix **A**, we say the new one is the *transpose* $\mathbf{A}^T$ of the old one. Thus

$$[a_{ji}] = [a_{ij}]^T \tag{4.44}$$

If the transpose of a matrix is the same as the original matrix, as it is for the stress matrix (because $\sigma_{21} = \sigma_{12}$, $\sigma_{23} = \sigma_{32}$, $\sigma_{31} = \sigma_{13}$), then we say the matrix is symmetrical. A special symmetrical square matrix is the unit matrix, **I**, for which all terms are zero except those forming the principal diagonal (for which $i = j$); these terms are all equal to 1.

Matrix addition is defined very simply:

$$\mathbf{C} = \mathbf{A} + \mathbf{B}$$

if and only if

$$c_{ij} = a_{ij} + b_{ij} \tag{4.45}$$

So two matrices can be added only if they have the same number of both rows and columns. Similar rules apply to subtraction.

Division by a matrix is not defined: the equivalent is multiplication by the *inverse matrix*. The inverse of a square matrix **A**, if it exists, is denoted $\mathbf{A}^{-1}$ and has the property that:

$$\mathbf{A}^{-1}\mathbf{A} = \mathbf{I} \tag{4.46}$$

Formally, the inverse matrix can be used to solve a set of linear equations. For example, the equation:

$$\Sigma \mathbf{l} = \mathbf{p}$$

can be solved for **l** (the vector of direction cosines, if they were unknown, given the stress matrix and stress vector) by multiplying both sides by the inverse of Σ:

$$\Sigma^{-1}\Sigma l = \Sigma^{-1}p$$

or

$$l = \Sigma^{-1}p$$

In practice, this method of solving the equations is not used, because numerically it is more difficult to obtain the inverse matrix than to solve the equations! But it is a useful notation, and inverse matrices do have important uses in other applications.

4.17.5 Tensor notation

Tensor notation does essentially the same work as matrix notation, but in a different way. Part of the notation was introduced in Section 2.16 but we review and extend it in this section.

Instead of writing a sum using summation notation, we use a repeated ("dummy") index to indicate the summation. For example, the sum

$$\sum_{j=1}^{3} \sigma_j \ell_j$$

is simply written $\sigma_j \ell_j$, where the range of summation, from j = 1 to j = 3 in a three-dimensional context, is understood as part of the convention.

If an index is not repeated in the same expression, then no summation is implied. In order to avoid ambiguity no index may be repeated more than once in the same expression (i.e., repeated indices always come in pairs).

So, returning to Equation (4.39), it can be written in tensor notation:

$$p_i = \sigma_{ij}\ell_j \tag{4.47}$$

We can see how useful this is if we write out the full set of equations (*Cauchy's formulae*) for three dimensions:

$$
\begin{aligned}
p_1 &= \sigma_{11}\ell_1 + \sigma_{12}\ell_2 + \sigma_{13}\ell_3 \\
p_2 &= \sigma_{21}\ell_1 + \sigma_{22}\ell_2 + \sigma_{23}\ell_3 \\
p_3 &= \sigma_{31}\ell_1 + \sigma_{32}\ell_2 + \sigma_{33}\ell_3
\end{aligned}
\tag{4.48}
$$

p_i is the stress vector acting on a plane whose direction cosines are a vector ℓ_j of unit length, normal to the plane. Therefore we can interpret the stress tensor σ_{ij} in Cauchy's formulae as a way of relating the vector p_i to the vector ℓ_j.

A useful symbol that plays much the same role for tensors as the unit matrix does for matrices is *Kronecker's delta* δ_{ij}. It takes the value unity when $i = j$ but is zero otherwise. The equivalent of Equation (4.46) is therefore

$$a_{ik}^{-1} a_{kj} = \delta_{ij}$$

4.17.6 Stress invariance under rotation of axes

Suppose we want to transform the tensor components σ_{ij} of stress related to one set of orthogonal axes x_1, x_2, x_3 to the equivalent set of tensor components σ'_{ij} related to a different set of axes x'_1, x'_2, x'_3, produced by rotation of the original axes. The new set of axes can be related to the old set by a set of direction cosines ℓ_{ij}, where the first subscript indicates the new axis, and the second subscript indicates the old axes (see Section 2.16). Each direction cosine is a measure of the angle between one of the original (old) axes, and one of the rotated (new) axes.

It can be shown that (in tensor notation) the new stress components are related to the old stress components by the equation

$$\sigma'_{ij} = \ell_{ip} \ell_{jq} \sigma_{pq} \tag{4.49}$$

Note that we must use two dummy indices p and q to carry out the two summations. To make the meaning of Equation (4.49) clear, we write it out in full for the case of $i = 1, j = 1$:

$$\begin{aligned}
\sigma'_{11} = {} & \ell_{11}\ell_{11}\sigma_{11} + \ell_{11}\ell_{12}\sigma_{12} + \ell_{11}\ell_{13}\sigma_{13} && \text{(sum over } q, \text{for } p = 1) \\
& + \ell_{12}\ell_{11}\sigma_{21} + \ell_{12}\ell_{12}\sigma_{22} + \ell_{12}\ell_{13}\sigma_{23} && \text{(sum over } q, \text{for } p = 2) \\
& + \ell_{13}\ell_{11}\sigma_{31} + \ell_{13}\ell_{12}\sigma_{32} + \ell_{13}\ell_{13}\sigma_{33} && \text{(sum over } q, \text{for } p = 3)
\end{aligned}$$

To obtain the whole of Equation (4.49) we would have to repeat the summation for all nine components of the stress tensor, which would take up a lot of space, and be very confusing to the reader. We now see the advantage of the compact tensor notation.

In matrix notation we could write the equation as

$$\begin{aligned}
[\sigma'_{ij}] &= [\ell_{ij}][\sigma_{ij}][\ell_{ji}] \\
&= [\ell_{ij}][\sigma_{ij}][\ell_{ij}]^T
\end{aligned}$$

or as

$$\boldsymbol{\Sigma}' = \mathbf{L}\boldsymbol{\Sigma}\mathbf{L}^T$$

To simplify matters, let us consider only two dimensions (see Figure 4.29). Equation (4.47) solves the problem of finding the unknown stress components on a plane of known orientation: that is, given normal and shear stresses $\sigma_{11}, \sigma_{12}, \sigma_{22}, \sigma_{21}$ acting on two orthogonal planes, what are the corresponding stress components acting on a different plane of known orientation? The answer is given by Cauchy's formula.

Now, consider the problem in a different way: given the stress components acting on the two orthogonal planes, what is the orientation of a plane on which a particular, but different, set of stress components act? The only case that we will consider is the one of most interest: what is the orientation of the plane normal to a principal stress (i.e. a principal plane)? That is, given the stress components σ_{ij} on the two orthogonal planes, as in Figure 4.29, what is the orientation (given by the direction cosines l_1, l_2) of a plane on which the total stress is a normal stress, i.e., on which the shear stresses are equal to zero, and what is the magnitude σ of this normal stress? For such a plane, the components of **p** are given by

$$p_1 = \sigma \ell_1 \quad \text{and} \quad p_2 = \sigma \ell_2$$

These relations provide the additional constraint needed to find ℓ_1, ℓ_2, and σ. Actually, there will be two such planes, perpendicular to each other since, as we shall see, the normal stress magnitude σ can have two values, σ_1 and σ_2, which are, of course, the two principal stresses. Combining the relations above with Cauchy's formula, it follows that

$$p_1 = \sigma \ell_1 = \sigma_{11}\ell_1 + \sigma_{12}\ell_2$$
$$p_2 = \sigma \ell_2 = \sigma_{21}\ell_1 + \sigma_{22}\ell_2$$

We can rewrite these equations as:

$$(\sigma_{11} - \sigma)\ell_1 + \sigma_{12}\ell_2 = 0$$
$$\sigma_{21}\ell_1 + (\sigma_{22} - \sigma)\ell_2 = 0$$

(4.50)

or, in matrix notation:

$$(\Sigma - \sigma \mathbf{I})\mathbf{l} = \mathbf{0}$$

The theory of equations tells us that this has a solution (for ℓ_1, ℓ_2 nonzero) only if

$$|\Sigma - \sigma \mathbf{I}| = 0 \qquad (4.51)$$

that is, if

$$\begin{vmatrix} (\sigma_{11} - \sigma) & \sigma_{12} \\ \sigma_{21} & (\sigma_{22} - \sigma) \end{vmatrix} = 0 \qquad (4.52)$$

Writing out the determinant gives

$$(\sigma_{11} - \sigma)(\sigma_{22} - \sigma) - \sigma_{12}\sigma_{21} = 0$$

or

$$\sigma^2 - (\sigma_{11} + \sigma_{22})\sigma + (\sigma_{11}\sigma_{22} - \sigma_{12}^2) = 0 \qquad (4.53)$$

This is a quadratic equation in σ with two roots σ_1 and σ_2, which are the two principal stresses. The theory of equations tells us that for a quadratic equation of the type

$$ax^2 + bx + c = 0$$

the two roots are given by

$$x = \frac{-b \pm \sqrt{(b^2 - 4ac)}}{2a}$$

The sum of the roots is $-b/a$, and the product of the roots is c/a; these are both invariants (they do not vary with x).

In the case of the quadratic equation in σ, $a = 1$, $b = -(\sigma_{11} + \sigma_{22})$ and $c = (\sigma_{11}\sigma_{22} - \sigma_{12}^2)$, so that the solutions σ_1 and σ_2 are related by

$$\sigma_1 + \sigma_2 = \sigma_{11} + \sigma_{22} = I \quad \text{(first stress invariant)}$$

$$\sigma_1\sigma_2 = \sigma_{11}\sigma_{22} - \sigma_{12}^2 = II \quad \text{(second stress invariant)}$$

The two principal stresses σ_1 and σ_2 are normal to two planes with direction cosines, say, ℓ_{11}, ℓ_{21} and ℓ_{12}, ℓ_{22}.

σ_1 and σ_2 are frequently called the *eigenvalues* of the matrix Σ and the two pairs of direction cosines are called the *eigenvectors* of the matrix. I is known as the *trace* of the stress matrix and II is the determinant of the stress matrix. Both of these may be determined directly from any known stress tensor, but since they are completely determined by the values of the principal stresses, they do not change (as the values in the stress tensor will) with rotation of the reference axes.

The eigenvectors are easily determined from σ_1 and σ_2 by using Equations (4.50), remembering to apply the condition that $\ell_1^2 + \ell_2^2 = 1$.

4.17.7 Numerical example

Let

$$[\sigma_{ij}] = \begin{bmatrix} 3 & -1 \\ -1 & 3 \end{bmatrix}$$

Then σ_1 and σ_2 are the roots of

$$\begin{vmatrix} (3-\sigma) & -1 \\ -1 & (3-\sigma) \end{vmatrix} = 0$$

i.e.

$$\sigma^2 - 6\sigma + 8 = 0$$

or

$$(\sigma - 2)(\sigma - 4) = 0$$

So the solutions are:

$$\sigma_1 = 4, \qquad \sigma_2 = 2$$

This means that, by a suitable change of axes we can transform our original stress matrix into the matrix whose only nonzero terms are the normal stresses along the principal diagonal:

$$\begin{bmatrix} 3 & -1 \\ -1 & 3 \end{bmatrix} \Rightarrow \begin{bmatrix} 4 & 0 \\ 0 & 2 \end{bmatrix}$$

We can easily confirm that the two invariants are the same for the two stress matrices:

$$\begin{array}{ccccc} & & \text{first matrix} & \text{second matrix} & \\ I & = & (3+3) & = & (4+2) & = 6 \\ II & = & (3 \times 3 - 1) & = & (4 \times 2 - 0) & = 8 \end{array}$$

- For the first root, $\sigma = 4$, and Equation (4.50) is:

$$-1\ell_1 - 1\ell_2 = 0$$
$$-1\ell_1 - 1\ell_2 = 0$$

so $\ell_1 = -\ell_2$ and because $\ell_1^2 + \ell_2^2 = 1$, $\ell_1 = 1/\sqrt{2}$, and $\ell_2 = -1/\sqrt{2}$ (or vice versa).

- For the second root, $\sigma = 2$, and

$$1\ell_1 - 1\ell_2 = 0$$
$$-1\ell_1 + 1\ell_2 = 0$$

so $\ell_1 = \ell_2 = \pm 1/\sqrt{2}$.

4.17.8 Stress invariants

In three dimensions, there are three principal stresses. The equivalent to Equation (4.50) is a system of three equations, which yields a cubic equation with three roots. Three sets of direction cosines define the three planes to which the three principal stresses are normal. The arithmetic is a little clumsier, but still straightforward.

In three dimensions there are three (not two) stress invariants:

First stress invariant:

$$I = \sigma_{ii}$$
$$= \sigma_{11} + \sigma_{22} + \sigma_{33} = \sigma_1 + \sigma_2 + \sigma_3 \tag{4.54}$$

i.e., the sum of the normal stresses (also called the *trace* of the stress matrix) is a constant, and equal to the sum of the three principal stresses.

Second stress invariant:

$$II = \sigma_{ii}\sigma_{jj} - \sigma_{ij}\sigma_{ij} \tag{4.55}$$

Third stress invariant:

$$III = |\sigma_{ij}| \tag{4.56}$$

i.e., the determinant of the stress tensor is a constant.

The second and third stress invariants, as written concisely above in tensor notation, are actually long, clumsy sums of products. However, we can easily determine what they are, if we remember that they are the same for all σ_{ij}, including the special case

$$[\sigma_{ij}] = \begin{bmatrix} \sigma_1 & 0 & 0 \\ 0 & \sigma_2 & 0 \\ 0 & 0 & \sigma_3 \end{bmatrix}$$

so we have

$$II = \sigma_2\sigma_3 + \sigma_3\sigma_1 + \sigma_1\sigma_2 \tag{4.57}$$
$$III = \sigma_1\sigma_2\sigma_3 \tag{4.58}$$

4.18 Why stress is a tensor

This is really just a matter of definition. Mathematicians define several different varieties of tensors. Stress can be described as a second-order Cartesian tensor because:

(i) it has two free indices over which summation can take place;

(ii) two sets (i.e., matrices) of direction cosines are required to transform a stress tensor from one Cartesian reference frame (i.e., one set of axes) to another (Equation 4.49).

A vector may be described as a first-order Cartesian tensor, because (i) it has one free index, and (ii) it needs only one matrix of direction cosines to transform it from one frame to another. This is the correct way to define a vector mathematically, i.e., that it transforms from one set of axes to another by means of an equation of the type that we first met in Chapter 2:

$$v'_i = l_{ij}v_j \qquad (4.59)$$

Not all quantities that have both direction and magnitude are vectors, though the exceptions are rather obscure (Borisenko and Tarapov, 1968, pp. 4–5). For a fuller discussion of tensors see Aris (1962), Borisenko and Tarapov (1968), or a text on continuum mechanics, such as Spencer (1980).

A simple description of a tensor, comparable to the (imprecise) statement that a vector is a quantity with both direction and magnitude, is that a tensor is a quantity with two or more directions associated with each of its components. This can be seen by considering the definition of a stress vector: it is the force per unit area acting on a plane with a particular orientation. To define a stress vector fully, one must specify not only the direction of the vector, but also the direction of the normal to the surface on which the vector is acting. Stress can be considered as a vector only when it is understood what plane it is acting on: if we have to consider not only the stress vector but also the plane it acts on as variable quantities, then we are not dealing with a vector but with a second-order tensor.

4.19 Review problems

(1) **Problem:** The two-dimensional state of stress at a point P in the $x_1 x_2$-plane is given by

$$\text{stress normal to } x_1 = 5 \text{ kPa}$$
$$\text{stress normal to } x_2 = 3 \text{ kPa}$$
$$\text{shear stress} = 2 \text{ kPa}$$

(a) In comparison with our everyday experience of stresses, are these large or small? On a slope of 0.001, how deep a river would be required to produce a shear stress of 2 kPa on the bed? (Remember that slope is expressed as the tangent of the angle, which for low slopes is equal to the sine of the angle.)

(b) Write the above data as a stress tensor (or matrix).

(c) Write the deviatoric stress matrix.
(d) What are the components of stress (in the x_1- and x_2-directions) acting on a surface inclined at an angle of 60 degrees to the x_1-axis?

Answer:

(a) The stresses are small, compared with e.g., atmospheric pressure ($\approx$ 100 kPa). They are not negligible, however. A column of water 1 m high produces a pressure on its base of about 10 kPa. On a slope of 0.001, therefore, the downslope shear stress produced by a river 1 m deep would be 0.01 kPa: so to produce 2 kPa, we would require a very large flow, 200 m deep. The stress would certainly be large enough to move very coarse sediment on the bed. This shows that what is large or small depends on the context. Often it is the difference between applied and resisting stress that is important, rather than the absolute magnitude.

(b) The stress tensor is

$$[\sigma_{ij}] = \begin{bmatrix} 3 & 2 \\ 2 & 5 \end{bmatrix}$$

The mean stress is $(\sigma_1 + \sigma_2)/2 = (\sigma_x + \sigma_y)/2$ (i.e., half the first stress invariant). In the example its value is 4 kPa.

(c) By definition, the deviatoric stress matrix is obtained by subtracting the mean stress from each term in the principal diagonal of the stress matrix (the shear-stress terms are not affected):

$$\begin{bmatrix} -1 & 2 \\ 2 & 1 \end{bmatrix}$$

(d) The surface inclined at an angle of 60 degrees to x_1 has the following direction cosines:

$$\begin{bmatrix} \ell_1 \\ \ell_2 \end{bmatrix} = \begin{bmatrix} \cos 30 \\ \cos 60 \end{bmatrix} = \begin{bmatrix} 0.866 \\ 0.500 \end{bmatrix}$$

So, the state of stress is given by Cauchy's formula (Equation 4.39):

$$\begin{bmatrix} p_1 \\ p_2 \end{bmatrix} = \begin{bmatrix} 3 & 2 \\ 2 & 5 \end{bmatrix} \begin{bmatrix} 0.866 \\ 0.500 \end{bmatrix} = \begin{bmatrix} 3.598 \\ 4.232 \end{bmatrix}$$

(2) **Problem:** Refer to the figure showing the stress ellipse (Figure 4.15). The point P (σ_x, σ_y) on the ellipse represents the stress on a plane that is not normal to one of the principal stresses. Let it be connected to the origin of the stress ellipse by a line that makes an angle θ with the σ_x-axis.

Derive an equation showing how θ is related to α, where $\ell = \cos\alpha$ and $m = \sin\alpha$ are the direction cosines of the plane that has the stresses represented by the point P.

Answer: It is easy to have the mistaken impression that θ and α are the same angle. But we can see from inspection of the stress ellipse diagram that

$$\cos\theta = \sigma_x/R, \quad \sin\theta = \sigma_y/R$$

where

$$R = \sqrt{\sigma_x^2 + \sigma_y^2}$$

whereas

$$\ell = \cos\alpha = \sigma_x/\sigma_1$$
$$m = \sin\alpha = \sigma_y/\sigma_2$$

From these equations $\tan\theta = \sigma_y/\sigma_x$ and $\tan\alpha = (\sigma_y/\sigma_x)(\sigma_1/\sigma_2)$. So the equation we require is

$$\tan\alpha = \frac{\sigma_1}{\sigma_2}\tan\theta$$

Note that this equation applies only in two dimensions.

(3) **Problem:** The following are typical values for a granite: angle of internal friction $\phi = 50$ degrees, cohesion $C = 55$ MPa, tensile strength $\sigma_o = -12$ MPa. Use these to determine the minimum shear stress at which failure by shearing can take place and the value of the maximum principal stress for this condition.

Answer: The problem can be solved quite simply using the Mohr circle, if we remember that the minimum shear stress that can produce a shear will correspond to the state of stress where the rock is just about to break by tensile fracture. This will happen when $\sigma_2 = -12$ MPa. First we construct the failure envelope, using the Navier–Coulomb criterion ($\tau_c = C + \sigma\tan\phi$) – it is the line with a slope of 50 degrees that passes through $(0, C)$. Then we construct the circle that has a left intersection of -12 MPa with the σ-axis (abscissa) and just touches the failure envelope.

The result can also be obtained numerically using the following three equations derived in the text:

$$\sigma = \sigma_1\cos^2\alpha + \sigma_2\sin^2\alpha$$
$$\tau = (\sigma_1 - \sigma_2)\sin\alpha\cos\alpha$$
$$\tau = C + \sigma\tan\phi$$

Bear in mind that $\alpha = 45° + (\phi/2)$. (We shall understand the symbols σ and ϕ as referring to the critical values for shear fracture.) Inserting the numerical values gives

$$\sigma = 0.117\sigma_1 - 10.60$$

$$\tau = 0.321\sigma_1 + 3.852$$

$$\tau = 1.192\sigma + 55$$

From these three equations in three unknowns it can be determined that $\sigma_1 = 211.6$, $\sigma = 14.2$, $\tau = 71.8$.

(4) **Problem:** Suppose that as the depth d within the crust increases, the ratio σ_1/σ_2 remains constant at $3/2$, and $\sigma_2 = \sigma_3$. Assume the density of the crust is 2500 kg m^{-3}.

(a) What depth of burial is necessary to produce a maximum shear stress of 10^7 N m^{-2}? What will be the orientation of the planes on which this shear stress acts?

(b) If the rock has an angle of internal friction of 60 degrees, and breaks under the influence of the shear stresses at a depth of 2 km to form a fault, what will be the orientation of the fault plane, the normal and shear stresses on the fault surface and the type of fault (normal, reverse, or strike-slip) that is formed?

Answer:

(a) Use the equation for τ derived for the Mohr circle:

$$\tau = \left(\frac{\sigma_1 - \sigma_2}{2}\right)\sin 2\alpha$$

It follows that there is a maximum value of τ for angles of 45 degrees, i.e., for all planes that make an angle of 45 degrees with the horizontal. For these planes

$$\tau_{max} = \left(\frac{\sigma_1 - \sigma_2}{2}\right)$$

and as $\sigma_1/\sigma_2 = 3/2$, we have $\sigma_2 = 2\sigma_1/3$ and $\tau_{max} = \sigma_1/6$. For $\tau_{max} = 10^7$ N m^{-2} and $\sigma_1 = 2500 \times 9.8 \times d$ where d is the depth in metres, we obtain $d = 2.45 \times 10^3$ m, or 2.45 km.

(b) Use the same Mohr circle equation. The rock breaks at an angle $2\alpha = 90 + 60$ degrees. The sine of this angle is $1/2$, and $(\sigma_1 - \sigma_2)/2$ is $\sigma_1/6$ as before, so the value for the shear stress at rupture is $\tau = \sigma_1/12$. At 2 km, $\sigma_1 = 2500 \times 9.8 \times 2000 = 49$ MPa, so the required value of τ is 4.08 MPa. Then we use the equation for the normal stress

$$\sigma = \left(\frac{\sigma_1 + \sigma_2}{2}\right) + \left(\frac{\sigma_1 - \sigma_2}{2}\right) \cos 2\alpha$$

to obtain $\sigma = 0.689\sigma_1 = 33.76$ MPa. Note that the shear stress is quite a bit smaller than either the normal stress on the fault plane or the principal stresses.

The orientation of the fault plane is such that the normal to the plane makes an angle of $45 + (\phi/2) = 75$ degrees with the direction of σ_1, i.e., with the vertical. So the plane itself dips at 75 degrees to the horizontal. Such a steeply dipping plane must be either a normal or reverse fault – it is clear in this case that it must be a normal fault because movement of the faulted block (between two such conjugate planes) must be downwards in response to the larger normal stress.

(5) **Problem:** Consider the following matrices:

$$\mathbf{M} = \begin{bmatrix} 2 & 1 \\ 1 & 4 \end{bmatrix}, \quad \mathbf{N} = \begin{bmatrix} 3 & 2 \\ 4 & 1 \end{bmatrix}$$

(a) Which could be a stress matrix (explain why) ?

(b) What is $\mathbf{M} + \mathbf{N}$?

(c) What is $\mathbf{MN}$?

(d) What is $\mathbf{NM}$?

(e) What is the inverse of $\mathbf{N}$?

Do the calculations by hand or pocket calculator, then check your answers using the program MATLIB.

Answer:

(a) Only matrix $\mathbf{M}$ *could* be a stress matrix, because $\mathbf{N}$ is not symmetric. Of course, to be a stress matrix, the elements of $\mathbf{M}$ would have to be stresses measured on orthogonal planes.

(b), (c), (d) The answers to these three calculations are given by the program.

(e) To obtain the inverse, go back to the definition:

$$\mathbf{A}^{-1}\mathbf{A} = \mathbf{I}$$

If we simplify the notation and call the inverse matrix $\mathbf{B}$, it is easy to obtain the following four equations:

$$b_{11}a_{11} + b_{12}a_{21} = 1$$

$$b_{11}a_{12} + b_{12}a_{22} = 0$$
$$b_{21}a_{11} + b_{22}a_{21} = 0$$
$$b_{21}a_{12} + b_{22}a_{22} = 1$$

Solve for b_{11} by multiplying the first equation by a_{22} and the second by a_{21} and subtracting to eliminate b_{12}. This gives

$$a_{11}a_{22}b_{11} - a_{12}a_{21}b_{11} = a_{22}$$

Using det **A** for the determinant of **A**, this can be written as

$$(\text{det } \mathbf{A})b_{11} = a_{22}$$

In a similar way we can find the set of equations

$$b_{11} = a_{22}/\text{det } \mathbf{A}$$
$$b_{12} = -a_{12}/\text{det } \mathbf{A}$$
$$b_{21} = -a_{21}/\text{det } \mathbf{A}$$
$$b_{22} = a_{11}/\text{det } \mathbf{A}$$

In the example, the determinant of **N** is -5 and the inverse is

$$\begin{bmatrix} -1/5 & 2/5 \\ 4/5 & -3/5 \end{bmatrix}$$

(6) **Problem:** Consider the stress matrix

$$[\sigma_{ij}] = \begin{bmatrix} 4 & 2 & 0 \\ 2 & -6 & 0 \\ 0 & 0 & 8 \end{bmatrix}$$

(a) What are the principal stresses?

(b) What are the stress invariants?

(c) What are the direction cosines of the planes on which these principal stresses act?

Note that the values in the stress matrix have been carefully chosen so that the calculation is easy, and we do not need to resort to special numerical methods.

Answer: The principal stresses are obtained by solving the determinant

$$\begin{vmatrix} (4-\sigma) & 2 & 0 \\ 2 & -(6+\sigma) & 0 \\ 0 & 0 & (8-\sigma) \end{vmatrix} = 0$$

For details of the method see Appendix C. This gives

$$(4 - \sigma)[-(6 + \sigma)(8 - \sigma)] + 2[-2(8 - \sigma)] = 0$$

which can be reduced to

$$(8 - \sigma)(\sigma^2 + 2\sigma - 28) = 0$$

So one root is $\sigma = 8$, and the others are the solutions of a quadratic equation and are given by

$$\sigma = \frac{-2 \pm \sqrt{4 + 4(28)}}{2} = -1 \pm 5.385$$

Therefore $\sigma_1 = 8$, $\sigma_2 = 4.385$ and $\sigma_3 = -6.385$. The stress invariants can now be calculated from the equations in Subsection 4.17.8. To find the direction cosines of the planes on which these act, we find the eigenvectors corresponding to each eigenvalue (principal stress), using the three-dimensional form of Equation (4.50). For $\sigma_1 = 8$,

$$(4 - 8)\ell_1 + 2\ell_2 + 0\ell_3 = 0$$
$$2\ell_1 - (6 + 8)\ell_2 + 0\ell_3 = 0$$
$$0\ell_1 + 0\ell_2 + (8 - 8)\ell_3 = 0$$

The first two equations give $\ell_1 = \ell_2/2$ and $\ell_1 = 7\ell_2$. This can only be true if both ℓ_1 and ℓ_2 are equal to zero. Then $\ell_3 = 1$, because the sum of squares of the direction cosines must be equal to one. For $\sigma_2 = 4.385$,

$$-0.385\ell_1 + 2\ell_2 = 0$$
$$2\ell_1 - 10.385\ell_2 = 0$$
$$3.615\ell_3 = 0$$

The last equation gives $\ell_3 = 0$. The first two equations are identical, giving $\ell_1 = 5.1926\ell_2$. To solve for both ℓ_1 and ℓ_2, we need another constraint, which is that the sum of squares of the direction cosines must be equal to one, $\ell_1^2 + \ell_2^2 = 1$. Combining the two expressions for ℓ_1 and ℓ_2 gives

$$\ell_1 = 0.982, \quad \ell_2 = 0.189$$

For $\sigma_3 = -6.385$,

$$10.385\ell_1 + 2\ell_2 = 0$$
$$2\ell_1 + 0.385\ell_2 = 0$$
$$16\ell_3 = 0$$

Once again $\ell_3 = 0$, and using the same method as before we obtain

$$\ell_1 = 0.189, \quad \ell_2 = -0.982$$

4.20 Suggested reading

Texts that give an elementary introduction to the concept of stress include the following:

Bayly, B., 1992. *Mechanics in Structural Geology*. New York, Springer-Verlag, 253 pp. (QE601.B36)

Carson, M. A., 1971. *The Mechanics of Erosion*. London, Pion, 174 pp. (QE471.C364)

Goodman, R. E., 1980. *Introduction to Rock Mechanics*. New York, John Wiley and Sons, 478 pp. (TA706.G65)

Jaeger, J. C., 1971. *Elasticity, Fracture, and Flow: with Engineering and Geological Applications*, third edition. New York, John Wiley and Sons, 268 pp. (QA931.J22)

Jaeger, J. C. and N. G. W. Cook, 1979. *Fundamentals of Rock Mechanics*, third edition. London, Methuen, 593 pp. (TA706.J32)

Johnson, A. M., 1970. *Physical Processes in Geology*. San Francisco, Freeman, Cooper, and Co. 577 pp. (QE33.J58)

Means, W. D., 1976. *Stress and Strain. Basic Continuum Mechanics for Geologists*. New York, Springer-Verlag, 339 pp. (QE604.M4)

Spencer, A. J. M., 1980. *Continuum Mechanics*. New York, John Wiley and Sons, 183 pp. (QA808.2S63)

Truesdell, C., 1968. The creation and unfolding of the concept of stress, Chapter IV in *Essays in the History of Mechanics*. New York, Springer-Verlag, pp. 184–238. (QC122.T7)

For rock friction:

Byerlee, J. D., 1978. Friction of rocks. Pure and Applied Geophysics, v.116, pp. 615–26.

Scholz, C. H., 1990. *The Mechanics of Earthquakes and Faulting*. Cambridge University Press, 439 pp. (QE534.2.S37)

For internal friction, dispersive pressure, and angle of repose in sands:

Allen, J. R. L., 1985. *Principles of Physical Sedimentology*. London, Allen and Unwin, 272 pp. (QE471.A5597)

Bagnold, R. A., 1956. The flow of cohesionless grains in fluids. Phil. Trans. Roy. Soc. London, v.249 A, pp. 235–97 (Allen's book gives a discussion of Bagnold's ideas that is considerably easier to read than the original.)

Bak, P., C. Tang and K. Wiesenfeld, 1988. Self-organized criticality. Physical Review, v.38 A, pp. 364–74 (also see Scientific American, January 1991, pp. 46–53.)

Campbell, C. S., 1990. Rapid granular flows. Ann. Rev. Fluid Mechanics, v.22, pp. 57–92.

Statham, I., 1977. *Earth Surface Sediment Transport*. Oxford, Clarendon Press, 184 pp. (QE471.2.S73)

For application to fracture and faulting:

Paterson, M. S., 1978. *Experimental Rock Deformation – the Brittle Field*. New York, Springer-Verlag, 254 pp. (QE604.P37)

Rutter, E. H., 1986. On the nomenclature of mode of failure transitions in rocks. Tectonophysics, v.122, pp. 381–7.

Suppe, J., 1985. *Principles of Structural Geology*. Englewood Cliffs NJ, Prentice-Hall, 537 pp. (QE601.S94)

Turcotte, D. L. and G. Schubert, 1982. *Geodynamics, Applications of Continuum Physics to Geological Problems*. New York, John Wiley and Sons, 450 pp. (QE501.T83)

Applications to soil mechanics, landslides, and rock avalanches are discussed by:

Carson, M. A. and M. J. Kirkby, 1972. *Hillslope Form and Process*. Cambridge University Press, 475 pp. (See especially Chapter 7; GB406.C3)

Cruden, D. M., 1976. Major rock slides in the Rockies. Canadian Geotechnical Journal, v.13, pp. 8–20.

Cruden, D. M. and J. Krahn, 1973. A reexamination of the geology of the Frank Slide. Canadian Geotechnical Jour., v.10, pp. 581–91.

McEwen, A. S., 1989. Mobility of large rock avalanches: evidence from Valles Marineris, Mars. Geology, v.17, pp. 1111–4.

Scheidegger, A. E., 1975. *Physical Aspects of Natural Catastrophes*. New York, Elsevier, 289 pp. (GB70.S33)

Terzaghi, K., 1943. *Theoretical Soil Mechanics*. New York, John Wiley and Sons, 510 pp. (TA710.T4)

5

Pressure, buoyancy, and consolidation

5.1 Concept of pressure in a fluid

A fluid is defined as a substance that deforms permanently and continuously in response to application of a shear stress, no matter how small. The rate of response depends on the viscosity, but some deformation begins as soon as the stress is applied.

In a fluid at rest, therefore, all shear stresses must be identically equal to zero. Only normal stresses are nonzero. We showed in Chapter 4 (Equation 4.21), however, that the shear stress on a plane parallel to the principal stress σ_3 and inclined at an angle α to the plane of σ_2 and σ_3 is given by

$$\tau = \frac{(\sigma_1 - \sigma_2)}{2} \sin 2\alpha \qquad (5.1)$$

It is clear that this shear stress can only be equal to zero, for all values of α, provided that

$$\sigma_1 = \sigma_2$$

The result can easily be generalized to the third principal stress.

For a fluid at rest it follows that the three principal stresses are equal. Thus, on all planes through a given point in a fluid at rest, no matter what the orientation of the plane, the normal stresses are equal. The magnitude of the stress is called the (static) pressure in the fluid at that point. Although it is generally the convention for tensional normal stresses to be positive, compressive pressures are universally regarded as positive, so that

$$\sigma_1 = \sigma_2 = \sigma_3 = -p \qquad (5.2)$$

Note that the shear stresses are not all zero in a fluid which is undergoing deformation so we cannot assume in this case that the three principal stresses are all equal. It is, however, still useful to employ the concept of pressure,

162

meaning a stress normal to a given surface, and it is common practice to consider such a pressure as made up of two parts: the static pressure (the pressure that the fluid would have at that point if it were at rest), and the dynamic pressure (the difference between the real normal stress and the static pressure).

The origin of the static pressure, for most practical examples of fluids at rest on the surface of the earth, is the gravitational body force, g (per unit mass of fluid) or γ (per unit volume of fluid).

As g and γ can often be taken to be constant, the pressure in a lake or sea depends on the distance z, below the free surface. If we arbitrarily regard the atmospheric pressure as zero, and neglect the compressibility of water (we consider compressible fluids in the next section), the hydrostatic pressure (sometimes called the *gauge pressure*) is therefore given by

$$p = -\gamma z = -\rho g z \tag{5.3}$$

Equation (5.3) is actually a complete statement of the equations of motion (Newton's second law) for a fluid at rest. The inertia terms are all zero, and so the pressure must balance the gravity force per unit area (see Chapter 9 for further discussion). In a fluid at rest, therefore, planes of equal pressure are horizontal and pressure increases in direct proportion with the distance from the (horizontal) free surface. The rate of increase with depth depends on the density of the fluid: for fresh water it is about 9.8 kPa m^{-1}, and for a saturated salt solution it can reach 11.9 kPa m^{-1}. Most surface waters, groundwaters and subsurface formation fluids lie between these extremes.

Note that in a fluid at rest the pressure at a point is the same in any direction: therefore we can treat it as if it were a scalar – though strictly speaking it is still a tensor quantity. It is, of course, possible for the pressure to vary from point to point. Equation (5.3) tells us how p varies in the z-direction, and if the only force producing pressure is gravity, and if the density is constant, then the pressure should be constant in the x- and y-directions. However, if the density varies (as it does, for example, in the oceans because of changing temperature and salinity) then it possible for p to vary in the x- and y-directions also. We represent the rate of change of a variable, like p, that may vary in more than one direction by a special notation. The rates of change in the three different directions are written as *partial derivatives*:

$$\frac{\partial p}{\partial x}, \qquad \frac{\partial p}{\partial y}, \qquad \frac{\partial p}{\partial z}$$

Partial differentiation, and its complement, multiple integration, are very important for dealing with variables like pressure that vary continuously in space (or with time or temperature, etc.). It may be that the reader is already familar with these concepts (they are used extensively in thermodynamics) but in case this is not so, we will introduce them gradually in this and the next chapter, and summarize the main rules in Appendix D. In many cases, partial derivatives are the same as ordinary derivatives, e.g., from Equation (5.3) it is clear that

$$\frac{\partial p}{\partial z} = -\gamma$$

We will show later that the partial derivative of a pressure with respect to some direction (in this case z) is equal to the force per unit volume acting in the opposite direction. We might anticipate this result by using dimensional analysis: pressure is a force per unit area, so a pressure gradient has the units of a force per unit volume.

If a fluid has no internal friction (i.e., if its viscosity is zero) then there can be no shear stresses even when the fluid is in motion and is being sheared. In this case (the case of the "ideal" or "inviscid" fluid) the normal stresses at a point in the fluid must be the same no matter what the orientation of the plane on which the force acts. So the concept of the pressure at a point is unambiguous. If the fluid is in motion, however, the pressure is not necessarily equal to the hydrostatic pressure.

If a fluid has nonzero viscosity (as all real fluids do), then shear stresses can exist in the fluid while it is being sheared, and the concept of the "pressure at a point" becomes less clear. We come back to this point when we consider viscous fluids in Chapter 9.

5.2 Pressure in a solid (geostatic pressure)

The shear stresses acting on a free surface of a solid must be zero (if not, what is producing the stress?). Within the solid, however, shear stresses will not generally be zero, even in a solid at rest. Neglecting small elastic deformations, shear stresses within a solid do not produce deformation unless they exceed the strength of the solid. It also follows, as we have seen in Chapter 4, that the three principal stresses are not generally equal.

There is a sense, however, in which one may speak of a "geostatic" pressure analogous to a hydrostatic pressure. At sufficiently large depths of burial (or in sufficiently weak rocks, such as ice or salt) under conditions of large compressive normal stress ("confining pressure") and at sufficiently high temperatures, rocks deform easily in response to relatively low shear

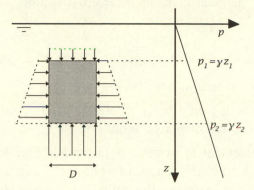

Fig. 5.1. Pressure forces acting on a submerged cylinder.

stresses, provided that the stresses are maintained for long periods of time. Under these conditions, therefore, shear stresses will tend to be reduced to low values and the three principal normal stresses will tend to become almost equal. Even though such rocks still behave as solids in response to stresses imposed over short time periods (e.g., seismic waves) they respond as fluids to stresses of long duration. Thus, averaged over long periods of time (and over large volumes of rock) shear stresses disappear and the magnitude of the three principal stresses becomes equal to the pressure exerted by a column of rock extending to the surface. If the bulk density of rock at depth z is ρ then the geostatic pressure is

$$p_g = \int_{-z}^{0} \rho g \, dz = g \int_{-z}^{0} \rho \, dz \tag{5.4}$$

(g remains almost constant over the depths being considered.) As we have seen in Chapter 4, the continental lithosphere seems to behave like a brittle solid to depths of the order of 15–20 km so it is not likely that the concept of lithostatic pressure is valid above these depths. Where the oceanic lithosphere is relatively old, and therefore cold and rigid, it is probably brittle to even greater depths.

5.3 Buoyancy

Consider a solid cylinder enclosed within a static fluid, as shown in Figure 5.1. It is apparent from the figure that the pressure forces acting on the sides of the cylinder cancel out, and that the net pressure force (acting upwards) has magnitude

$$P = (p_2 - p_1)A \tag{5.5}$$

where $A = \pi D^2/4$, the area of the cylinder cross-section. Therefore

$$P = \frac{\pi}{4}(z_2 - z_1)D^2\,\gamma$$
$$= \text{volume of cylinder} \times \text{ specific weight of displaced fluid}$$

This is a particular example of *Archimedes' principle*:

The buoyant force acting on a body wholly or partly immersed in a fluid is equal to the weight of the displaced fluid.

A more general statement is as follows (after Hubbert and Rubey, 1959). Consider a solid of arbitrary shape, of volume V, and exterior surface A. The total pressure force acting on the body is simply the sum of the pressure forces acting on all infinitesimal directed elements $\delta \mathbf{A}$ of the surface:

$$\mathbf{F} = -\iint_\mathbf{A} p\,d\mathbf{A} \tag{5.6}$$

The double integral signs simply indicate that we must perform the integration over a surface, rather than over a line as in elementary integral calculus. Rather than write the equation in the vector form shown we might choose to write it in terms of the three components of force parallel to the x-, y-, and z-axes, and the three components of areas normal to these axes. For example, for the z-component we have

$$F_z = -\iint_{A_z} p\,dA_z \tag{5.7}$$

Even for bodies of arbitrary shape, it is apparent that the x- and y-components will sum to zero, since the pressure depends only on z, and the surface is a closed surface.

To integrate Equation (5.7), it may be converted into a volume integral. For each vertical column of cross-sectional area dA_z,

$$F_z = -\iint_{A_z} (p_2 - p_1)\,dA_z \tag{5.8}$$

where p_2 and p_1 are the pressures at z_2 and z_1, respectively. Therefore

$$F_z = -\iint_{A_z} \frac{\partial p}{\partial z}(z_2 - z_1)\,dA_z$$
$$= -\iiint_V \frac{\partial p}{\partial z}\,dV \tag{5.9}$$

where dV is the volume of the vertical column of cross-section dA. But $\partial p/\partial z = -\gamma$, so it follows that

$$F = F_z = \gamma V \tag{5.10}$$

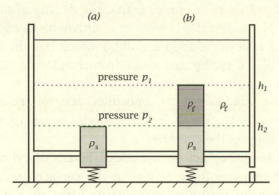

Fig. 5.2. Hubbert and Rubey's buoyancy experiment.

where V is the total volume of the immersed solid. Equation (5.10) is a more general version of Equation (5.5) and, therefore, a restatement of Archimedes' principle.

Note that we have shown that the net pressure force acting on the surface of a submerged body is equal, but opposite in direction, to the net body force (gravity force) that would have acted on the volume of fluid which the body has displaced. Thus, the buoyancy force acts *as though* it were a body force, even though it results from surface forces.

Equation (5.9) is a special case of a general theorem (*Green's theorem*) that relates surface to volume integrals. For further discussion see Aris (1962, Chapter 3).

5.4 Is buoyancy a surface or a body force?

Most texts of fluid mechanics treat buoyancy as though it were a surface force resulting from the pressure acting on the solid–fluid interface. But is it? Chapman (1981, p. 11) claimed that it is, in fact, a body force. His claim was based on his interpretation of Hubbert and Rubey's theory of buoyancy (Hubbert and Rubey, 1959, 1960, 1961). This theory was developed in order to extend the concept of buoyancy to porous solids, but the problem is perhaps best illustrated by the following example (after Hubbert and Rubey, 1960, pp. 623–6).

Consider a solid cylinder, which passes through the bottom of a liquid-filled tank (by way of an airtight, frictionless sleeve) and is supported by a spring balance, as shown in Figure 5.2(*a*). A supporting spring exerts an upward force equal to that of the weight of the cylinder plus that of the column of fluid above it. A second cylinder, shown in (*b*), has an extra section cemented to its top. We suppose that the density of this upper section is equal to that of the liquid ρ_l. Now it is apparent in each case that the spring supports a column of solid and fluid which has the same weight, so no compression of the spring can result from emplacing the second length of solid cylinder.

This is what we would expect if buoyancy acts on the added length of solid cylinder, because the weight of the cylinder is exactly balanced by the buoyant force acting on the cylinder.

But how does buoyancy act on the added length of cylinder? Obviously not directly as a pressure force acting on the surface, because there is no surface on which an upward directed pressure force can act.

Hubbert and Rubey regarded this as a demonstration that the concept of buoyancy could be generalized to apply to volumes whose boundaries were not necessarily everywhere solid–fluid boundaries. Does this further imply that buoyancy is a body force? Perhaps it can be so regarded by virtue of Equation (5.9), but it is surely not a body force in the sense that gravity forces or electromagnetic forces are body forces. The buoyancy force applies to a vessel enclosing a vacuum just as well as it does to solid bodies. It seems that, in general, it can best be regarded as a surface (pressure) force, which in some circumstances can be transmitted from the fluid into the interior of adjacent masses.

In the example shown in Figure 5.2, the equivalence of the forces exerted on the springs at (*a*) and (*b*) can be explained in terms of surface forces as follows. At (*a*) the force supported on the spring is the weight of the short cylinder plus the fluid pressure force $p_2 A = \rho_l g(h_1 + h_2)A$, where A is the cross-sectional area of the cylinder.

At (*b*), emplacing the second cylinder reduces the fluid pressure force on the upper surface to $p_1 A = \rho_l g h_1 A$, i.e., by $\rho_l g h_2 A$. But at the same time as the fluid pressure force is removed, an equal force due to the weight of the second cylinder is added. The total normal stress acting on the upper surface of the first cylinder is the same in both cases. In this interpretation, therefore, there is no buoyancy force acting on any part of the cylinder in (*b*).

As a further example consider the case of a cylinder which is gradually lowered to the bottom of a tank of liquid, as shown in Figure 5.3. The density of the solid is assumed to be greater than that of the liquid. The cylinder is clearly acted on by buoyancy in position (*a*) but does buoyancy act when the cylinder rests on the bottom in position (*b*)? It seems this question can be answered in two ways:

• yes, buoyancy still acts – in which case we must also assume that the fluid

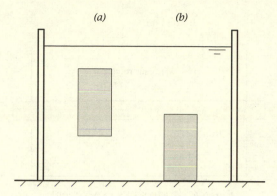

Fig. 5.3. Solid cylinder (*a*) submerged in fluid, and (*b*) resting on the bottom of the tank.

pressure still acts on the bottom of the tank below the cylinder (this is possible if there is a thin film of fluid between the cylinder and the bottom of the tank);

- no, buoyancy no longer is effective – and neither does the fluid pressure act directly on the bottom of the tank below the cylinder.

Consider what happens as the cylinder is gradually lowered into the tank, supported on a cable from above. While it is above the bottom it must be supported by a force equal to its weight less the force of buoyancy. But when it touches the bottom the support from above is rapidly reduced to zero, and the complete weight is supported by the base. The change is not instantaneous, because the pressure regime within the cylinder changes from one of tension to one of compression and therefore there will be a very small elastic shortening of the cylinder as it is lowered onto the bottom. Ultimately, however, the whole weight must be supported by the base of the tank (this, incidentally, is consistent with the analysis given by Hubbert and Rubey, 1961, p. 1591). We assume, of course, that there is no longer a film of water underneath the cylinder. If there were, then buoyancy would act upwards on the cylinder, and fluid pressure downwards. The net result would be the same.

One further point needs to be made: it is an unrealistic idealization to suppose that fluid pressures have no effect on solids enclosed within the fluid. A solid enclosed in a fluid is, in fact, subject to elastic compression transmitted across the solid–liquid boundary and into the interior of the solid. This is demonstrated by an experiment by Bridgman, reported by Hubbert and Rubey (1961, p. 1592). Consider a solid rod in a pressure vessel. If the rod is totally enclosed in the pressure vessel it will be compressed as the

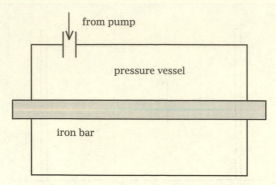

Fig. 5.4. Bridgman's experiment.

pressure increases. But if the ends protrude as shown in Figure 5.4, then compression in the middle of the rod and low pressures at the ends will tend to produce a lengthening of the rod reflecting tensional forces within the rod. If the pressure on the mid-section exceeds the tensile strength of the rod, it actually breaks, and the two parts are "shot" at considerable speeds out of the two ends of the pressure vessel. Note that the tensional forces producing fracture of the rod are produced entirely by pressures acting on the surface of the rod normal to the direction of tension – which is an impossible result unless the elasticity of the rod is taken into account.

5.5 Isostasy

One important application of buoyancy to the earth sciences is the concept of isostasy, named in 1889 by the American geologist Clarence E. Dutton (1841–1912). Though the concept is basically simple, its ramifications are still the topic of active research – for a historical review see Beckinsale and Chorley (1991). In its modern form it refers to the theory that relatively rigid lithospheric blocks tend to float in the underlying, relatively ductile asthenosphere. So any load imposed on, or removed from the lithosphere (e.g., a lake, an icecap, or a thrust sheet) tends to cause sinking or elevation of the surface, as the lithosphere responds by sinking or rising in the asthenosphere, until there is a balance between gravity and buoyancy.

As an example, consider the simplest possible isostatic model of a continent, shown in Figure 5.5. Subscripts w, o, c, and m refer to water, ocean, crust, and mantle respectively. We assume the depth of water in the oceans is t_w, the thickness of the oceanic crust is t_{co}, and the thickness of the continental crust is t_c. The density ρ, of each material is assumed to be constant in each block. To apply isostasy to this model we must further

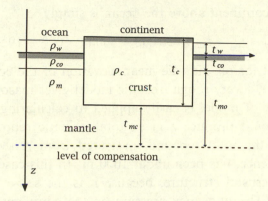

Fig. 5.5. Simple isostatic model of a continental block.

assume that continental and oceanic blocks start to float in the mantle at some level, generally called the *level of compensation*, lying a distance t_{mo} below the oceanic crust and a distance t_{mc} below the continental crust. This is the level z_c at which the mantle begins to flow readily in response to stress, so that here the principal stresses are all equal to the lithostatic pressure p. It is generally assumed to coincide with the top of the asthenosphere – in fact, the terms "lithosphere" and "asthenosphere" were introduced by the American geologist Joseph Barrell (1869–1919) in the context of isostatic theory long before the development of plate tectonics.

When gravity and buoyancy forces are in equilibrium, there will be no relative vertical movement of the continental and oceanic blocks. Then the pressure at the level of compensation must be the same under both blocks, and must be equal to the weight of a column of unit cross-sectional area extending to the surface. The integral of Equation (5.4) may be replaced by two sums:

$$\left(\sum t_i \rho_i\right)_{\text{ocean}} = \left(\sum t_i \rho_i\right)_{\text{cont}} = p/g \qquad (5.11)$$

Writing out the two sums:

$$[(t\rho)_w + (t\rho)_c + (t\rho)_m]_{\text{ocean}} = [(t\rho)_c + (t\rho)_m]_{\text{cont}}$$

Using reasonable estimates of thicknesses and densities (in SI units, see Appendix B):

$$5000 \times 1000 + 5000 \times 2900 + t_{mo} \times 3300 = 30\,000 \times 2800 + t_{mc} \times 3300$$

or $t_{mo} - t_{mc} = 19\,545$. From inspection of Figure 5.5 we can see that the elevation of the continent above the ocean is simply

$$t_c + t_{mc} - t_w - t_{co} - t_{mo} = 30\,000 - 5000 - 5000 - 19\,545 = 455 \text{ m}$$

This figure is not far off the true mean elevation of the continents (840 m) showing that even a very rough isostatic model gives reasonable results.

The same type of model is easily applied to calculating the thickness of ice sheet that would produce a given total isostatic rebound. In the case of Scandinavia, the total rebound, measured from raised shorelines and geomorphic evidence, has been about 1000 m. In this case we do not need to consider the crustal structure, because it is the same both during and after glaciation. The buoyancy pressure on the continent must have been equal to the pressure previously imposed by the ice sheet. The latter is proportional to the thickness of ice, t, times its density (≈ 1000 kg m^{-3}). The buoyancy pressure is, by Archimedes' principle, equal to the weight of the column of asthenosphere that was originally displaced by the weight of the ice. Assuming that the postglacial rise is complete, this column was 1000 m thick, with density ≈ 3300 kg m^{-3}. So

$$t \times 1000 = 1000 \times 3300, \qquad t \approx 3 \text{ km}$$

The *rate* of glacial rebound depends on the rate at which the asthenosphere can flow, and therefore on its viscosity. Rates of isostatic rise are one of our main sources of information about the viscosity of the mantle (Chapter 9, Subsection 9.10.4).

5.6 Rise and intrusion of magma in the crust

The most common volcanic material is basalt, which is denser than most sedimentary rocks. How is it that molten basalt is intruded into sedimentary rocks, or rises through them to be extruded as lava flows? Two simple models for the intrusion of molton rock material (*magma*) have been proposed (Figure 5.6).

- If a fissure is formed that connects a deep magma chamber with the surface, then the magma pressure may drive the magma to the surface. The pressure must be large enough to support a column of magma extending from the chamber to the surface. How does such a pressure arise? The pressure in the magma chamber must be approximately equal to the lithostatic pressure at that depth: if it were much larger, the magma would break through the walls of the chamber, if it were much smaller the chamber would collapse. The lithostatic pressure is produced

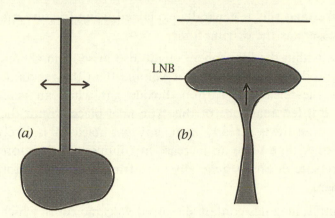

Fig. 5.6. Two mechanisms of igneous intrusion: (*a*) magma pressure, (*b*) buoyant rise. LNB is the level of neutral buoyancy.

by a column of rocks (generally sedimentary or volcanic) extending to the surface. Experimental studies have shown that the density of magma does not increase as rapidly with pressure as the density of most rocks does. Near the surface the density of the surrounding rocks may be less than the magma, but at depths of several kilometres it is probably greater. Magma is less dense than most compacted sedimentary and volcanic rocks, because of the expansion produced by heating and melting. So it is quite possible to get pressures capable of driving magma to the surface.

- Because magma is less dense than deeply buried sedimentary and volcanic rocks, it may rise through them by buoyancy, in much the same way that salt rises through sedimentary rocks. In this case, the rise should stop, and the magma should tend to spread laterally, or form a magma chamber, when it reaches the level at which it has the same density of the surrounding rocks. This level has been called the *level of neutral buoyancy* (Ryan, 1987; Walker, 1989).

There are, of course, many other factors that complicate the mechanics of igneous intrusion (Wilson *et al.*, 1987). Magma pressures must be large enough to fracture the rocks that are intruded, and to overcome the viscous forces of resistance opposing flow. If flow takes place too slowly, the magma cools, which increases its viscosity by several orders of magnitude, and ultimately produces crystallization, which ends the intrusive episode. Studies of some modern volcanoes, such as Hawaii, give considerable support for the buoyancy theory of magma rise, but the formation of smaller intrusions such as dykes and sills is generally explained using fracture mechanics with magma pressure as the driving force.

Pressures within the lithosphere might also arise from chemical reactions, where the products occupy a larger volume than the reactants. A familiar example is the release of carbon dioxide gas when an acid reacts with limestone. But few reactions of this type take place within the lithosphere, and if the pressure is already high, any gas released is likely to go into solution, rather than force an increase in volume and therefore in pressure. Magma pressure does not generally arise from chemical factors, but rather controls them.

For example, magmas contain dissolved volcanic gases, which are released only when the magma rises to about two kilometres below the surface. Magma pressure does not generally result from generation of gases; rather gas release results from the decrease in pressure as the magma rises up the vent. Alternatively, steam may be generated rapidly as magma rises into water-saturated sediments or rocks. If the pressure is no longer large enough to prevent volume expansion due to gas release, a violent explosion may result. A violent eruption may also result from a sudden decrease in lithostatic pressure produced by landsliding: this was the cause of the 1980 lateral blast at Mount St Helens, which released pressure from a magma chamber only 650 m below the top of the volcano (Kieffer, 1984).

If gas is released into a rising magma by pressure decrease, the density of the magma decreases dramatically. The pressure produced by a column of magma-plus-gas is much less than the pressure of a column of magma, so the excess driving pressure is much greater, and the eruption is more violent. "Lava fountains" may form where gas is released from relatively low-viscosity basaltic magma, or spectacular explosive eruptions may result if the magma is too viscous to permit easy separation of gas (Williams and McBirney, 1979, Chapter 4). This is the probable mechanism for the "plinian" eruption that sent volcanic gases and ash high into the atmosphere, following the lateral blast at Mount St Helens. The magma probably rose up from a reservoir 7 km below the surface, after the sudden release of pressure at the surface.

Though the pressures released by explosive volcanism are large, the fact that they are derived from lithostatic pressure limits their value to about 1 GPa, which is much less than the short lived pressures, up to 100 GPa, that may result from meteorite impacts (Gratz *et al.*, 1992). These very high pressures produce minerals (e.g., coesite), and mineral textures, that cannot be produced at the lower pressures characteristic of volcanic eruptions.

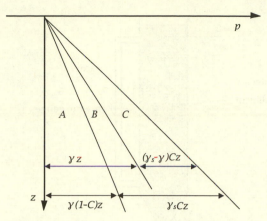

Fig. 5.7. Pressure versus depth for a dispersion of grains, as given by Equation (5.13). The width of region A gives the pressure of the fluid at depth z in the dispersion; the width of region B gives the pressure of the fluid that has been displaced by the grains; the total width of $A + B$ thus gives the pressure in the absence of grains; the width of C gives the added pressure due to the grains.

One possible example of pressures in excess of lithostatic produced by the release of gas results from the generation of methane and carbon dioxide by the thermal alteration ("maturation") of shales very rich in organic matter. In this case, gas may be produced in quantities too large to be held in solution, and at rates too high for the gas to leak away in rocks of such low permeability. In this case, fluid pressures may exceed lithostatic, producing natural expansion and fracturing of the source rocks (e.g., see the discussion of the Bakken shale in the Williston Basin given by Meissner, 1984).

5.7 Forces on settling grains

Consider a grain settling at a steady speed (called the *terminal settling velocity* or *fall velocity*) through a fluid. Because there is no acceleration of the grain, there must be a balance of the forces acting on it:

gravity force = buoyancy force + fluid drag force

So the fluid exerts a total force on the grain equal to the weight of the grain. The reverse is also true: the grain transmits a force to the fluid equal to its weight. This force causes the pressure to increase. Consider a body of static fluid containing grains dispersed through the fluid, with a uniform volume concentration, C. Let γ and γ_s be the unit weights of the fluid and the solid grains, respectively. The pressure at a point will be equal to the total weight of a column of unit cross-sectional area of both fluid and solids above the

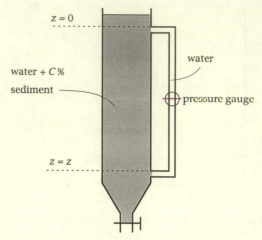

Fig. 5.8. Settling tube using a pressure gauge. The pressure difference is $(\gamma_s - \gamma)Cz$.

point:

$$p = \gamma(1 - C)z + \gamma_s Cz \tag{5.12}$$

We rewrite this equation as

$$p = \gamma(1 - C)z + (\gamma_s - \gamma)Cz + \gamma Cz \tag{5.13}$$

The last term, γCz, is the buoyancy force acting on the grains and the first and last terms sum to γz, the fluid pressure that would exist in the absence of solid grains. The second term is evidently the added pressure due to the presence of the grains. This is called the *buoyant weight* of the grains. Figure 5.7 shows the variation of these terms with z.

Figure 5.8 shows a settling tube which makes use of the difference in the pressures produced by a column of sediment dispersion and by a column of water of equal height. Sediment is introduced at the upper end, producing an initial average concentration C between levels 0 and z. The initial difference in pressure between level z in the settling tube and the same level in the small tube to the right is $(\gamma_s - \gamma)Cz$. The average concentration between levels 0 and z remains constant until some of the sediment settles past level z, when the average C begins to decrease, and therefore so does the pressure difference. The time interval δt taken for grains to settle through a distance δz depends on the settling velocity w:

$$\delta t = \delta z / w$$

If the specific weights and the distance z are known, the pressure difference against time after sediment release can be plotted. This plot is essentially a graph of C against settling velocity. Settling velocity can be directly related to grain size (Chapter 3, Problem 1). Suppose the change in concentration between two times t_1 and t_2 is $C_1 - C_2$: this corresponds to the settling past level z of those grains whose diameters lie between d_1 and d_2, and therefore the size distribution can be derived from the concentration–time graph.

5.8 Buoyancy in debris flows

Debris flows consist of concentrated mixtures of sediment grains and water, which are capable of flowing like very viscous "fluids". The concentration of grains, which reaches as much as 50 per cent by volume, increases both the apparent density and viscosity of the "fluid". There is generally a wide range of grain sizes carried by the flow, from boulders to mud. If there is a muddy matrix it may not be a true fluid but may have an appreciable strength, as well as a high viscosity.

As the motion of debris flows has often been observed to be laminar, not turbulent, the question arises as to what forces are preventing the large grains from settling out of the flow. Three forces appear to be important:

(1) the strength of the muddy matrix (we return to this in Section 10.5);
(2) dispersive pressure resulting from collisions between the larger grains (see subsection 4.2.2);
(3) buoyancy.

We consider here only the buoyancy force acting on the larger grains (gravel and boulders). It is useful to consider the sand and mud fraction, together with the water, as the "matrix" surrounding the larger grains. Grains settling in a fluid transfer their submerged weight to the fluid, thus causing an increase in fluid pressure. As shown above (Figure 5.7) this causes an increased pressure gradient with depth. So the high concentration of sand and mud in the matrix produces not only an increased pressure gradient but also an increased buoyancy effect on large clasts submerged in it.

For a typical boulder, the solid density ρ_s, is about 2500 kg m^{-3}, and for a typical debris flow the density of the matrix ρ_m may be as high as 2000 kg m^{-3}. So the buoyancy force can be twice as large as it would be for a boulder in pure water, and the residual weight, which must be supported by matrix strength or dispersive pressure, may be only one third of what it would be for a boulder in pure water. Hampton (1979) has demonstrated this effect experimentally on a small scale, using sand grains in a matrix of mud.

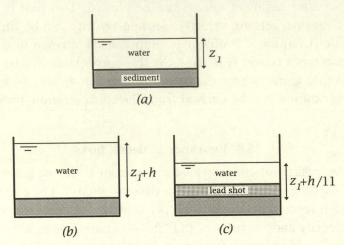

Fig. 5.9. Experiment on consolidation.

The small grains may settle so slowly that their downward movement is negligible during the lifetime of the debris flow (a few hours). In real flows, where the concentration of small grains in the matrix may be large, it appears that the small grains (especially the clay minerals) are supported partly by forming a transient structural network of sheet-like grains (like box-work, or a "house of cards"). Such a network can be compressed by the weight of overlying grains, but the rate of compaction is limited by the rate at which water can be squeezed out of the pores between the grains. The result is that, even though the matrix has strength, much of the weight of the small grains is still transferred to the pore fluids (see Section 5.9).

5.9 Effective and neutral stress

The response of porous materials to stress depends not only on the total state of stress but also on the pore fluid pressure. In fact, the stress effective in producing material deformation is exactly equal to the difference between the total stress and the pore pressure (also called the *neutral stress*). Although this principle was clearly stated by Charles Lyell (1797–1875) in his *Elements of Geology* (Skempton, in Terzaghi, 1960), the concept of effective stress was not widely used until it was defined in soil mechanics by Terzaghi (1943; see also Terzaghi and Peck, 1967; Terzaghi's earlier paper is reprinted in Terzaghi, 1960).

Terzaghi demonstrated the nature of effective stress with the following experiment (Figure 5.9). A clay layer is slowly deposited in the bottom of each of two identical beakers. Then additional water having a weight per unit area of $\delta\sigma$ is added to one beaker, and the same weight of lead shot is added to the other. Increasing the depth of fluid produces no consolidation but adding the layer of lead shot does produce consolidation of the clay. So it is clear that a simple increase in total stress does not produce consolidation.

In the first beaker, the increase in total stress was produced by increasing the depth of water by h, where $\delta\sigma = \gamma h$. This is simply the increase in pore pressure δu in the sediment at the bottom of the beaker. Increasing the pore pressure did not produce consolidation, so we can distinguish this component of the total stress from a component, called *effective stress* which does.

We can divide the total normal stress into two parts:

$$\text{total stress} = \text{effective stress} + \text{pore pressure}$$
$$\sigma \qquad = \qquad \sigma_e \qquad + \qquad u$$

In the second beaker, the total pressure increase is also equal to $\delta\sigma$. The unit weight of lead is about eleven times that of water, so adding the same weight of lead raises the water level by h/11. This increases the pore pressure by $\delta u = \gamma h/11$ and so the effective stress acting on the top of the clay layer is:

$$\sigma_e = \sigma - u \approx 10\gamma h/11$$

The total weight of the lead shot is supported partly by buoyancy, and partly by the strength of the framework of solid grains within the layer of shot and underlying it. The clay framework responds by deformation: the combination of deformation and the expulsion of pore fluids produced by deformation is called *consolidation*.

Finally, we consider the distribution of effective stress in a saturated porous solid, for example, a saturated soil below a level ground surface. Following the same line of reasoning that we used to construct Figure 5.7, we consider the various components contributing to the stress at some depth z. We use the porosity P (the volume concentration of pores), rather than the volume concentration of solids C. The effective normal stress is the difference between the total stress and the pore pressure. The total normal stress at depth z is the sum of the weights of the solids, $\gamma_s z(1 - P)$, and of the water, $\gamma z P$. If the pore fluids are not moving, the pressure is hydrostatic, γz. Therefore

$$\sigma_e = \gamma_s z(1 - P) + \gamma z P - \gamma z$$

which can be rearranged as

$$\sigma_e = (\gamma_s - \gamma)z(1 - P) \qquad (5.14)$$

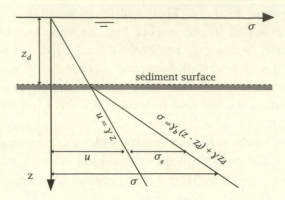

Fig. 5.10. Stresses in normally consolidated sediment. σ, σ_e and u are the total stress, effective stress, and pore pressure. γ_b is the bulk unit weight.

This equation states that, for a saturated soil under hydrostatic conditions, the effective stress is equal to the submerged weight of the solids.

5.10 Consolidation

If overlying sediment is added very slowly, so that water is progressively expelled during consolidation, the pore fluid pressure in the clay layer will remain essentially hydrostatic (Figure 5.10). The effective stress σ_e will remain equal to the submerged weight of the overlying solids, $(\gamma_s - \gamma)(z - z_d)(1 - P)$ (cf. Equation 5.14). The total stress σ is equal to the hydrostatic pressure at the sediment surface, γz_d, plus the total weight of the saturated sediment below this level. If the bulk unit weight is $\gamma_b = \gamma_s(1 - P) + \gamma P$, the bulk weight below the level z_d is $\gamma_b(z - z_d)$. Such a sediment is said to be *normally consolidated*. Since consolidation is generally irreversible, a normally consolidated material is one whose present effective stress is the largest stress experienced by the material.

Now suppose that instead of adding the overlying sediment slowly we add a layer suddenly. In nature, for example, coarse sand brought in by a storm surge or turbidity current may be suddenly deposited on top of a clay layer. Let the thickness of sand be δz (Figure 5.11). The total stress suddenly increases by an amount

$$\delta\sigma = (\gamma_s - \gamma)\delta z \qquad (5.15)$$

where γ_s is the specific weight of the sand.

Initially water can easily escape from the sand, so pore pressures in the sand are hydrostatic. Water cannot easily escape from the clay, however, so

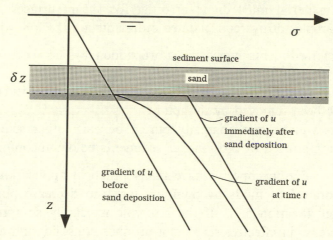

Fig. 5.11. Stresses in an underconsolidated sediment. After the sudden addition of sand of thickness δz, the sudden increase in total stress is $\delta\sigma = (\gamma_s - \gamma)\delta z$.

at first no consolidation takes place and all the additional stress is transferred to the fluid, producing an *excess pore pressure* (i.e., a neutral pressure that is larger than hydrostatic). Only at the very top of the clay layer is the effective pressure equal to the increment in total stress. As time goes by, the pore fluid, driven by the excess pore pressure, gradually escapes upwards to the sediment surface (Figure 5.11). The fluid flow reduces the excess pore pressure, the effective stress increases proportionately, and so the clay layer gradually consolidates.

At first, the effective pressure increases only near the top of the clay surface. So consolidation proceeds from the top down. As consolidation reduces the size of the pore spaces it also reduces the rate at which the pore fluids can escape, and makes it more difficult for fluid to escape from the lower clay layers. So long as sufficient time has not been allowed for the pore pressures to decline to hydrostatic, the clay layer is described as *underconsolidated*.

Another possibility is that, after sediments have been consolidated, the overlying effective stress is removed, for example, by erosion of some of the sediment. Though this may produce some elastic expansion of the remaining sediment, it does not restore the original pore structure. The sediment now has a lower porosity than it did during consolidation at the same effective stress, and is described as *overconsolidated*. Overconsolidated sediment tends to have a higher frictional strength than normally consolidated sediments. As we saw in Chapter 4, they may show a peak strength in triaxial tests, because the material must dilate in order for shear surfaces to develop. Common examples of overconsolidated sediments are the following:

(1) Pleistocene sediments which were overridden by ice sheets (that have now melted away);
(2) sediments originally deposited on a slope and covered by other sediments that have been removed by slumping;
(3) sediments which were originally buried beneath other sediments that have been removed by erosion, i.e., sediments below unconformities.

These consolidation concepts have many practical applications. For example, withdrawal of fluids by pumping tends to decrease pore pressure around a well (Section 6.7). If the reservoir is readily compressible, the resulting increase in effective stress can produce consolidation and settling of the ground surface: this has happened in Florida and Nevada due to withdrawal of water, and in California due to the withdrawal of oil.

Note that the effective stress is equal to the weight of the overlying sediment, less the buoyancy effect. Terzaghi originally argued that because buoyancy results from fluid pressure acting on the exposed lower surfaces of the grains, the full buoyancy force should act only so long as the contact between grains are very small. As the contact between grains enlarges, due to compaction and decrease in the porosity, the buoyancy force should be decreased by a factor f_b, called the "boundary porosity", equal to the proportion of the surface across which the fluid pressure is applied (see Hubbert and Rubey, 1959, p. 135).

Actual tests, however, have shown that f_b is always close to unity, indicating that the full buoyancy effect predicted by Archimedes' principle still acts on the solid framework in porous solids, and that this is so even for porosities as low as one per cent, provided the pores are interconnected (Brace, 1968). A more complete theoretical analysis by Nur and Byerlee (1971) taking into account not only the fluid pressures, but also the elastic properties of the grains, indicates that $f_b = 1 - (k/k_s)$ where k and k_s are the bulk elastic moduli for the whole rock and the mineral grains respectively (see Chapter 8). As $k \ll k_s$, for most porous rocks f_b is almost equal to unity. This is another example of the fact that it is not always possible to analyze the effect of fluid pressures on solids without taking account of the elastic deformation of the solid that results from pressure.

Consolidation depends on the rate at which fluids can escape from porous rocks. Effective pressure is needed to deform the rock framework, but porosity cannot be reduced if the pore fluids cannot escape. So a more complete discussion of the theory of consolidation is deferred until the next

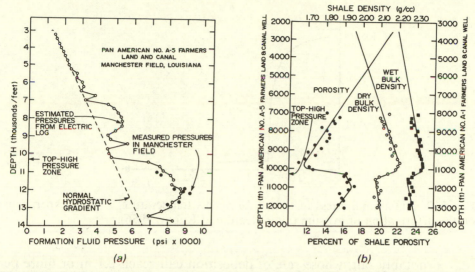

Fig. 5.12. Geopressured zone in a Gulf Coast well (after Schmidt, 1973; reprinted by permission). (*a*) shows the normal hydrostatic pressure gradient extrapolated from the upper to the lower part of well A-5, together with the actual pressures measured by drill stem tests (solid circles) and those estimated from electric logs (open circles). (*b*) shows data on shale density and porosity from two wells: the depths in A-5 are plotted on the left and the depths in A-1 are plotted on the right. The depth scales are adjusted using the top of the high-pressure zone as a datum.

chapter, where we discuss the laws governing the flow of fluids through rocks.

5.11 Geopressured zones

It is observed in thick sedimentary sequences that there are certain zones where pore pressures are greatly in excess of hydrostatic, and may approach the total load. These zones are called geopressured zones. Such high pore pressures indicate, either that load has been added to the top of the section faster than fluid can be driven out of the sediment column by consolidation, or that the volume of fluid within the sediment has been increased faster than it can escape by compaction. Figure 5.12 shows a typical example of a geopressured zone, from a well in the Gulf Coast region.

The possible causes of excess pressure therefore include the following.

(1) Rapid loading, relative to the permeability of the zone:

    (a) sedimentary loading due to rapid deposition at the surface; some sediments of low permeability seem to be deposited very rapidly,

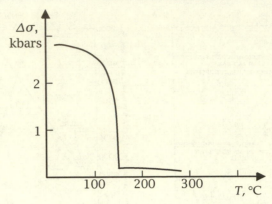

Fig. 5.13. Differential stress at failure $\Delta\sigma$ versus temperature T from experiments on the deformation of gypsum (after Heard and Rubey, 1966).

notably salt, whose rate of deposition can reach 0.1 m or more per year;

(b) sudden emplacement of large load, e.g. glacier advance, landslide, volcanic ash deposit, etc.;

(c) tectonic loading, e.g., due to the advance of thrust sheets, and also tectonic build-up of lateral stresses during the pre-thrusting stage.

(2) Expansion of water due to increasing temperature with depth of burial. Magara (1975) found that the actual increase in pressure due to thermal expansion in the Gulf Coast was about 984 Pa m^{-1} (1.4 psi ft^{-1}) of burial. He calculated that if a formation is "sealed off" at 2438 m (8000 ft), then burial at 6096 m (20 000 ft) will produce a pore pressure equal to the total overburden load.

(3) Release of water due to phases changes: e.g., gypsum to anhydrite plus water, or smectite clays to illite plus water. Such phase changes take place at critical temperatures that are likely to be reached after a few kilometres of burial.

In certain fold belts it is clear that there has been slip (décollement) over evaporite layers. It is possible that this is due to high pore pressures resulting from dehydration of gypsum.

Figure 5.13 shows results obtained experimentally by Heard and Rubey (1966). $\Delta\sigma$ is the differential stress $\sigma_1 - \sigma_2$ at yield for samples of gypsum. The stress required drops radically at temperatures above 150 °C, clearly indicating that a transition from gypsum to anhydrite is involved.

Geopressures are a matter of considerable concern in the petroleum industry. They pose problems for drilling oil wells: in the early days of the industry, oil "gushed" to the surface (where it generally caught fire) when a well was drilled into an unexpected geopressured zone. Modern wells are drilled with large valves (*blow-out preventers*) installed to prevent such disasters. The generation of oil and natural gas may itself produce geopressures, or it may assist in creating high-pressure gradients and forming subsurface fractures. These features may facilitate the primary migration of petroleum out of its source rocks into the reservoir rocks in which it accumulates. For examples and discussion see Sahay and Fertl (1988), Mudford and Best (1989) and Hunt (1990).

5.12 Effect of neutral stresses on shear strength

The effective stress represents the portion of the total stress acting on the solid structure of a soil or rock. It is not surprising, therefore, that extensive experimentation has shown that it is the effective stress that determines the frictional strength. The Coulomb failure criterion should therefore be revised to the form

$$\tau_c = C + \sigma_e \tan \phi \qquad (5.16)$$

i.e., the normal pressure is the effective stress σ_e, not the total stress σ. The concept of effective stress applies to all the normal stresses, not just the largest. In the two-dimensional case displayed as a Mohr circle, the effect is to move the whole circle toward lower values, without changing its size. The pore pressure is subtracted from both σ_1 and σ_2, but the radius of the circle, and the values of shear stress, depend only on the difference between the normal stresses, so are not affected.

Because increasing the neutral stress (pore pressure) moves the Mohr circle to the left, where it may intersect the failure envelope, it can lead to failure, even though the principal (total) normal stresses remain the same (Figure 5.14). This is probably the most common cause of landslides in soils and rocks, where increase in pore pressure is caused by rainfall or snowmelt.

In many underconsolidated clays, the pore pressures may be as high as 90 per cent or more of the total stress, so the shear stresses required for failure are very low, and may be achieved even on very low slopes (Morgenstern, 1967). Note that some authors express this by saying that the angle of internal friction is very low: but in fact Terzaghi showed that if the angle of internal friction is measured under "drained" conditions (where enough time is allowed for water to escape, after a normal load has been applied, see Section 4.4) then the angle ϕ is generally quite large, even for clays with a very high porosity. Under "undrained" conditions (where excess

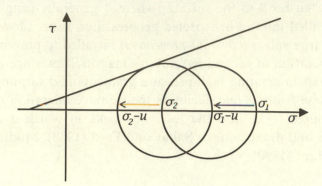

Fig. 5.14. Failure due to increase in pore pressure.

pore pressures develop in response to applied normal loads) the angle of internal friction appears to become very low. Clays with a large fraction of smectite (a clay mineral with large amounts of interlayer water, also called montmorillonite) may, however, have low values of internal friction ($\phi =$ 4–6 degrees) even under drained conditions (Goodman, 1980, p. 76).

In the limit where all the applied stress acts only to increase the pore pressure, effective stress and strength do not increase at all. It follows from Equation (5.16) that the angle of friction is zero for these completely undrained conditions.

5.13 Review problems

(1) **Problem:**

    (a) An iceberg has a density of 950 kg m^{-3}. What fraction of its volume is submerged in sea water with a density of 1025 kg m^{-3}?

    (b) A nearly spherical boulder with a specific weight of 27 000 N m^{-3} and a diameter of 2 m is observed to have half of its volume projecting out of a debris flow. Matrix taken from the active debris flow had a specific weight of 20 000 N m^{-3}. What part of the total weight of the boulder is supported by buoyancy, and what part must be supported by some other mechanism?

  Answer:

    (a) In order for it to float, the weight of the iceberg must be equal to the weight of the displaced sea water. If V_i is the total volume of the iceberg, and V_w is the volume of the water displaced, then

$$\rho_i g V_i = \rho_w g V_w$$

It follows that

$$\frac{V_w}{V_i} = \frac{\rho_i}{\rho_w} = \frac{950}{1025} = 0.927$$

The volume of submerged ice is equal to the volume of displaced water, so the proportion submerged is 92.7 per cent.

(b) The total weight of the boulder is

$$\frac{\pi d^3}{6}\gamma_s = \frac{3.142 \times 8}{6} \times 27\,000 = 113.1 \text{ kN}$$

Half is submerged in the matrix, so the weight of the displaced matrix is

$$\frac{\pi d^3}{12}\gamma_m = \frac{3.142 \times 8}{12} \times 20\,000 = 41.9 \text{ kN}$$

So roughly 37 per cent is supported by buoyancy and the remainder must be supported by some other force.

(2) **Problem:** A shield volcano builds up to the surface of the ocean, in a time that is short compared with isostatic subsidence. The volcano is composed of rock of density ρ_v. When its height is about equal to the depth of the ocean, $d = 5000$ m, volcanism ceases.

(a) What do we expect the total subsidence to be, near the centre of the volcano? Assume that $\rho_v = 2900$ kg m^{-3} and that the density of the asthenosphere $\rho_m = 3300$ kg m^{-3}.

(b) Use dimensional analysis to compare the subsidence accompanying the growth of a large shield volcano, perhaps 100 km across, with the isostatic rebound following the melting of a small ice sheet, of about the same width, but half the thickness. Would you expect the time required for this isostatic response to be shorter, longer, or about the same as that for the subsidence of the volcano?

Answer:

(a) Essentially 5000 m of sea water have been replaced by 5000 m of volcanic rock. Subsidence continues until the top of the volcano is a distance h below the surface, which is possible only if the base of the lithosphere sinks a distance h into the asthenosphere. The isostatic balance is given by

$$d\gamma_w + h\gamma_m = d\gamma_v + h\gamma_w$$

Note that we ignore the thickness and density of the lithosphere beneath the ocean, because we assume it was not changed by the emplacement of the volcano (this assumption is not as likely to be true as the corresponding assumption for growth of an ice cap, discussed in the text). Solving for h gives

$$h = \frac{d(\gamma_v - \gamma_w)}{(\gamma_m - \gamma_w)}$$

Inserting numerical values gives $h \approx 4130$ m.

(b) The difference in specific weight between ice and mantle is much larger than between volcanic rock and mantle. So for equal thickness, more subsidence occurs for volcanoes. On the other hand, the driving pressure difference $\delta\gamma = \gamma_v - \gamma_w = 1900g$ is also larger than it would be for ice ($\delta\gamma = \gamma_i \approx 1000g$). A dimensional analysis, similar to those given in Chapter 3, suggests that

$$\frac{t\,\delta\gamma\,d}{\mu} = \text{constant}$$

where μ is the viscosity of the asthenosphere. In this case t should be much shorter because both $\delta\gamma$ and d are almost twice as large for the volcano, while μ is the same in the two cases.

(3) **Problem:**

(a) In a series of tests on the strength of a sandstone, the following stresses produced failure:

Test	σ_2 (MPa)	σ_1 (MPa)
1	1.0	9.6
2	5.0	28.0
3	9.5	48.7
4	15.0	74.0

Find the Navier–Coulomb strength criterion.

(b) In the field, the state of stress in this sandstone is $\sigma_3 = 9$ MPa, $\sigma_1 = 34.5$ MPa. What *pore pressure* is necessary to cause failure?

Answer:

(a) A graphical answer is obtained by constructing the Mohr circles (Figure 5.15) and then drawing in the envelope, which has the following equation:

$$\tau_c = 1.4 + \sigma \tan 39.6 \qquad (5.17)$$

It is also possible to obtain an algebraic solution. Suppose the con-

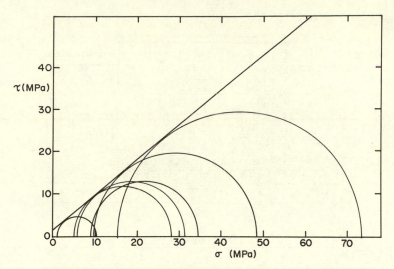

Fig. 5.15. Failure envelope for the data of Problem 3.

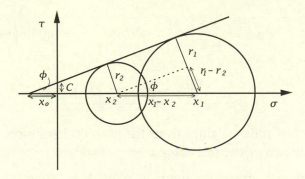

Fig. 5.16. Algebraic solution for failure envelope.

ditions for failure are known for two cases, as shown in Figure 5.16. From the figure,

$$\sin \phi = (r_1 - r_2)/(x_1 - x_2) \qquad (5.18)$$

where, for each circle, $x = (\sigma_1 + \sigma_2)/2$ and $r = (\sigma_1 - \sigma_2)/2$. From Figure 5.16

$$\sin \phi = r_2/(x_2 + x_0), \quad \text{so} \quad x_0 = (r_2/\sin \phi) - x_2 \qquad (5.19)$$

and

$$\tan \phi = C/x_0, \quad \text{so} \quad C = x_0 \tan \phi \qquad (5.20)$$

Applying Equations (5.18), (5.19) and (5.20) to the results of tests 2

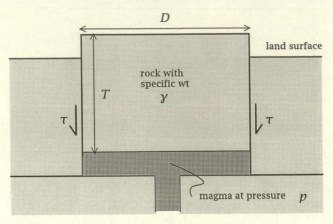

Fig. 5.17. A piston model of a laccolith.

and 4 gives

$$\sin \phi = \frac{29.5 - 11.5}{44.5 - 16.5} = 0.643, \quad \phi = 40 \text{ degrees}$$

$$x_0 = 1.39 \text{ MPa}, \quad C = 1.17 \text{ MPa}$$

The results differ slightly from the graphical solution, because they use only two of the test results.

(b) The original stress condition is plotted as a circle on the Mohr diagram, then *both* σ_1 and σ_2 are decreased until the Mohr circle just touches the failure envelope. The left shift is the pore pressure required to cause failure: in the example it is $\sim$ 3.1 MPa.

(4) **Problem:** G. K. Gilbert suggested that a laccolith could be represented by the "piston" model shown in Figure 5.17. In order to form a laccolith, the magma pressure, p must be large enough to overcome the combined effects of the weight of the overlying cylinder of rock (of height T and diameter D) and the shear strength, τ_c, of the rock walls of this cylinder. Derive an equation for the diameter of the laccolith. Use γ for the specific weight of the rock. Comment on the geological signifance of the equation. (This problem is taken from Arvid Johnson's book *Physical Processes in Geology* which may be consulted for further discussion of the model.)

Answer: The balance of forces is:

$$\begin{matrix} \text{pressure force} & & \text{weight} & & \text{strength} \\ \text{on base of} & = & \text{of} & + & \text{of} \\ \text{cylinder} & & \text{cylinder} & & \text{rock walls} \end{matrix}$$

$$p \times \text{base area} = \text{volume} \times \gamma + \tau_c \times \text{wall area}$$

$$p(\pi D^2/4) = \pi D^2 T/4 \times \gamma + \tau_c \times \pi D T$$

$$p = T\gamma + \tau_c(4T/D)$$

Solving for D:

$$D = \frac{4\tau_c T}{(p - T\gamma)} \tag{5.21}$$

$(p - T\gamma)$ is the excess of magma pressure over lithostatic pressure; the larger this excess, the smaller the diameter of the laccolith. (Note that in Johnson's book, the excess magma pressure is simply called the pressure.) If we assume that the excess pressure is constant, and the shear strength of the walls is also a constant, then the diameter of the laccolith should be proportional to the depth T at the time of emplacement, i.e., deeper laccoliths should be larger.

If we were to attempt a dimensional analysis of this problem using the list of variables D, T, p, τ_c, and γ, we would run into the problem that two pairs of variables have the same dimensions (D, T and p, τ_c). The result is that we can define only two repeating variables, so we end up with three dimensionless products rather than the two we might expect, and the result is not very helpful. We could obtain better results, however, by remembering that it is really the excess pressure $p - T\gamma$ that interests us, not the separate contributions of pressure and rock weight. If we call this Δp then the list becomes: D, T, τ_c, and Δp. Dimensional analysis would give

$$\left(\frac{D}{T}, \frac{\tau_c}{\Delta p} \right) = 0$$

This is not as useful a result as we obtain from the balance of forces but it does get us part of the way towards a solution of the problem – in particular, it does suggest that there might be a direct proportionality between size and depth for laccoliths.

5.14 Suggested reading

Goodman, R. E., 1980. *Introduction to Rock Mechanics.* New York, John Wiley and Sons, 478 pp. (A good general introduction. Table 3.3 on p. 78 lists some typical values of internal friction and cohesion for different rock types. TA706.G65)

Gretener, P. E., 1977. *Pore Pressure: Fundamentals, General Ramifications, and Implications for Structural Geology.* Amer. Assoc. Petroleum Geol. Continuing Education Course Notes, No. 4, 87 pp. (QE601.2G74)

Hubbert, M. K., and W. W. Rubey, 1959, 1960, 1961. Role of fluid pressure in mechanics of overthrust faulting. Geol. Soc. America Bull., v.70, pp. 115–60; v.71, pp. 617–28; v.72, pp. 1587–94. (A classic series of papers)

Hunt, J. M., 1990. Generation and migration of petroleum from abnormally pressured fluid compartments. Amer. Assoc. Petroleum Geol. Bull., v.74, pp. 1–12. (See also an interesting discussion of the paper, *ibid.* v.75, pp. 326–38, 1991)

Terzaghi, K., 1943. *Theoretical Soil Mechanics.* New York, John Wiley and Sons, 510 pp. (A classic work. TA710.T4)

Terzaghi, K., 1960. *From Theory to Practice in Soil Mechanics.* New York, John Wiley and Sons, 425 pp. (Includes a good article on the "effective stress concept" by A. W. Skempton, pp. 42–53, TA710.T33F)

6

Flow through porous media

6.1 Darcy's law

We have already seen that pore fluid pressure is important in determining the criterion for failure, and therefore for sliding and faulting. It is also an important control on the consolidation of soils and sediments. Flow of fluids through soils and rocks is important also for the recharge and movement of groundwater and for the migration of petroleum. In this chapter we can only discuss a few aspects of this large topic: we concentrate on the flow of a single fluid, water, through a fully saturated porous medium. It turns out that the fundamental equation, Darcy's law, has much in common with many other equations, such as the heat equation, which describe the diffusion of one "substance" through another. The mathematical properties of diffusion equations are well understood, and the equations are amenable to numerical solution, so the study of flow through porous media has been dominated recently by mathematical models. To simplify the models, drastic assumptions are often made about the porous medium, which ignore the complex nature of real soils and rocks. In this chapter, we cannot avoid making such assumptions – but it is essential for earth scientists to retain some skepticism about the results of models, until they have been verified by field observations.

The fundamental empirical law for the movement of any kind of fluid (water, oil, gas) through porous media (such as soils or rocks) was originally published in 1856 by a French engineer, Henry Darcy (1803–58) in a study entitled "Les Fontaines Publiques de la Ville de Dijon". The original paper has been reprinted in Hubbert (1969) and is discussed extensively by Hubbert and by Chapman (1981, Chapter 3). A good discussion is given in Freeze and Cherry (1979, Chapter 2); see also Freeze and Back (1983).

The following discussion follows Hubbert, rather than attempting to follow

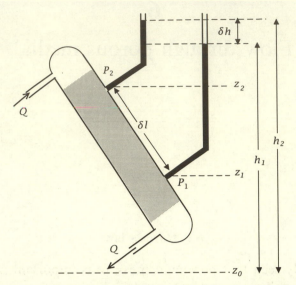

Fig. 6.1. Darcy's experiment (Hubbert's version).

Darcy exactly. Figure 6.1 shows a tube containing sand. The (interior) cross-sectional area of the tube is A, and water is pumped in at the top and out at the bottom at a volume rate of flow Q. Two manometers, separated by a length $\delta \ell$ along the tube, indicate heads of h_1 and h_2, measured above some arbitrary datum, $z_0 = 0$.

The discharge per unit area, or *unit discharge*, is

$$q = Q/A \qquad (6.1)$$

q therefore has dimensions of velocity, $[LT^{-1}]$. Darcy found that the unit discharge was proportional to the gradient in head along the tube:

$$q = -K(\delta h/\delta \ell) \qquad (6.2)$$

or, in general, for flow in the x-direction:

$$q_x = -K_x(dh/dx) \qquad (6.3)$$

The coefficient K is called the *hydraulic conductivity*, and also has the dimensions of a velocity, $[LT^{-1}]$, since dh/dx is the hydraulic gradient, and is dimensionless (it is basically a slope). The negative sign on the right-hand side of the equation indicates that flow is from the region of higher head to one of lower head, i.e., in the direction opposite to an increase in head. Note that:

(1) The unit discharge is not the actual velocity at which fluid travels through

the pores of the rock: it is simply a measure of the volume per unit time of fluid traveling through unit cross-sectional area of the rock.

The proportion of the cross-section of a porous rock that is occupied by pores is given by the fractional porosity, P (i.e., the volume of pores divided by the total, or bulk, volume). So the actual average velocity of the fluid moving through the pores is $q^* = q/P$. As the porosity is often only 10–20 per cent, the actual velocity v is frequently 5–10 times larger than the Darcy unit discharge q.

(2) Changing the angle of tilt of the tube produces no effect on the relation between q and $\delta h/\delta \ell$.

(3) The pressure at P_1, $p_1 = \gamma(h_1 - z_1)$, is larger than the pressure at P_2, $p_2 = \gamma(h_2 - z_2)$, so flow in the experiment illustrated is from a region of low fluid pressure to one of high fluid pressure.

This last point illustrates nicely that

$$dp/dx \neq dh/dx \tag{6.4}$$

i.e., the hydraulic gradient is not the same thing as the pressure gradient. Darcy's law relates the flow to the gradient in hydraulic head, not to the pressure gradient. Hydraulic head is related to the energy of the pore fluids. It is a fundamental concept that is frequently misunderstood, and so we discuss it in detail in the next section.

6.2 Hydraulic head

Figure 6.2 shows a cross-section through a water-bearing rock, i.e., an *aquifer*. Groundwater flows from a source in the mountains, on the left of the diagram, to a discharge point, where the aquifer outcrops on the plains and loses water, on the right of the diagram. Schematically this could be a cross-section from west to east across the western Canada sedimentary basin.

At any point along the cross-section, we can drill a well for the measurement of hydraulic head. Such a well, called a *piezometer*, differs from a normal well in that it is a pipe open only at the top and bottom. The bottom is the point at which the head is determined by observing the level to which the water rises within the piezometer. In other words, the piezometer is a kind of enormous manometer.

In many cases, the level to which water rises is above the ground surface, so the well drilled at that place will flow freely without pumping (i.e., it is an *artesian* well).

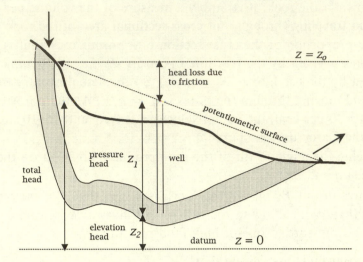

Fig. 6.2. Variation in hydraulic head across an artesian basin.

We define a surface, called the *piezometric surface* or *potentiometric surface*: this is an imaginary surface connecting all the points to which water would rise if the appropriate piezometers were installed. It is not the same as the *water table*, which is the level to which water will rise in a well that penetrates the zone, generally not far from the surface, where the soil or rock becomes saturated with water. The water table is a subdued replica of the topography. It is determined mainly by rates of recharge, discharge and flow near the topographic surface, whereas the piezometric surface is determined by the conditions within a particular aquifer, and may differ greatly for different aquifers (or even for different levels in the same aquifer).

The elevation of the piezometric surface at any point is the *total head*, which can be considered to be composed of two parts: the *elevation head*, z_2 (the elevation of the base of the piezometer), and the *pressure head*, z_1. We know that the pressure is given by

$$p = \gamma z_1 \tag{6.5}$$

So, we can write the total head h as

$$h = (p/\gamma) + z_2 \tag{6.6}$$

Note that in Figure 6.2, as in Figure 6.1, flow proceeds from a region of higher total head to a region of lower total head. Darcy's law states simply that the unit discharge is directly proportional to the gradient in total head.

The total head may be interpreted as the total potential energy per unit weight of the fluid. To see why, consider the movement of a *unit mass* of fluid from some standard state (density ρ_0, pressure p_0 = atmospheric pressure) at elevation $z = 0$, to some final state (density ρ, pressure p') at elevation z.

Work is done on this mass of fluid in four ways.

(1) The mass is lifted against gravity: in Chapter 2 (Section 2.9) we demonstrated that the work required to raise a mass, m by a vertical distance z was mgz. Expressed in units of energy per unit mass $[L^2T^{-2}]$, this is

$$w_1 = gz \tag{6.7}$$

(2) The fluid pressure is increased from p_0 to p':

$$w_2 = \int_{p_0}^{p'} V \, dp = \int_{p_0}^{p'} (1/\rho) dp \tag{6.8}$$

V is the volume occupied by unit mass of fluid, so $V = 1/\rho$. Most of the liquids we will be concerned with, such as water, are almost incompressible, so we may assume ρ is a constant, independent of p, and write

$$w_2 = (p' - p_0)/\rho \tag{6.9}$$

$p' - p_0$ is the gauge pressure, which we have previously simply denoted as p.

(3) The mass is accelerated from rest to its final speed u: in Chapter 2 we showed that for a mass m, this work is $mu^2/2$. Expressed in units of energy per unit mass this is

$$w_3 = u^2/2 \tag{6.10}$$

but as u is very small (of the order of 10^{-3} m s^{-1} or less) w_1 and w_2 are generally many orders of magnitude larger, and w_3 can be neglected.

(4) Friction must be overcome.

If we neglect energy loss to friction (and other possible sources or sinks of energy) mechanical energy must be conserved, that is, the other three components listed above must sum to a constant, which we call the *fluid potential* ϕ:

$$gz + \frac{p}{\rho} + \frac{u^2}{2} = \phi \tag{6.11}$$

Equation (6.11) is *Bernoulli's equation* (see Chapter 9, Section 9.5, for a more complete derivation and discussion). It can also be written in terms of energy per unit weight (rather than mass) simply by dividing through by g:

$$z + \frac{p}{\rho g} + \frac{u^2}{2g} = \frac{\phi}{g} \tag{6.12}$$

It holds exactly when there is no friction. We can take friction into account formally by rewriting Equation (6.12) as:

$$z + \frac{p}{\rho g} + \frac{u^2}{2g} + h_f = \frac{\phi}{g} \tag{6.13}$$

where h_f is a term representing lost energy, called the *head loss due to friction*. One advantage of this form of the equation is that each component of energy (per unit weight) is expressed as a component of head (with dimension [L]): z is the elevation head, $p/\rho g$ is the pressure head, and $u^2/2g$ is called the *dynamic head*.

If we neglect the dynamic head and the friction loss, we see by comparison with Figure 6.1 that the remaining energy components are equal to the head that drives flow. The sum of these terms has been defined as the total head h. Returning to units of energy per unit mass we define the *fluid potential* at a point as:

$$\phi = hg = gz + \frac{p}{\rho} \tag{6.14}$$

Note the use of the term *potential*. If there exists a conservative force field such that work must be done to bring a unit "object" (mass, charge, etc.) from some reference state to a particular position in the field, then a potential exists from which the field can be derived. We remember from Section 2.9 that, in a conservative force field, the work done against the field is independent of the path taken by the moving object. This is true of gravity fields, provided that we ignore friction.

We discuss potentials more completely in Section 6.5, once we have developed the concept of the gradient of a vector field. In the discussion given above, we have assumed that the pores are filled with a single fluid. In soils above the water table, air is present as well as water. In petroleum reservoirs, oil or gas is present as well as water. The concept of potential can be extended to these cases also (e.g., see a text on soil physics, such as Jury *et al.*, 1991).

6.3 Hydraulic conductivity

From experiments (see Chapman, 1981, for a review) it can be concluded that the hydraulic conductivity, K is a function of several variables:

(1) Gravity is the force driving the flow, so we expect that the hydraulic conductivity will be directly proportional to the specific weight of the fluid, γ.
(2) Viscosity opposes the flow, so we expect that the hydraulic conductivity will be inversely proportional to the viscosity of the fluid, μ.
(3) Large pores should offer less resistance than small pores. For a granular medium, the pore size should be proportional to the grain size. Experiments performed with well-sorted spherical glass beads show that hydraulic conductivity is directly proportional to the square of the diameter of the beads, D.

From a dimensional analysis of Darcy's law (Equation 6.3) we conclude that the dimensions of K are $[LT^{-1}]$. Using the known dimensions of the other three variables listed above (see Chapter 3) we can easily verify that the variables can be reduced to one dimensionless product, which must therefore equal a constant, C:

$$\frac{K\mu}{\gamma D^2} = C \tag{6.15}$$

Rearranging, we obtain an expression for K,

$$K = CD^2 \frac{\gamma}{\mu} \tag{6.16}$$

which confirms the generality of the experimental results.

It is clear, therefore, that K is a function of two different types of variable: those such as specific weight and viscosity, which are a function only of the properties of the fluid (and the gravity field driving its flow), and those such as grain size, which are somehow related to the geometry of the pore system. There is some advantage to reformulating Darcy's law in a form that separates out these different components of the hydraulic conductivity:

$$q = -k \left(\frac{\gamma}{\mu}\right) \frac{dh}{dx} \tag{6.17}$$

k in this equation is the *coefficient of permeability*. From our dimensional analysis we know that its dimensions are $[L^2]$. Experiments show that it is related only to the geometry of the pore system, and not to the properties of the fluid in the pores.

There is no named SI unit of permeability but the unit in common use is the *darcy*. Unfortunately, the darcy was defined in a peculiar way. We can rewrite Equation (6.17) as

$$q = -\left(\frac{k}{\mu}\right) \frac{dp'}{dx} \tag{6.18}$$

where $p' = \gamma h = p + \gamma z$. If z can be neglected, as is generally the case in experimental determinations of the permeability, then dp'/dx is the pressure gradient across the specimen. If q is in cm s^{-1}, μ is in centipoise (i.e., poise/100), and dp'/dx is in atmospheres per centimetre, then k is in darcies.

It follows that one darcy is 0.987×10^{-8} cm^2, or roughly 1 μm^2 (one square micrometre). The reasons for this peculiar choice of units were:

(1) water has a viscosity of about 1 centipoise;
(2) permeability is generally measured by establishing a strong pressure gradient across a small specimen;
(3) a loose sand, composed of grains with a diameter of about 0.5 mm, has a permeability of about 1 darcy.

Permeability is actually measured either in the laboratory or indirectly in the field by pumping tests (see Section 6.7). In the laboratory, the permeability of soils is measured by measuring the flow rate at a known hydraulic head (Jury *et al.*, 1991). The permeability of rocks is often measured by forcing a fluid (generally air or mercury) through a dried sample (Monicard, 1980).

A permeability of 1 darcy corresponds roughly to a hydraulic conductivity (for water, at near-surface temperatures) of about 10^{-5} m s^{-1}, or about 10 m per day. A few aquifers, composed of well-sorted gravels or coarse sands, have permeabilities of several darcies, but most have permeabilities of less than a darcy. In the petroleum industry the common unit used is the millidarcy (darcy/1000) and many good reservoir rocks have permeabilities of only a few tens of millidarcies. "Impermeable" rocks have permeabilities much less than a millidarcy, probably ranging down to less than 10^{-9} darcies. It is very difficult to measure very low permeabilities realistically: low values are sensitive to confining pressure (which tends to narrow the pores), and the real permeability of "impermeable" rocks (e.g., igneous rocks) is mainly related to fractures which are poorly represented in laboratory samples (Neuman and Neretnieks, 1990).

Many authors have tried to derive Darcy's law from the fundamental equations governing the slow flow of viscous fluid (the Navier–Stokes equations, see Chapter 9). But in order to do this, it has always been necessary to make use of a simplified "model" of a porous medium. Real porous media are very complicated pore networks. Thus it is best to regard Darcy's law simply as a well-established empirical law. What we know about viscous fluids, however, does lead us to expect that Darcy's law will break down when the flow in the pores changes from laminar to turbulent. We know that this generally takes place at a critical value of the Reynolds number (see Chapter 3, Subsection 3.3.3). For porous media, the Reynolds number may be defined as $Re = \rho q D/\mu$, where, for a sand, D is the grain size (or roughly, the largest common pore diameter). For sands it has been found experimentally that Darcy's law breaks down at Reynolds numbers greater than about 4. Values this large are rarely found in the flow of groundwater.

6.4 Nature of permeability: relation to porosity, tortuosity, and specific surface area

Several attempts have been made to relate the coefficient of permeability to measurable geometrical properties of a porous rock. Such properties include the following.

(1) Porosity. There is no necessary relationship between porosity and permeability. A rock may have only a little pore space, but if the pores are large and freely interconnected, the permeability may be high. Nevertheless, we expect, and observe, a general statistical trend within certain classes of rock (sandstones, for example): those rocks with higher porosities generally have higher permeabilities.

(2) Tortuosity. This is a measure of the length that a fluid particle must travel in order to move unit distance in the average flow direction (the x-direction). It is not easy to measure directly, and is usually estimated from the electrical resistance of the permeable solid saturated with an electrolyte (e.g., a saline formation fluid). This resistance is equal to the resistivity, R_0, multiplied by the length of the block of saturated solid and divided by its cross-sectional area.

Archie (1942, 1950) related the electrical resistivity of the saturated porous solid, R_0 to the resistivity of the pore fluid, R_w by a "formation factor", F:

$$R_0/R_w = F \tag{6.19}$$

Consider the simple case of a straight capillary tube, with circular cross-section and diameter D, inclined at an angle α to the bulk flow direction (Figure 6.3). We assume that the solid part of the rock has a very high electrical resistivity and that the electric current flows from one side of the block of porous rock to the other entirely through the liquid in its pores. In the model shown, therefore, the resistance r should depend inversely on the cross-sectional area of the capillary a and directly on the length of the capillary ℓ_t: $r = \ell_t R_w/a$. If r_0 is the resistance of a block of rock saturated with water, and r_w is the resistance of a similar block composed wholly of water, then:

$$F = R_0/R_w = r_0/r_w$$

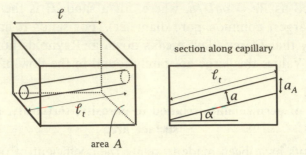

area A

Fig. 6.3. Capillary model of a porous rock.

$$= \frac{R_w \ell_t / a}{R_w \ell / A}$$
$$= (A/a)(\ell_t / \ell)$$
$$= (A/a)T \qquad (6.20)$$

where $T = \ell_t / \ell$ is the tortuosity. Now the ratio of the cross-sectional area of the tube a to the cross-sectional area of the block A is not equal to the porosity, because the first is measured normal to the orientation of the tube, and the second normal to the direction of bulk flow. In a porous rock the orientation of the pores (tubes) will, on average, differ from that of the bulk flow.

The porosity is given by

$$P = a_A / A \qquad (6.21)$$

where a_A is the cross-sectional area of a tube in the plane of A. It is apparent that the porosity is also given by:

$$P = \text{volume of tube/volume of rock}$$
$$= (\ell_t a)/(\ell A)$$

Equating these two expression for the porosity gives

$$a_A / a = \ell_t / \ell = T \qquad (6.22)$$

or

$$a = a_A / T$$

and we can write:

$$F = (A/a_A)T^2 = T^2/P \qquad (6.23)$$

Thus the formation factor, which can be measured in a borehole by electric logging techniques, can be used (when combined with porosity measurements) to estimate the value of the tortuosity. It is to be expected that tortuosity will depend on porosity: the larger the proportion of a rock that is made up of pores, the easier it is for the fluid to flow along almost straight paths through the pore space. For unconsolidated sands, the tortuosity is generally 1.3 or higher, and for sandstones it is generally in the range of 2 to 3.

For a given porosity, as the size of pores decreases, the number of pores must increase, and so must their surface area. It is to be expected that the surface area will increase as the inverse square of the pore size. There are techniques available for measuring the total surface area of unit bulk volume of a porous rock (specific surface area S) and this has suggested to some workers that it might be useful to try to related permeability to some combination of porosity and S.

Kozeny (1927, see Chapman, 1981, p. 206) suggested that k is proportional to P^3/S^2 and if a rock were composed of spherical grains of diameter, d then S is proportional to $(1-P)/d$ so that

$$k \propto P^3 d^2/(1-P)^2 \tag{6.24}$$

Various other forms of this basic Kozeny equation have been proposed by other authors. A useful review is given by Chapman (1981).

Finally, we should note that it is to be expected that the permeability of real sediments and rocks will be anisotropic, i.e. will depend on the direction of flow. We particularly expect that the permeability along the bedding will be much greater than the permeability across the bedding. Freeze and Cherry (1979, p. 154) cite measurements on sandstones which indicate that the horizontal permeability is generally about 1.5 times the vertical permeability and may be as much as three times as large. Minor variations in permeability have also been reported even within the plane of the bedding, due to variations in grain orientation (see Blatt *et al.*, 1980, p. 423 for discussion and references).

6.5 Gradient and curl and their application to Darcy's law and other diffusion equations

Suppose we have some scalar property $\phi(x, y)$ which is a function of position in space, as indicated by x- and y-coordinates. In the earth sciences, we generally represent that property by a contour map, showing lines of equal ϕ, plotted in the xy-plane.

The fluid potential ϕ, defined by Equation (6.14), is a scalar quantity. It actually varies in three, rather than two dimensions, but if we neglect the third dimension, its distribution in two-dimensional space can be represented by a contour map. Using Darcy's law (Equation 6.17), we can represent the components of flow in the x- and y-directions by the equations

$$q_x = -k_x \left(\frac{\rho}{\mu} \right) \frac{\partial \phi}{\partial x} \tag{6.25}$$

$$q_y = -k_y \left(\frac{\rho}{\mu} \right) \frac{\partial \phi}{\partial y} \tag{6.26}$$

These equations state that the unit discharges in the x- and y-directions are directly proportional to the rate of change of ϕ (i.e., the slope of the ϕ-contoured surface) in those directions. Partial derivatives ($\partial \phi/\partial x$ and $\partial \phi/\partial y$) are used because ϕ varies in both the x- and y-directions. The permeability may also be different in these two directions, as indicated by the subscripts to k. The real situation is actually even more complicated. If the permeability is not isotropic, then part of the flow in the x-direction may result from the hydraulic gradient in the y-direction (we will explain how in Section 6.11).

We could write a more general equation for the flow q_i in the ith direction using the Einstein summation convention:

$$q_i = -k_{ij} \left(\frac{\rho}{\mu} \right) \frac{\partial \phi}{\partial x_j} \tag{6.27}$$

In this form, we can begin to see that the permeability k is actually a tensor, very similar in its general form to the stress tensor. In many cases, however, we are willing to make the assumption that permeability is approximately isotropic, i.e., that $k_x = k_y = k$ or more generally that, in the permeability tensor, all the off-diagonal elements are equal to zero, and all the elements of the principal diagonal are equal to a single value, k. This effectively reduces the permeability tensor to a scalar.

The fluid potential is a scalar property. For a given depth z, it is defined by a single value, hg, at each map point. A map of ϕ, however, conveys much more information than a simple list of scalar values because, as we shall see, it also indicates the direction and magnitude of fluid flow, for an aquifer of constant isotropic permeability and for a fluid of constant density and viscosity.

We can indicate this by writing a vector form of Equation (6.27):

$$\mathbf{q} = -k \left(\frac{\rho}{\mu} \right) \text{grad } \phi = -k \left(\frac{\rho}{\mu} \right) \nabla \phi \tag{6.28}$$

In this equation, "grad" denotes the *gradient* of the function ϕ defined (for the two-dimensional case) by

$$\text{grad } \phi = \nabla \phi = \mathbf{i}\frac{\partial \phi}{\partial x} + \mathbf{j}\frac{\partial \phi}{\partial y} \tag{6.29}$$

∇ (the symbol "nabla") is called the *vector operator*, del. It converts the scalar ϕ into a vector $\nabla \phi$ which is directed up the steepest slope indicated by the contours of ϕ. Thus, $\nabla \phi$ is normal to the lines of equal potential. The magnitude of $\nabla \phi$ is equal to the slope itself (i.e., to what is, in fact, commonly called the "gradient" of the slope), just as in one dimension, the simple derivative of a function is equal to the slope of that function.

The concept of a potential gradient, however, is commonly generalized to three dimensions, in which case a graphical interpretation of ϕ using a relief map is no longer possible. However, just as one can draw equipotential lines on a cross-section through an aquifer, so one can draw equipotential surfaces on a three-dimensional block diagram of an aquifer.

The fluid potential is by no means the only one commonly used in the earth sciences. The following are some further examples:

(1) $\mathbf{q} = -(1/R) \text{ grad } \phi$

where $\mathbf{q}$ is the flow of electric current, R is the electrical resistance, and ϕ is the electrical potential.

(2) $\mathbf{q} = -k \text{ grad } \phi$

where $\mathbf{q}$ is the flow of heat, k is the thermal conductivity and ϕ is the temperature.

(3) $\mathbf{q} = -k \text{ grad } \phi$

where $\mathbf{q}$ is the flux of a chemical species, k is the diffusion coefficient and ϕ is the concentration of the chemical species.

(4) $\mathbf{g} = - \text{ grad } \phi$

where $\mathbf{g}$ is the gravitational acceleration and ϕ is the gravitational potential (there is no coefficient k in this case, because the units have been chosen to make it equal to one).

(5) $\mathbf{u} = -\text{grad } \phi$

where $\mathbf{u}$ is the fluid velocity and ϕ is the velocity potential.

All of these quantities (except gravity) represent the flux of some quantity from a region of higher to one of lower potential. The last example can be used to illustrate an important point: not all vector fields can be expressed as a potential gradient. In the case of the velocity field, the existence of a potential implies that the x- and y-components of velocity can be expressed as minus the gradients in the x- and y-directions:

$$u_x = -\frac{\partial \phi}{\partial x} \tag{6.30}$$

$$u_y = -\frac{\partial \phi}{\partial y} \tag{6.31}$$

Differentiating these two equations in the transverse direction gives

$$\frac{\partial u_x}{\partial y} = -\frac{\partial^2 \phi}{\partial x \partial y} \tag{6.32}$$

$$\frac{\partial u_y}{\partial x} = -\frac{\partial^2 \phi}{\partial x \partial y} \tag{6.33}$$

It follows that

$$\frac{\partial u_x}{\partial y} - \frac{\partial u_y}{\partial x} = 0 \tag{6.34}$$

i.e., fluid flow in a potential field requires that the transverse velocity gradients in the x- and y-directions must be equal. By no means all velocity fields have this property (see Chapter 9, Section 9.6). The ones that do are called *irrotational*. Equation (6.34), therefore, expresses the condition that a velocity field $\mathbf{u}$ must satisfy if it is to be expressed as the gradient of a velocity potential. This condition can be generalized to three dimensions and written

$$\text{curl } \mathbf{u} = \nabla \times \mathbf{u} = 0 \tag{6.35}$$

where the vector operator, "curl" is defined by

$$\text{curl } \mathbf{u} = \nabla \times \mathbf{u} = \left(\frac{\partial u_z}{\partial y} - \frac{\partial u_y}{\partial z}\right)\mathbf{i} + \left(\frac{\partial u_x}{\partial z} - \frac{\partial u_z}{\partial x}\right)\mathbf{j}$$
$$+ \left(\frac{\partial u_y}{\partial x} - \frac{\partial u_x}{\partial y}\right)\mathbf{k} \tag{6.36}$$

We now see that Equation (6.34) states that the component of the curl in the z-direction (indicated by the unit vector $\mathbf{k}$) is equal to zero. Note that the curl of a vector is a vector also, and can also be written in the form of a determinant (see Appendix C):

$$\nabla \times \mathbf{u} = \begin{vmatrix} \mathbf{i} & \mathbf{j} & \mathbf{k} \\ \frac{\partial}{\partial x} & \frac{\partial}{\partial y} & \frac{\partial}{\partial z} \\ u_x & u_y & u_z \end{vmatrix} \tag{6.37}$$

This is not a true determinant, since elements in the first row are unit vectors, those in the second row are differential operators, and those in the third row are velocity components. But it serves as a useful mnemonic.

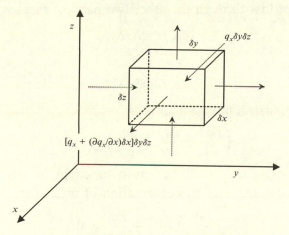

Fig. 6.4. Flow into and out of a small volume.

6.6 Divergence and the conservation of mass

Darcy's law tells us how **q** varies with ϕ. But how do **q** and ϕ vary in space? If we measured ϕ at a large number of locations, we could construct a contour map and then use Darcy's law to calculate **q**. Such an observational program, however, is rarely possible. Often, no direct observations of ϕ are available, although values of ϕ are known along certain surfaces, e.g., along impermeable surfaces, for which $\mathbf{q} = -K \nabla\phi = 0$. Such surfaces provide boundary conditions that we can use to solve for the flow field if we have some expression for the manner in which ϕ varies in space.

To derive such an equation we consider the formulation, for groundwater flow, of the condition of conservation of mass. To do this, consider flow into and out of the small volume of porous rock shown in Figure 6.4. We assume that the volume of pore space does not change (though in fact it might, due to compression or expansion of the reservoir, or deposition of cement in the pores) and that the fluid is incompressible. These assumptions mean that conservation of mass implies conservation of fluid volume.

In unit time, the total volume flowing through the two faces normal to the x-axis is

$$q_x \delta y \delta z \qquad \text{into the 'rear' face and}$$
$$[q_x + (\partial q_x/\partial x)\delta x]\delta y \delta z \quad \text{out of the 'front' face.}$$

The net loss to the volume through these two faces is therefore

$$(\partial q_x/\partial x)\delta x \delta y \delta z$$

Similarly the net loss through the other two pairs of faces is

$$(\partial q_y/\partial y)\delta x\delta y\delta z$$

and

$$(\partial q_z/\partial z)\delta x\delta y\delta z$$

So the total volumetric loss of fluid is

$$\left(\frac{\partial q_x}{\partial x} + \frac{\partial q_y}{\partial y} + \frac{\partial q_z}{\partial z}\right)\delta x\delta y\delta z$$

and for an *incompressible* fluid, this must be equal to zero. Therefore, we have the condition, imposed by conservation of mass, that

$$\frac{\partial q_x}{\partial x} + \frac{\partial q_y}{\partial y} + \frac{\partial q_z}{\partial z} = 0 \tag{6.38}$$

This can also be written as

$$\text{div}\,\mathbf{q} = \nabla\cdot\mathbf{q} = 0 \tag{6.39}$$

The *divergence* (div) operator is defined by Equations (6.38) and (6.39). Thus, the divergence of a vector is the scalar product of the del operator and the vector.

Though derived here for flow in porous media, there was nothing in the derivation that was peculiar to porous media, except the assumption that the porosity of the small volume shown in Figure 6.4 does not change with time. The equation applies to any kind of flow of an incompressible fluid, where q_x, q_y, and q_z are the components of flow in the three coordinate directions. Though the equation expresses the conservation of volume (and mass, if the fluid is incompressible), it is commonly called the *continuity equation*, because it also expresses the assumption that the fluid is continuously distributed in space (e.g., it assumes that there are no bubbles or empty pores).

If there is a change in the volume of the pores (perhaps due to compaction), the divergence does not sum to zero. Similarly, if more than one fluid fills the pore space, e.g., air or petroleum besides water, then this necessitates a new set of equations to express the conservation of mass of each of the fluids involved.

If we apply the continuity equation $\nabla\cdot\mathbf{q} = 0$ to Darcy's law $\mathbf{q} = -k(\rho/\mu)\nabla\phi$, we obtain

$$\nabla\cdot\nabla\phi = 0 \tag{6.40}$$

or

$$\frac{\partial^2\phi}{\partial x^2} + \frac{\partial^2\phi}{\partial y^2} + \frac{\partial^2\phi}{\partial z^2} = 0 \tag{6.41}$$

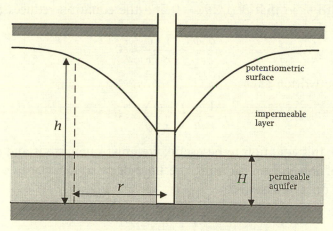

Fig. 6.5. Flow to a well in a confined aquifer.

This is known as *Laplace's equation.* It is a famous partial differential equation whose solutions are well known because they are important not only for the flow of fluids through porous media, but also in many other applications in science (see Appendix D). Laplace's equation tells us the potential field, i.e., how ϕ varies in space. Knowing this, we can use Darcy's law to determine the flow field, i.e., how **q** varies in space.

6.7 Flow to a well

We give here only a single example of the analytical solution of Laplace's equation. Consider flow to a well which produces water at a rate Q from an aquifer of thickness H confined between two impermeable layers (Figure 6.5).

The basic equation of flow is the two-dimensional Laplace equation:

$$\frac{\partial^2 \phi}{\partial x^2} + \frac{\partial^2 \phi}{\partial y^2} = 0 \tag{6.42}$$

We assume that the aquifer has isotropic permeability, and that there is no regional groundwater flow. Therefore, flow will be radially in towards the well, and it is better to write Laplace's equation in cylindrical coordinates:

$$\frac{\partial^2 \phi}{\partial r^2} + \frac{1}{r}\frac{\partial \phi}{\partial r} + \frac{1}{r^2}\frac{\partial^2 \phi}{\partial \theta^2} = 0 \tag{6.43}$$

(the derivation is not difficult, but is rather tedious). For radial flow, ϕ does not vary with θ, so that $\partial^2\phi/\partial\theta^2 = 0$ and the equation reduces to

$$\frac{\partial^2\phi}{\partial r^2} + \frac{1}{r}\frac{\partial\phi}{\partial r} = 0 \tag{6.44}$$

This may be written as

$$\frac{1}{r}\frac{\partial}{\partial r}\left(r\frac{\partial\phi}{\partial r}\right) = 0 \tag{6.45}$$

In order for this equation to be useful, we must integrate it and evaluate the constants of integration by using the boundary conditions (see Chapter 2). Integrating once, we have

$$r\frac{\partial\phi}{\partial r} = C_1 \tag{6.46}$$

where C_1 is a constant of integration. We cannot evaluate C_1, however, because the boundary conditions are given in terms of r and ϕ, not $\partial\phi/\partial r$. So we must integrate again:

$$\phi = C_1 \ln r + C_2 \tag{6.47}$$

To determine the constants of integration, consider the boundary conditions. Let $h = h_1$ at $r = r_1$ and $h = h_2$ at $r = r_2$. Along the piezometric surface $\phi = gh$, so we can write

$$gh_1 = C_1 \ln r_1 + C_2 \tag{6.48}$$
$$gh_2 = C_1 \ln r_2 + C_2 \tag{6.49}$$

or

$$g(h_1 - h_2) = C_1 \ln(r_1/r_2) \tag{6.50}$$

so that

$$C_1 = \frac{g(h_1 - h_2)}{\ln(r_1/r_2)} \tag{6.51}$$

From Darcy's law, we have the following expression for flow in the radial direction:

$$q_r = -K\left(\frac{\partial h}{\partial r}\right) = -\frac{K}{g}\frac{\partial\phi}{\partial r} \tag{6.52}$$

and using Equation (6.46) for $\partial\phi/\partial r = C_1/r$ we obtain

$$q_r = -\frac{KC_1}{gr} \tag{6.53}$$

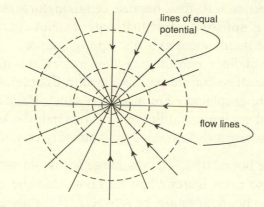

Fig. 6.6. Flow net for radial flow to a well in a confined aquifer.

So, if the borehole radius is r_0, the total *discharge* Q (defined as the volume rate of flow) from the well is given by

$$Q = 2\pi r_0 H q_0 \tag{6.54}$$
$$= -2\pi K H C_1/g \tag{6.55}$$
$$= -2\pi K H \frac{(h_1 - h_2)}{\ln(r_1/r_2)} \tag{6.56}$$

With this equation we can use piezometer wells to measure h at two locations, for a given rate of pumping Q, and determine K. H is known from the stratigraphy, which is determined when the well is drilled. It is important to have ways to measure the 'bulk' permeability of an aquifer, because laboratory determinations, made only on a few small samples, may be quite misleading in some aquifers, where much of the flow takes place along fractures, or in a few large pore spaces.

6.8 Flow nets

A solution to the two-dimensional Laplace's equation can be represented as a contour map of ϕ. The solution to the well-flow problem, given by Equation (6.47) is shown in Figure 6.6 by a set of circular lines of equal potential (*equipotential lines*). Superimposed on the equipotential lines are theoretical flow lines: as mentioned in Section 6.5, these are everywhere normal to the equipotential lines and point in the direction of decreasing ϕ; therefore they are everywhere parallel to the velocity vectors. If the flow is steady (does not change with time), flow is indeed along a flow line. A tubular part of the flow field bounded by flow lines is called a *stream tube*.

We will see in Section 6.10 that, because of mixing processes induced by the pore network, it is not strictly true that water cannot cross flow lines, but let us accept this for the moment as a first approximation.

The two sets of lines constitute a *flow net*. Flow nets have much in common with the stress trajectories discussed in Chapter 4, and the rules for constructing them graphically are much the same. Besides the requirement that the flow and equipotential lines be orthogonal, the key to constructing a flow net is the correct definition of the boundary conditions. These are:

(1) Impermeable boundaries. The discharge across an impermeable boundary is zero, so from Darcy's law we know that the gradient of h or ϕ normal to the boundary must be zero (e.g., if the plane formed by the x- and y-axes is impermeable, then $\partial h/\partial z = \partial \phi/\partial z = 0$ along that plane). Because flow lines are perpendicular to the gradient of ϕ, this boundary condition also requires that an impermeable boundary be a flow line.

(2) Constant-head boundaries. A line of constant head (e.g., along the horizontal bottom of a body of standing water) is a line along which ϕ is a constant, i.e., an equipotential line. Flow lines must be perpendicular to such a boundary.

(3) Water table. The water table is the boundary between unsaturated and saturated zones in the earth (neglecting effects due to surface tension). The (gauge) pressure along the water table is zero, so from Equation (6.14) we have $\phi = hg = gz$. If there is no flow across the water table (no recharge) and its position does not vary with time, then the water table is also a flow line. If there is flow across the water table (as when percolating rainwater recharges the groundwater), the water table is neither a flow line nor an equipotential line.

A flow net is a graphical solution to the Laplace equation. In the next section, we will discuss numerical solutions. Numerical methods are now more commonly used than graphical methods to solve the Laplace equation, although the solution in either case is often presented as a flow net. The boundary conditions used in either case are the same. Further details on flow nets may be found in Freeze and Cherry (1979, pp. 168–89).

One of the interesting and useful properties of the solutions to Laplace's equation is that it is possible to superpose solutions (add them together) and obtain another solution (see Appendix D). The following is an example.

Suppose that there is a regional flow of groundwater within a confined aquifer. Then the flow net would be as shown in Figure 6.7: in the flow shown, a steady east-to-west flow is produced by a constant (negative) gradient of fluid potential towards the west. If we now drill a well into the aquifer, and

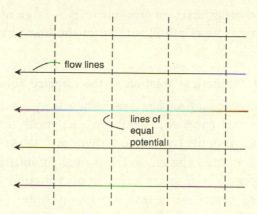

Fig. 6.7. Flow net for regional flow within a confined aquifer.

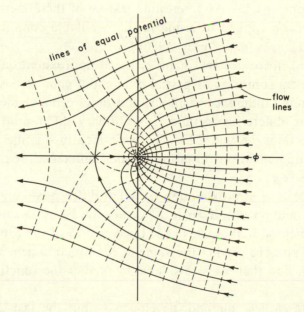

Fig. 6.8. Flow to a well in a confined aquifer, with a regional flow to the left (after Jacob, 1950).

begin to withdraw water at a rate Q, the flow net that results (Figure 6.8) can be obtained by superposing the two flow nets shown in Figures 6.6 and 6.7. The flow net shown in Figure 6.8 illustrates an important practical fact about such flows: a well producing water in a regional flow draws its supply of water from a large "catchment area" around and upstream from the well.

Other examples of flow nets are given by Hubbert (1969), Freeze and

Cherry (1979) and other texts on groundwater. A large number of analytical solutions of potential flows are illustrated in the book by Kirchhoff (1985).

6.9 Numerical solution of the Laplace equation

Though the Laplace equation has been extensively studied by mathematicians, the number of cases for which an analytical solution is known is limited. For most practical problems, therefore, it is necessary either to resort to analogue models (based on the formal similarity between electrical and hydraulic potentials) or to numerical approximation techniques. Numerical techniques are described in texts on groundwater (Freeze and Cherry, 1979, Chapter 5; De Marsily, 1986; Verruijt, 1982; Wang and Anderson, 1982; Bear and Verruijt, 1987) and in texts and articles on numerical analysis (see Appendix D). An interesting feature of these methods is that some of them can be applied by using readily available commercial spreadsheet software packages (e.g., Ousey, 1986).

In general, there are two main types of approximation method used to give numerical solutions of partial differential equations such as Laplace's equation: finite difference methods, and finite element methods. Examples of the finite difference method (for ordinary differential equations) were given in Sections 2.8 and 2.13. The basic idea in this method is to express the differential equations in terms of small differences (e.g., δx) rather than differentials (e.g., ∂x).

Details of how approximation by small differences can be used to solve Laplace's equation are given in Appendix D. Basically, however, the *finite difference* method makes use of the known form of the function at initial or boundary values to predict the value of the function at a small distance (or time) away, and then uses this value to predict the function again, and so on.

The *finite element* method, developed within the last 20 years (initially to solve problems in mechanical engineering) is now widely applied to the solution of partial differential equations, including those that describe the flow of groundwater. The approach taken differs from that used in the finite difference method. In the finite element method, the region of interest (e.g., part of the flow field) is divided into a number of small elements, within which the true function (e.g., Laplace's equation) may be replaced by a simpler function (e.g., a linear approximation to Laplace's equation). The problem is to find a set of these functions, such that they all fit together at the boundaries of the elements, and at the same time minimize errors of approximation. Though the techniques are now widely used, they are still considered relatively advanced and will not be discussed further in this text. For an introduction to both finite differences and finite elements and their application to groundwater problems see Wang and Anderson (1982) and for a more advanced treatment see Bear and Verruijt (1987) and Istok (1989).

6.10 Dispersion by flow

One of the main practical reasons for the study of groundwater is to ensure that water pumped from aquifers is not contaminated by wastes from disposal sites. We consider first the general question of how tracers are carried by a flow, and then some aspects peculiar to flow through porous media. We restrict the discussion to the flow of single-phase fluids, for example, to material dissolved in water, and do not consider the more difficult problem of the flow of two immiscible fluids. The flow of two or more immiscible fluids is of great practical importance, both in the petroleum industry (oil or gas and water) and in environmental studies (e.g., organic compounds denser than water), but we do not have the space to consider such questions in this book.

Some materials carried along by fluid flow can be considered to be conservative tracers, that is, they are not easily lost by decay (as are short-lived radioactive materials) or by adsorption onto clays, or by chemical reaction with porous rock. Generally, pore fluids do react with the rocks they flow through, so materials dissolved in groundwater are not conserved but may be lost or gained to the host rocks (Phillips, 1991). But we restrict our attention to materials, such as chloride ion, where this is not the case, and conservation of mass applies to the dissolved material as well as the fluid which carries it.

If the material were simply carried along passively (i.e., *advected*) by the flow, we might expect that its path could be followed by moving along flow lines leading from the source. Indeed, this would be the case if the flow were steady and if we were dealing with the flow of an inviscid fluid, or the flow of a viscous fluid in the laminar regime. But it is not true of the flow of groundwater, even though almost all flow through pores takes place in the laminar regime. Generally we think of the diffusion of dissolved tracers as being a feature diagnostic of turbulent flow as contrasted with laminar flow, yet diffusion is particularly characteristic of flow through porous media. Why is this?

The answer is that groundwater must move through an extremely complex network of pores. Two small parcels of fluid, originating in two adjacent pores, are constrained by the pore geometry to split, spread out and mix

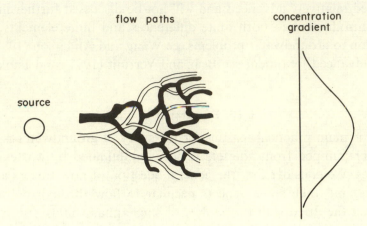

flow paths

concentration
gradient

source

Fig. 6.9. Dispersion and mixing of pore fluids (from Cherry *et al.*, 1975).

in the downflow direction (Figure 6.9). This tendency towards dispersion in directions both parallel to and transverse to the flow is also produced by molecular and turbulent motions, but in groundwater it is mainly a response to the geometry of the pore network.

Dispersion is generally expressed quantitatively by *Fick's law*. A one-dimensional form appropriate to porous media is

$$J_y = -PD_y \frac{\partial C}{\partial y} \tag{6.57}$$

J_y is the mass of tracer moving in the y-direction per unit area per unit time, so has dimensions $[ML^{-2}T^{-1}]$. C is the mass concentration of the tracer, with dimensions $[ML^{-3}]$, D_y is the diffusion coefficient for the y-direction, and P is the porosity. We assume that the y-direction is transverse to the flow. The quantity PC is the mass of tracer per unit volume of the porous medium. It follows from Equation (6.57) that the dimensions of D_y are $[L^2T^{-1}]$.

For molecular diffusion of a particular chemical species at a particular temperature, it may be assumed that the diffusion coefficient is a constant. From experiments on the flow of tracers through uniform porous media, however, it is known that D_y is proportional to the mean velocity of the flow within the pores, $q_x^* = q_x/P$. So we have

$$D_y = \alpha_y q_x^* \tag{6.58}$$

where α_y has dimensions of length, and is called the *dispersivity* of the tracer in the y-direction. It is often assumed that the dispersivity is a constant, generally measured on scales of about one metre. But observations

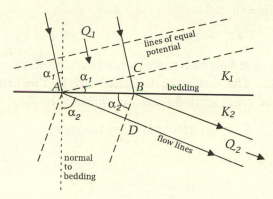

Fig. 6.10. Refraction of groundwater flow across a boundary between two beds with different permeability.

of the dispersion of tracers in the field have shown that it is actually scale dependent: for flow over length scales of the order of 1–10 km, the dispersivity may be as much as 100 m (Schwartz *et al.*, 1990). The reason for this is that natural porous media are not homogeneous, and cannot be fully characterized by a single "pore size". The permeability varies on many different scales, due to sedimentary structures, stratigraphic variation, and variations in cementation or compaction. Such variations contribute to the irregular paths taken by fluid particles moving through the medium, and therefore to the diffusivity. We will see in Section 11.5 that diffusion by fluid turbulence is also characterized by diffusion coefficients that are scale (and position) dependent.

The net result is that tracers in groundwater do not simply move along the "flow lines" that may be computed from measured or theoretical potential fields. Instead, they are rapidly spread out by diffusion, both laterally and in the flow direction. Though the rate of diffusion can in theory be predicted from the diffusion equation, in practice this is not easy to do, because of the difficulty of determining appropriate values of the diffusion coefficient or the diffusivity. For a detailed discussion, with many case histories, see Domenico and Schwartz (1990, Chapters 16–18).

6.11 Refraction of groundwater flow

Consider the flow of groundwater from an aquifer with hydraulic conductivity K_1 to one with a different (larger) hydraulic conductivity K_2. Let us consider the flow between the two flow lines shown in Figure 6.10. The cross-sectional areas AC (times unit length in the third dimension) and BD

are related by the expressions

$$AC = AB \cos \alpha_1, \quad BD = AB \cos \alpha_2 \tag{6.59}$$

From continuity, the volumetric discharge must be the same in each stream tube ($Q_1 = Q_2$). The discharge is calculated as the product of the unit discharge and the cross-sectional area normal to the flow, so

$$q_1 AB \cos \alpha_1 = q_2 AB \cos \alpha_2$$

or

$$q_1/q_2 = \cos \alpha_2 / \cos \alpha_1 \tag{6.60}$$

Points A and C, and B and D lie along two equipotential lines. Therefore the change in potential $\delta\phi$, from C to B, must equal the change from A to D. From geometry:

$$CB = AB \sin \alpha_1, \quad AD = AB \sin \alpha_2 \tag{6.61}$$

Applying Darcy's law to CB and AD:

$$q_1 = -K_1 \frac{\delta\phi}{AB \sin \alpha_1} \tag{6.62}$$

$$q_2 = -K_2 \frac{\delta\phi}{AB \sin \alpha_2} \tag{6.63}$$

Recalling that $\delta\phi$ is the same in both equations, we can use them to give the ratio q_1/q_2; combining the result with Equation (6.60) gives

$$\frac{K_1}{K_2} = \frac{\tan \alpha_1}{\tan \alpha_2} \tag{6.64}$$

Groundwater flow is therefore refracted as water moves from layers of one permeability to those of another, but unlike the refraction of light or seismic waves, the refraction law is not a sine law, but a tangent law.

In many stratified sequences, for example, interbedded sandstone (or sand) and mudrock (or clay) the permeabilities of the sandy layers may be several orders of magnitude larger than those of the muddy layers. It follows from Equation (6.64) that, within the sand, the flow will be almost parallel to the stratification but flow within the mud will be almost normal to the stratification, as shown in Figure 6.11. If we were to average such a hydraulic gradient, oriented obliquely across a sequence of thin sandstones interbedded with mudrocks, we would find that the direction of average flow across the sequence would not be the same as the direction of the average hydraulic gradient, because of the strong refraction of the flow along the more permeable beds. An extreme example of the same type of effect can

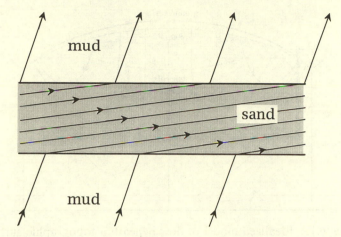

Fig. 6.11. Flow through strata of different permeability.

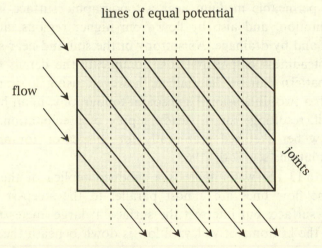

Fig. 6.12. Flow through a set of joints, shown as solid lines.

be seen by considering flow through a rock that is completely impermeable except for a set of fractures (joints). In this case (Figure 6.12) it is clear that flow can take place only in the direction of the joints, no matter what the direction of the hydraulic gradient. This is an example of the phenomenon of permeability anisotropy referred to in Section 6.5. Most simple groundwater theory ignores such anisotropy, but it is possible to take it into account in some numerical models.

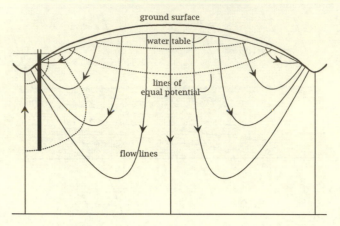

Fig. 6.13. Idealized model of flow beneath a topographic surface.

6.12 The water table: flow near a topographic surface

Water in a permeable medium near a topographic surface is replenished from precipitation, and also by flow from higher regions, and is lost by evaporation and by drainage. Anisotropy of the soil and near-surface rocks, and the unsteadiness of precipitation, complicate the details of real flows. The main pattern can be determined, however, using a simple model of flow beneath a two-dimensional sinusoidal topography, in an homogeneous, isotropic soil, resulting from a steady supply of precipitation. Figure 6.13 shows a flow net sketched graphically for this case: for analytical and numerical solutions see Toth (1963).

In Figure 6.13 the water table is a subdued replica of the topographic surface. The flow lines are almost parallel to the steepest parts of the topographic surface, but intersect the surface at large angles near the tops of hills and the bottoms of valleys. Flow is down beneath the hills, and up beneath the valleys. As required in a homogeneous, isotropic material, the equipotential surfaces are everywhere normal to the flow lines.

Suppose we drill a well beneath the valley, as shown in the figure. What height will the water rise to? The answer is obtained by noting the value of ϕ at the bottom of the well. From Equation (6.14), $\phi = hg = gz + (p/\rho)$, so the pore pressure at the bottom of the well is equal to $p = \rho(hg - gz)$, and the water will rise in the well to an elevation h above the bottom of the well. This will also equal the elevation found by tracing the equipotential line at the well tip to the water table, where $p = 0$, so $z = h$ (Figure 6.13).

As we can see from the figure, this elevation is generally somewhat above the ground surface, for wells drilled in valleys. Therefore, these wells will be

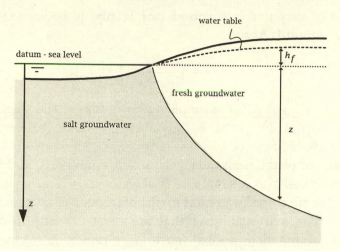

Fig. 6.14. The freshwater (Ghyben–Herzberg) lens.

artesian wells. Not all artesian wells, however, are to be explained in terms of confined aquifers with recharge areas that are topographically high (this was the provisional explanation given in Section 6.2).

6.13 Fluids of different densities: the freshwater lens

Consider a lens of freshwater, floating above (and not mixing with) saltwater in a homogeneous, isotropic aquifer underlying an oceanic island, as shown in Figure 6.14. The fluid potential in each fluid depends on the density:

$$\phi_1 = gz + \frac{p}{\rho_1}, \quad \phi_2 = gz + \frac{p}{\rho_2} \tag{6.65}$$

Along the interface between the two fluids, the pressure and elevation must be the same. So, eliminating p,

$$\phi_2 = gz + \frac{(\phi_1 - gz)}{\rho_2}\rho_1 \tag{6.66}$$

$$z = \frac{1}{g}\left(\frac{\rho_2}{\rho_2 - \rho_1}\phi_2 - \frac{\rho_1}{\rho_2 - \rho_1}\phi_1\right) \tag{6.67}$$

Letting $\phi = gh$, and using subscripts s to refer to saltwater and f to refer to freshwater,

$$z = \left(\frac{\rho_s}{\rho_s - \rho_f}\right)h_s - \left(\frac{\rho_f}{\rho_s - \rho_f}\right)h_f \tag{6.68}$$

If the head of saltwater, h_s is chosen (arbitrarily) to be zero and elevations are positive upwards, then

$$z = -\left(\frac{\rho_f}{\rho_s - \rho_f}\right) h_f \qquad (6.69)$$

For $\rho_f = 1$, $\rho_s = 1.025$ this implies $z = -40h_f$. The thickness of the freshwater lens, at any point, is thus about 40 times the elevation of the water table above sea level.

The existence of the freshwater lens is well established, and is important not only practically (as a supply of freshwater in oceanic islands) but also because of the effect freshwater has on the diagenesis of aragonitic carbonate sediments. It is well established that aragonite converts to calcite (by a dissolution–precipitation process) much more rapidly in freshwater than in saltwater.

A further complication that is important in diagenesis is that mixing of freshwater and saltwater may take place along the lower boundary of the freshwater lens. This is especially likely when there are fluctuations in sea level (produced by tides, storm surges, etc.) and fluctuations in the water-table elevation produced by variations in precipitation. The mixing of freshwater and saltwater may produce a fluid that is undersaturated with respect to calcite, but saturated with respect to dolomite. Alteration by such mixed fluids is one possible way in which dolomitization of carbonates may take place (to form "schizohaline" or "Dorag" dolomites; see Blatt *et al.*, 1980, p. 523).

6.14 Effect of groundwater on sliding on an infinite slope

In Section 4.16 we derived an expression for the conditions of incipient failure along a planar surface below an infinite slope (at an angle α; see Figure 4.24). We did this by equating an expression for the shear force acting along a surface parallel to the slope ($T = \gamma z l \cos \alpha \sin \alpha$) to the sliding resistance offered by the soil ($S = Cl + N \tan \psi$), where z is the vertical direction (not y as in Chapter 4) and we use ψ for the angle of internal friction, in order to avoid confusion with the fluid potential. As before, γ is the unit weight of the soil, l is a length parallel to the slope, and C is the cohesion. The normal force N is the component of the soil weight resolved normal to the surface ($N = \gamma z l \cos \alpha \cos \alpha$).

In Chapter 5, we demonstrated that the soil strength is actually a function of the effective normal stress $\sigma - u$, where u is the pore pressure in the soil. Now, we can estimate a value of u and combine these results to examine

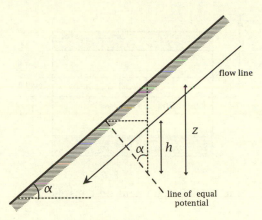

Fig. 6.15. Pore pressure distribution for the condition of slope-parallel flow on an infinite slope, for which $h = z \cos^2 \alpha$.

the conditions for sliding on an infinite slope containing groundwater. First, we define the resistance force from the soil strength in terms of the effective normal stress:

$$S = l[C + (\gamma z \cos\alpha\cos\alpha - u)\tan\psi]$$

and the solution becomes

$$\tan\alpha = \frac{C}{\gamma z \cos^2\alpha} + \left(1 - \frac{u}{\gamma z \cos^2\alpha}\right)\tan\psi \qquad (6.70)$$

For $u = 0$, this equation reduces directly to Equation (4.30) and for both u and C equal to zero, the particular result $\alpha = \psi$ directly follows.

A groundwater condition commonly assumed to represent the critical conditions for slope stability over a long period is that the water table is at the ground surface and that the flow is parallel to the slope (Figure 6.15). In a noninfinite slope, this assumption tends to overestimate the pore pressures below the slope top, where the flow is downwards, and underestimate the pore pressures below the slope toe, where the flow is upwards. For slope-parallel flow, the equipotential lines must be normal to the slope surface. From the slope geometry in Figure 6.15, the pore pressure is given by

$$u = \gamma_w h = \gamma_w z \cos^2\alpha \qquad (6.71)$$

where γ_w is the unit weight of the water, as distinguished from the total unit weight of the soil γ. For the case where $C = 0$, and using Equation (6.71) to substitute for u in Equation (6.70) we obtain

$$\tan\alpha = (1 - \gamma_w/\gamma)\tan\psi \qquad (6.72)$$

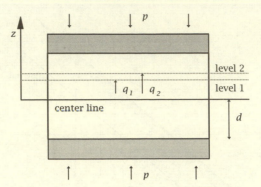

Fig. 6.16. Experiment on consolidation.

The saturated unit weight of soils varies between roughly 15 kN m^{-3} and 22 kN m^{-3}, whereas that of water is about 10 kN m^{-3}. Using $\gamma \approx 20$ kN m^{-3}, Equation (6.72) gives

$$\tan \alpha \approx \tfrac{1}{2} \tan \psi$$

In other words, incorporating the pore pressures in an effective-stress strength criterion reduces by half the friction coefficient for an infinite slope in cohesionless material. This applies only to cohesionless soils and an extreme groundwater condition. Because C is not affected by u, the reduction in critical slope angle for materials with some cohesive strength is smaller than that for cohesionless materials, but still significant.

6.15 Theory of consolidation

In our previous discussion of consolidation, we were not able to carry the theory very far, because we lacked an expression for the rate at which fluids could flow out of compacting sediments. We now have that expression (Darcy's law) and can indicate the general lines of development of a theory of consolidation. Consider the artificial consolidation of a sediment layer of thickness $2d$ between two porous plates, or (if you prefer) the natural consolidation of a clay layer between two porous aquifers (Figure 6.16). We apply a total normal stress of p. Within the porous plates the pore pressure is essentially zero (in natural porous aquifers it would be hydrostatic), but within the consolidating layer of clay it is $u(z, t)$, i.e., u varies with z because of the low permeability of the clay, and it varies with t because it take time for the fluid to flow out the pores and for the pore pressure to reach an equilibrium value. At the instant we apply p, u will equal the hydrostatic pressure plus p. This creates a gradient in u towards the porous plates. In the plates and in the clay immediately adjacent to them, u is essentially zero. u increases with depth into the clay. The rate of change of u with z is $\partial u/\partial z$. We will assume that both z and the flow rate q are positive in the upward direction.

We remember that the total head h is composed of two parts: a part due to pressure, u/γ, and a part due to the elevation z. The hydraulic gradient at z due to the variation in pressure alone is

$$i_1 = \frac{1}{\gamma} \frac{\partial u}{\partial z} \tag{6.73}$$

This will be negative in the upward direction because the pore pressure, u is decreasing in that direction. We assume that i_1 is large compared with the variation in h due to z.

At depth $z + \delta z$, the pressure is $u + (\partial u/\partial z)\delta z$ and the gradient in pressure head is

$$i_2 = \frac{1}{\gamma} \frac{\partial}{\partial z} \left(u + \frac{\partial u}{\partial z} \delta z \right) \tag{6.74}$$

$$= \frac{1}{\gamma} \left(\frac{\partial u}{\partial z} + \frac{\partial^2 u}{\partial z^2} \delta z \right) \tag{6.75}$$

By Darcy's law, the flow rate across level 1 (q_1) and the flow rate across level 2 (q_2) are given by $q_1 = -Ki_1$ and $q_2 = -Ki_2$ where we have assumed that the change in permeability between levels 1 and 2 is negligible. Conservation of fluid volume requires that the change in fluid volume between levels 1 and 2 be $\delta q = q_2 - q_1$. The water lost between levels 1 and 2 is therefore

$$\delta q = -K(i_2 - i_1) = -\frac{K}{\gamma} \frac{\partial^2 u}{\partial z^2} \delta z \tag{6.76}$$

Assuming that both the fluid and the solid part of the clay are incompressible, this must be equal to the rate of reduction of the volume of pore water in the element of clay between levels 1 and 2. Because the specimen is saturated, the rate of volume reduction of pore water must also equal the rate of volume reduction of the whole clay element. Let the latter rate be $\partial V/\partial t$. We have

$$\frac{\partial V}{\partial t} = \delta q \delta A \tag{6.77}$$

where δA is the cross-sectional area of the specimen $(V = \delta A \delta z)$, and δq is given by Equation (6.76). If we define the volume per unit area as $V_A = V/\delta A$, we arrive at the interim result that the time rate of volume change per unit area is equal to the flow rate per unit area out of the specimen:

$$\frac{\partial V_A}{\partial t} = \delta q \tag{6.78}$$

Now the rate of volume reduction must also be related to inherent properties of the clay specimen (e.g., its structure). We represent the compressibility of the specimen by a *coefficient of compressibility* m_v, which is defined as the fractional change in specimen volume, i.e., a volumetric strain $\delta V/V$, divided by the change in effective pressure $\delta(p - u)$. Because we assume there are no volume changes in the lateral direction, the volumetric strain is exactly equal to the proportional change in specimen thickness. Expressing the change in specimen thickness as the change in volume per unit area δV_A, the volumetric strain is $\delta V_A/\delta z$, and the definition of m_v is given by

$$m_v = \frac{\delta V_A/\delta z}{\delta(p - u)} \tag{6.79}$$

or

$$\delta V_A = m_v \delta(p - u)\delta z \tag{6.80}$$

We are looking for an expression for the time rate of change of specimen volume, so we divide Equation (6.80) by δt and take the limit as δt goes to zero:

$$\lim_{\delta t \to 0} \frac{\delta V_A}{\delta t} = \lim_{\delta t \to 0} m_v \frac{\delta(p - u)}{\delta t}\delta z$$

or

$$\frac{\partial V_A}{\partial t} = m_v \frac{\partial(p - u)}{\partial t}\delta z$$

Because p is a constant, this may be written as

$$\frac{\partial V_A}{\partial t} = -m_v \frac{\partial u}{\partial t}\delta z \tag{6.81}$$

We now have two expressions for $\partial V_A/\partial t$, one derived from considering the flow of pore water out of the specimen (Equation 6.78), the other defined using the coefficient of compressibility (Equation 6.81). If we equate these two expressions (using Equation 6.76 to define δq) and solve for $\partial u/\partial t$, we obtain

$$\frac{\partial u}{\partial t} = \frac{K}{\gamma m_v}\frac{\partial^2 u}{\partial z^2} = C_v \frac{\partial^2 u}{\partial z^2} \tag{6.82}$$

where $C_v = K/\gamma m_v$ is called the *coefficient of consolidation*.

Equation (6.82) is the basic (one-dimensional) partial differential equation of consolidation, originally derived by Terzaghi. It states that the rate

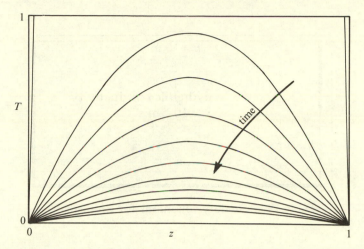

Fig. 6.17. Solution of the consolidation equation, obtained using program EHEAT. z is the distance across the layer ($z = 0$ at the base, $z = 1$ at the top). T, which is the dimensionless temperature difference for a heat-flow problem, is here the *consolidation ratio*, that is, the fractional difference between the ultimate thickness decrease and the decrease at time t (so T is initially unity for all z and is ultimately zero for all z). The 10 curves are for 10 equal increments of time.

of change of fluid pressure with respect to time is proportional to the second derivative of the fluid pressure with respect to distance (in the flow direction). It shows how fluid diffuses out of an infinitesimal volume because of a pressure gradient, and so it is an example of a *diffusion equation*.

A similar equation applies to heat flow:

$$\frac{\partial T}{\partial t} = \kappa \frac{\partial^2 T}{\partial z^2} \tag{6.83}$$

where T is the temperature, and κ is the *thermal diffusivity*, which is equal to the thermal conductivity divided by the specific heat times the density (see Chapter 12: details of the derivation are given in Turcotte and Schubert, 1982, Chapter 4). It is perhaps no surprise that this is so, because there is a close analogy between Darcy's law and the law of heat conduction (Section 6.5). The theory of heat flow was developed earlier than the theory of flow through porous media, so many of the mathematical methods used for groundwater flow were developed first for heat flow.

For the case of vertical consolidation, as shown in Figure 6.16, the consolidation ratio is as shown in Figure 6.17.

The theory of consolidation outlined above is adequate for small consolidations, such as those resulting from construction of buildings or dams. It assumes that the changes in permeability produced by compaction can be

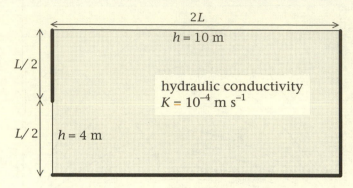

Fig. 6.18. Boundary conditions for Problem 1.

neglected, so it is not adequate for describing the extensive consolidation of sediments produced by burial over long periods of time in sedimentary basins. For this type of consolidation a more complex treatment is required: Audet and Fowler (1992) give a review, and some numerical solutions.

6.16 Review problems

(1) **Problem:** Consider the groundwater flow problem shown in Figure 6.18. Though the geometry is somewhat artificial, this might represent flow under a dam (on the left), where the head to the right of the dam is 10 m and that to the left is 4 m.

(a) Sketch in the flow net by hand, using six streamtubes (i.e., draw five flow lines and five lines of equal potential). Remember that

   (i) flow is parallel to impermeable boundaries;
   (ii) flow is normal to surfaces of equal potential (i.e., of equal head);
   (iii) flow and potential lines are normal to each other.

   Draw your sketch in pencil on a sheet of squared paper, and be prepared to do a lot of erasing and revision!

(b) Use the program LAPLACE (or a spreadsheet) to compute the flow potentials at grid points. For best results use a 13 × 7 grid or larger. The largest grid that can readily be printed out on computer tractor paper (sidewise) is 15 × 29. Draw contour lines of equal potential, and use them to draw flow lines – then compare with your previous sketch (LAPLACE offers a computer contouring routine, but as an exercise it is better to ignore it and draw contours by hand).

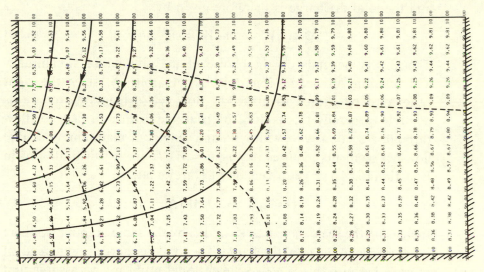

Fig. 6.19. Flow net for Problem 1, from potential values obtained using LAPLACE.

(c) Calculate the total discharge through the system, if $L = 10$ m and $K = 10^{-4}$ m s^{-1}. Use the equation

$$Q = mK\,\Delta h/n$$

where m is the number of streamtubes, n is the number of divisions of head (both set to 6 in the first part of the question) and Δh is the difference in head from surface 1 to surface 2. Notice that the answer does not depend on the scale, L. Explain in words why not.

Answer:

(a) See (b).

(b) A typical flow net obtained by using LAPLACE to obtain grid values of the potential, and then sketching in potential and flow lines, is shown in Figure 6.19.

(c) The equation for Q is derived as follows: if the total head difference is Δh and there are n divisions of head, then the head difference across a streamtube is $\delta h = \Delta h/n$. If the total cross-sectional width of flow is W at some section, and there are m flow tubes, then the width of the streamtube is $\delta W = W/m$. The flow through one section of the streamtube, which has length δW (since the sections of the flow net are nearly square), is given by Darcy's law,

$$q\delta W = K(\delta h/\delta W)\delta W = K\Delta h/n$$

and the total discharge is therefore given by

$$Q = qW = qm\delta W = mK\,\Delta h/n \qquad (6.84)$$

The answer depends on the choice of the number of streamtubes and divisions of head – but this is not arbitrary, since the right ratio must be chosen to make the flow-net segments as square as possible (generally this is done by trial and error). In fact, for flow through a section with width W and length L, maintaining square flow-net segments requires that $m/n = W/L$. If we substitute this in Equation (6.84) we get a version of Darcy's law:

$$Q = qW = -KW(\Delta h/L)$$

Increasing the scale increases W and L by the same amount, so there is no change to the total discharge. The total discharge is independent of scale because this is a two-dimensional problem: we assume unit length in the third dimension. If we consider the geometry in the third dimension, an increase in scale increases the discharge because the cross-sectional area A increases as L^2.

(2) **Problem:** What is the gradient of:

(a) $\phi = x^2 + y^2$,

(b) $\phi = e^x \sin 2y$.

Answer: By simple partial differentiation:

(a)

$$\frac{\partial \phi}{\partial x} = 2x, \quad \frac{\partial \phi}{\partial y} = 2y$$

so

$$\text{grad } \phi = 2x\mathbf{i} + 2x\mathbf{j}$$

(b) Similarly

$$\text{grad } \phi = \sin 2y\, e^x \mathbf{i} + 2\cos 2y\, e^x \mathbf{j}$$

(3) **Problem:** If the fluid potential is given by:

$$\phi = x^2 + xy + yz$$

then at the point (2,1,4) find

(a) the coordinates of the unit vector pointing in the direction of maximum rate of change of the potential,

(b) the derivative of the potential in the x-direction.

Answer:

(a) The vector pointing in the direction of maximum rate of change of the potential is the gradient,

$$\text{grad } \phi = (2x + y)\mathbf{i} + (x + z)\mathbf{j} + y\mathbf{k}$$

and at (2,1,4) this is given by

$$\text{grad } \phi = 5\mathbf{i} + 6\mathbf{j} + \mathbf{k}$$

This is not a unit vector, however, because it has length

$$L = \sqrt{25 + 36 + 1} = \sqrt{62}$$

We divide by this quantity to obtain the unit gradient vector:

$$\mathbf{r} = 0.635\mathbf{i} + 0.762\mathbf{j} + 0.127\mathbf{k}$$

(b)

$$\frac{\partial \phi}{\partial x} = 2x + y$$

and at (2,1,4) this equals 5.

(4) **Problem:** The existence of a hydraulic gradient implies that potential energy is being lost to friction as fluid moves through a porous medium. Because of frictional losses, the total mechanical energy decreases in the direction of flow, which implies that the fluid exerts a force, the *seepage force*, on the porous medium. Show that, if the flow is uniform, i.e., does not change in the flow direction x, this force per unit volume F_s is given by

$$F_s = \gamma(dh/dx)$$

Answer: The potential energy per unit mass is $\phi = hg$. Its rate of change in the x-direction is $d\phi/dx = g(dh/dx)$. There are no other forces to balance the drop in mechanical energy in the flow direction, so the seepage force is exactly equal to the drop in potential. Multiplying by density, the force per unit volume exerted on the porous medium is $\rho(d\phi/dx) = \gamma(dh/dx)$.

(5) **Problem:** When flow in a porous medium is directed upward, the vertical component of the seepage force acts in the direction opposite to the gravitational force. If the upward gradient is strong enough, the seepage force can be large enough to support the weight of the solid grains. These are said to be *quick* conditions, for which the medium loses all

frictional strength (the effective stress goes to zero). If the material is cohesionless, the material under this condition is said to be *fluidized*. Show that the "critical" upward hydraulic gradient necessary for quick conditions is given by

$$\frac{\partial h}{\partial z} = \frac{(\gamma_s - \gamma)}{\gamma}(1 - P)$$

and that the corresponding discharge is

$$q = -K\frac{(\gamma_s - \gamma)}{\gamma}(1 - P)$$

where γ_s and γ are the specific weights of the solids and fluid, respectively, and P is the porosity.

Answer: Consider a thickness H of a porous medium. From Problem 5, the seepage force per unit area is $\gamma H(\partial h/\partial z)$, where z is the vertical direction. The buoyant weight (per unit area) of the solid medium is $(\gamma_s - \gamma)H(1 - P)$. These are the only forces acting in the vertical direction, and we balance for them for quick conditions. Equating the two and solving for $\partial h/\partial x$ gives

$$\frac{\partial h}{\partial z} = \frac{\gamma_s - \gamma}{\gamma}(1 - P)$$

The same solution can be obtained by starting with the definition that quick conditions occur when the effective stress equals zero. The pore pressure at the base of H is given by $u = \gamma(H + \delta h)$, where δh is the change in head across H. We see that the pore pressure is different from the hydrostatic pressure γH by the amount $\gamma \delta h$. The total stress at the base of H is given by the total weight of solids and fluid:

$$\sigma = \gamma_s H(1 - P) + \gamma H P$$

Equating σ with u and noting that the gradient in head across H is given by $\delta h/H$ gives the same solution as above. For values of γ_s and γ representative of mineral grains and water, the critical upward gradient generally falls within the range of 0.84 for loose granular material (P = 0.5) and 1.12 for dense granular material (P = 0.33). From Darcy's law, the gradient in head is equal to $-q/K$. Substituting this into the expression for the critical gradient and solving for q gives

$$q = -K\frac{\gamma_s - \gamma}{\gamma}(1 - P)$$

6.17 Suggested reading

Chapman, R. E., 1981. *Geology and Water: An Introduction to Fluid Mechanics for Geologists*. The Hague, Martinus Nijhoff, 228 pp. (Includes a good treatment of permeability and tortuosity. QC809.F5C44)

Domenico, P. A. and F. W. Schwartz, 1990. *Physical and Chemical Hydrogeology*. New York, John Wiley and Sons, 824 pp. (This and Freeze and Cherry (1979) have become the standard textbooks on the subject. GB1003.2.D66)

Freeze, R. A. and W. Back, eds., 1983. *Physical Hydrogeology. Benchmark Papers in Geology*, v.72. Stroudburg PA, Hutchinson Ross, 431 pp. (Reprints of classic papers, with comments on their historical significance. GB1003.2.P49)

Freeze, R. A. and J. A. Cherry, 1979. *Groundwater*. Englewood Cliffs NJ, Prentice-Hall, 604 pp. (GB1003.2.F73)

Heath, R. C., 1983. Basic Ground-Water Hydrology. US Geol. Survey Water-Supply Paper 2220, 84 pp. (Good, elementary summary)

Hubbert, M. K., 1969. *The Theory of Ground-water Motion and Related Papers*. New York, Hafner, 310 pp. (Reprints of Darcy's and Hubbert's classic papers. TC176.H83)

Jury, W. A., W. R. Gardner and W. H. Gardner, 1991. *Soil Physics*. New York, John Wiley and Sons, fifth edition, 328 pp. (An introduction to the special problems of unsaturated porous media. S592.3.J86)

Schwartz, F. W. *et al.*, 1990. *Ground Water Models: Scientific and Regulatory Applications*. Washington DC, National Academy Press, 303 pp. (Readable discussion of the problems and achievements of mathematical models in groundwater hydrology. TC176.G76)

Verruijt, A., 1982. *Theory of Groundwater Flow*. London, Macmillan, second edition, 144 pp. (Clear and concise. TC176.V36)

Wang, H. F. and M. P. Anderson, 1982. *Introduction to Groundwater Modelling, Finite Difference and Finite Element Methods*. San Francisco, W. H. Freeman, 237 pp. (Elementary introduction to numerical methods, with programs in BASIC. TC176.W36)

7

Strain

7.1 Definitions

In this chapter we begin to describe the effects produced by forces or stresses acting on continua. A *continuum* consists of material smoothly distributed through space (Chapter 1). The smallest part of a body considered by continuum mechanics is generally called a *particle* or element of the continuum. The bulk properties of the continuum, such as density or stress, are assumed to vary smoothly from one element to the next.

Forces acting on bodies of continua may produce accelerations, or changes in the motion of the body, just as they do for discrete particles or rigid bodies of matter. In Chapter 2 we considered the use of vector methods in Cartesian and other types of coordinate systems to describe such motions. Now we would like to generalize those methods so that they can be applied to bodies that experience not just rigid-body translation and rotation, but also change of shape, or distortion.

A *deformation* (see, for example, Means, 1976) can be described in terms of a displacement field. A *displacement field* is an array of displacement vectors connecting the positions of particles before and after deformation.

In this sense, deformation includes:

(1) rigid-body translation;
(2) rigid-body rotation;
(3) strain (= distortion).

A deformation is *homogeneous* if a straight line (of any orientation) in the original body remains a straight line after the deformation. All the different types of infinitesimal strain discussed in this chapter are homogeneous.

234

Nonhomogeneous deformations result when strains are finite, and larger in some places than in others. Many different terms are used to describe strain, but we will see that all strains can be described by combining an elongation (or, in more than one dimension, a dilatation) and a shear strain.

Geologists (particularly structural geologists) are frequently much concerned with large degrees of deformation, including large strains. A good discussion is given by Means (1976) and especially by Ramsay (1967) and Ramsay and Huber (1983). In this book, however, we will be concerned mainly with infinitesimal strains. Infinitesimal strains are those which are so small that only linear relationships between stresses and the elongation and shear strain need be considered. Or, to put it another way, we can neglect all squares and products of the two quantities, ε and γ, which measure elongation and shear strain (i.e., we can ignore second- and higher-order terms, see below).

Although strains larger than about one per cent cannot be considered infinitesimal, the infinitesimal strain taking place in a solid during an infinitesimal increment of time can always be related to the instantaneous stress. So a description of infinitesimal strains is important even for the case of a body which is undergoing much larger strains. Furthermore, in solid rocks, stresses are transmitted by small elastic deformations of the rock that can indeed be regarded as infinitesimal strains in most cases.

In fluids, a shear stress produces an indefinitely large strain; for this reason, stress is not related to strain, but to rate of strain. To evaluate rates of strain, however, we must consider small increments of strain. So it turns out that infinitesimal strains are important practically, and their consideration forms the basis of both elasticity theory (Chapter 8) and the theory of the flow of viscous fluids (Chapter 9).

In this chapter, we are concerned only with methods for describing deformation or strain. In later chapters, we will relate strain to the stresses that produce it. Newton's laws do not tell us how to do this, and therefore it is necessary to develop a new set of relationships, called *constitutive equations*, which do provide this information. With rare exceptions, constitutive equations cannot be derived from first principles (e.g., from statistical mechanics); therefore, they remain empirical and approximate in character. Nevertheless, elaborate theories have been developed for certain simple constitutive equations, for example, those that define a linear elastic body, or a Newtonian viscous fluid, and it has been found that these theories give good descriptions of some of the important ways in which real rocks and fluids behave.

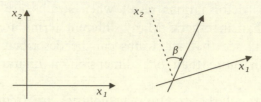

Fig. 7.1. Change in angle between two lines, that were originally perpendicular.

7.2 Measures of strain

Several measures are described more fully by Means (1976). In this book we will be using mainly the elongation ε and the shear strain γ.

7.2.1 Elongation

The elongation ε is given by

$$\varepsilon = \frac{\text{length after deformation} - \text{undeformed length}}{\text{undeformed length}}$$

$$= \frac{\delta\ell}{\ell} \tag{7.1}$$

Note that, if the strain is infinitesimal, and if we ignore solid-body translations, we do not have to specify that ℓ is the undeformed length, because $\delta\ell$ is very small compared with ℓ, i.e.,

$$\varepsilon = \frac{\delta\ell}{\ell} \approx \frac{\delta\ell}{\ell + \delta\ell} \tag{7.2}$$

The Greek symbol *epsilon*, which is used here to represent elongation, comes in two forms: ϵ and ε. Some authors distinguish their use (as we do, see below), but there is no universal convention. Similarly, some authors prefer to use the term "extension" for what we have called elongation. Note that a negative elongation is a contraction.

7.2.2 Shear strain

Shear strain is measured as the change in angle between two lines in the material that were initially perpendicular. In the example shown in Figure 7.1, note that shear strain has been accompanied by translation (from left to right across the page) and by an anticlockwise rotation of both lines. We ignore this translation and rotation by allowing the reference frame (x_1, x_2) to translate and rotate also.

The shear strain γ is defined as follows:

$$\gamma = \tan \beta \qquad (7.3)$$

where the angle β is called the *angular shear*, and is generally measured in radians. It is the angle by which the original $\pi/2$ radians (90 degrees) between the line segments, has been reduced. Thus angular shear does not include the effect of rigid-body rotations that affect both line segments equally. We note that the particular component of shear strain shown is γ_{12}, i.e., the strain in the x_1x_2-plane. Note also that $\gamma_{12} = \gamma_{21}$ and that in general there will be three components of shear strain, $\gamma_{12}, \gamma_{13}, \gamma_{23}$, corresponding to the three planes that each include two of the coordinate axes. The axes, and therefore the planes of strain, are indicated by the subscripts.

The plane that shows strain is not the plane along which movement is taking place within the body. If, in the example shown, x_1 and x_2 lay in the plane of the bedding, then the shear would be taking place by movement along planes *normal* to the bedding.

7.2.3 Displacement gradients

Elongation and shear strain are ways of measuring the two main categories of strain:

Normal strain: the change in length of lines within a strained body.

Shear strain: the change in angle between intersecting lines within a strained body.

Note at once the analogy with stress: we can already appreciate that the full representation of strain involves a strain tensor, consisting of three normal strain components and three shear strain components.

The definitions of ε and γ given above were expressed in terms of the finite increments $\delta\ell$ and β. In another commonly used notation, ε and γ are expressed in terms of displacements with components u, v, and w (or u_1, u_2 and u_3) in the x-, y- and z- (or x_1-, x_2-, and x_3-) directions. For example, the normal strain might be measured by the displacement $u = \delta\ell$, over a distance $x = \ell$. At first, it is somewhat confusing to have two different symbols that refer to distances measured in the x-direction. It may help to remember that x refers to the dimension of the body (in the x-direction) and u to a displacement of the body in that direction. If the dimension of the body is small, we refer to it as δx, and if the (small) body suffers a (very) small displacement, the displacement is δu. In general, $\delta u \ll \delta x$. Often we are concerned with the strain at a point, so we must take limits. For example,

$$
\begin{aligned}
\varepsilon &= \lim_{\ell \to 0} \frac{\delta \ell}{\ell} \\
&= \lim_{\delta x \to 0} \frac{\delta u}{\delta x} \\
&= \frac{\partial u}{\partial x}
\end{aligned}
\tag{7.4}
$$

where, in the second line, we simply change the notation: $\delta x = \ell$, $\delta u = \delta \ell$. Unfortunately, in works on the mechanics of solids, u is commonly taken as the notation for a *displacement*, whereas in works on the mechanics of fluids, u is generally the notation for a velocity. We follow this convention. In this chapter and Chapter 8, u is a displacement, but in Chapter 9 it will be a velocity. $\partial u / \partial x$ is a partial derivative, because u may vary in the y- (x_2-) and z- (x_3-) directions also, even though, in the present chapter, it is the component of displacement in the x-direction. Such partial derivatives are called *displacement gradients*. We will see below that shear strains can also be expressed in terms of displacement gradients.

7.3 Plane strain

In this section, we consider the case where there is no strain in the x_3-direction. For example, if a body is long in the x_3-direction, and if the main forces act across the length, but do not vary in the x_3-direction, then it may be assumed that all $x_1 x_2$ cross-sections are similar. There may be strain in the x_1- and x_2-directions, but not in the x_3-direction.

Consider the deformation of a small element, rectangular in the $x_1 x_2$-section, as shown in Figure 7.2. We will demonstrate that such a deformation can be expressed as the sum of four different types of movement:

(1) translation;
(2) rotation;
(3) elongation (or contraction);
(4) pure shear.

The German scientist H. L. F. von Helmholz (1821–94) demonstrated in 1858 that this is true of any deformation (not only of plane strains). We will first examine the case of plane strain (i.e., strain in two dimensions only) and then provide a generalization, using tensor notation, to three dimensions. Note that each type of movement is independent of the others. For example, elongation does not necessarily involve net translation, and pure shear (though it involves rotation of some lines within the deformed

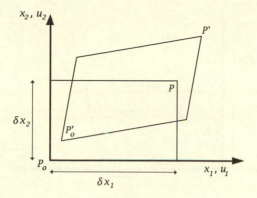

Fig. 7.2. Deformation of a small area element. $P_0P_0' = \mathbf{u}_0$, $PP' = \mathbf{u}$; $\mathbf{u} - \mathbf{u}_0 = \delta\mathbf{u}$ is the displacement of P due to rotation, elongation, and shear strain.

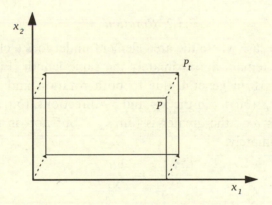

Fig. 7.3. Translation of a small area element. Point P is moved to P_t by the displacement $\mathbf{u}_0$.

body) does not involve net rotation, because the average rotation over all lines is zero. To make this distinction clear, translation and rotation are sometimes called "solid body" movements.

7.3.1 Translation

The components x_1, x_2 of any point in the element (Figure 7.3) are given after a translation of $\mathbf{u}_0$ by

$$x_1' = x_1 + (u_1)_0, \quad x_2' = x_2 + (u_2)_0 \tag{7.5}$$

where $(u_1)_0$ and $(u_2)_0$ are the displacements in the x_1- and x_2-directions. The notation $(u_1)_0$, $(u_2)_0$ indicates that these are displacements due only to translation.

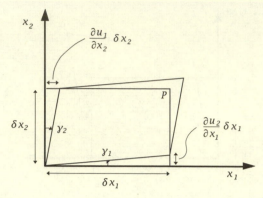

Fig. 7.4. Rotation and shear strain of a small area element. If $\gamma_2 = -\gamma_1$, we have pure (solid body) rotation; if $\gamma_2 = +\gamma_1$, we have pure shear strain.

7.3.2 Rotation

Let us consider the case where the area element undergoes a change in shape although the sides remain approximately the same length (Figure 7.4). The displacement of P is, in general, due to both rotation and shear and has components $(\delta u_1)_{r,s}$, $(\delta u_2)_{r,s}$ in the x_1- and x_2-directions. $(\delta u_1)_{r,s}$ depends on δx_2 – the longer is δx_2, the greater is $(\delta u_1)_{r,s}$ – but not on δx_1. Thus, we may write, approximately,

$$\frac{(\delta u_1)_{r,s}}{\delta x_2} = \frac{\partial u_1}{\partial x_2}$$

and, therefore,

$$(\delta u_1)_{r,s} = \frac{\partial u_1}{\partial x_2}\delta x_2$$

as shown in Figure 7.4. In a similar way,

$$(\delta u_2)_{r,s} = \frac{\partial u_2}{\partial x_1}\delta x_1$$

The horizontal sides of the element have been rotated by γ_1 and the vertical sides by $-\gamma_2$. The average rotation ω is given by

$$\omega = (\gamma_1 - \gamma_2)/2$$

We assume that the angle of rotation is small: if so, the angle (in radians) is almost equal to the tangent. For example, one degree is 0.01745 radians, and the tangent of one degree is 0.01746. Thus, in Figure 7.4 γ_1 in radians is almost equal to $(\partial u_2/\partial x_1)\delta x_1/\delta x_1$ and γ_2 in radians is almost equal to

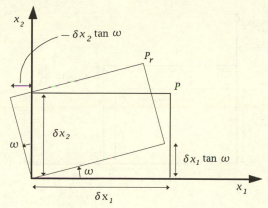

Fig. 7.5. Pure rotation of a small area element. Point P is moved to P_r.

$(\partial u_1/\partial x_2)\delta x_2/\delta x_2$ so that the above equation becomes

$$\omega = \frac{1}{2}\left(\frac{\partial u_2}{\partial x_1} - \frac{\partial u_1}{\partial x_2}\right) \tag{7.6}$$

We have used the convention that anticlockwise rotations are positive.

A small displacement of P produced by *rotation alone* thus has components (see Figure 7.5)

$$(\delta u_1)_r = -\delta x_2 \ \tan \ \omega = -\frac{1}{2}\left(\frac{\partial u_2}{\partial x_1} - \frac{\partial u_1}{\partial x_2}\right)\delta x_2 \tag{7.7}$$

$$(\delta u_2)_r = \delta x_1 \ \tan \ \omega = \frac{1}{2}\left(\frac{\partial u_2}{\partial x_1} - \frac{\partial u_1}{\partial x_2}\right)\delta x_1 \tag{7.8}$$

We will use this result in Section 7.4. The x_1-component is negative because a positive rotation moves P in the negative x_1-direction, whereas the x_2-component is positive because a positive rotation moves P in the positive x_2-direction. From Equation (7.6), the condition for *no rotation* of the element (i.e., $\gamma_1 = \gamma_2$) is that

$$\frac{\partial u_2}{\partial x_1} = \frac{\partial u_1}{\partial x_2} \tag{7.9}$$

7.3.3 Elongation or contraction (dilatation or compression)

In the case where the area element undergoes elongation or contraction but the sides remain perpendicular (Figure 7.6), we have (Equation 7.4)

$$\varepsilon_{11} = \lim_{\delta x_1 \to 0} \frac{(\delta u_1)_d}{\delta x_1} = \frac{\partial u_1}{\partial x_1} \tag{7.10}$$

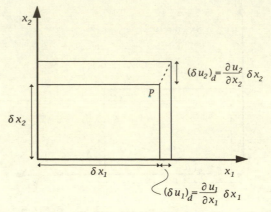

Fig. 7.6. Dilatation of a small area element. The sides are elongated by $(\delta u_1)_d, (\delta u_2)_d$.

$$\varepsilon_{22} = \lim_{\delta x_2 \to 0} \frac{(\delta u_2)_d}{\delta x_2} = \frac{\partial u_2}{\partial x_2} \tag{7.11}$$

These are measures of elongation or contraction in the x_1- and x_2-direction. From Equations (7.10) and (7.11), the displacement of P has x_1- and x_2-components that are approximately given by

$$(\delta u_1)_d = \frac{\partial u_1}{\partial x_1} \delta x_1$$

$$(\delta u_2)_d = \frac{\partial u_2}{\partial x_2} \delta x_2$$

The *dilatation* θ is the fractional change in area of the element (Figure 7.6) $\delta A / A$, where

$$A = \delta x_1 \delta x_2$$

and

$$\delta A = \left(\delta x_1 + \frac{\partial u_1}{\partial x_1} \delta x_1 \right) \left(\delta x_2 + \frac{\partial u_2}{\partial x_2} \delta x_2 \right) - \delta x_1 \delta x_2$$

So, if we use the linear approximation and neglect second-order terms:

$$\delta A = \frac{\partial u_1}{\partial x_1} \delta x_1 \delta x_2 + \frac{\partial u_2}{\partial x_2} \delta x_1 \delta x_2$$

and

$$\theta = \frac{\delta A}{A} = \frac{\partial u_1}{\partial x_1} + \frac{\partial u_2}{\partial x_2} = \varepsilon_{11} + \varepsilon_{22} \tag{7.12}$$

Note that we are here defining dilatation to be positive, and compression to be negative.

In three dimensions it can be shown that the dilatation is

$$\theta = \varepsilon_{11} + \varepsilon_{22} + \varepsilon_{33} \tag{7.13}$$

By analogy with the discussion of stress invariants given in Chapter 4, we can see that the dilatation is the *first invariant of the strain tensor*. In other words, the dilatation is not altered by choice of a different coordinate system – a property that we consider essential for any measure of volumetric change.

7.3.4 Shear strain

The shear strain ε_{12} of the element shown in Figure 7.4 can be defined as the average shear strain of the two sides, i.e., the average angular change from the original right angle between the two sides.

$$\varepsilon_{12} = \frac{1}{2}(\gamma_1 + \gamma_2) \tag{7.14}$$

As before, we use the figure to obtain

$$\varepsilon_{12} = \frac{1}{2}\left[\frac{(\partial u_2/\partial x_1)\delta x_1}{\delta x_1} + \frac{(\partial u_1/\partial x_2)\delta x_2}{\delta x_2}\right]$$

$$= \frac{1}{2}\left(\frac{\partial u_2}{\partial x_1} + \frac{\partial u_1}{\partial x_2}\right) \tag{7.15}$$

ε_{12} is called the *tensor component of shear strain* in the $x_1 x_2$-plane. Note that it is one-half the shear strain as previously described in Subsection 7.2.2: that *total* or *engineering component* ϵ_{12} is defined as twice ε_{12}:

$$\epsilon_{12} = 2\varepsilon_{12} = \left(\frac{\partial u_2}{\partial x_1} + \frac{\partial u_1}{\partial x_2}\right) \tag{7.16}$$

This is essentially the measure shown in Figure 7.1, and there called the tangent of angular shear. Note the use of a different form of epsilon to represent the engineering component of shear strain. In this book, we generally use the tensor component rather than the engineering component. From the definition given above it is clear that $\varepsilon_{12} = \varepsilon_{21}$.

As for the other types of strain, the displacement of P due to shear strain may be split into x_1- and x_2-components:

$$(\delta u_1)_s = \frac{1}{2}\left(\frac{\partial u_2}{\partial x_1} + \frac{\partial u_1}{\partial x_2}\right)\delta x_2$$

$$(\delta u_2)_s = \frac{1}{2}\left(\frac{\partial u_2}{\partial x_1} + \frac{\partial u_1}{\partial x_2}\right)\delta x_1$$

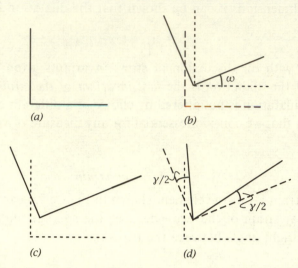

Fig. 7.7. Deformation shown by two line segments: (*a*) original position; (*b*) after translation and rotation by an angle ω; (*c*) after elongation of one of the segments; (*d*) after an angular shear γ.

7.3.5 *Summary*

Figure 7.7 illustrates graphically how a deformation of two line segments may be produced by a sequence of four operations: translation, rotation, elongation, and pure shear. At each stage of the operation the preceding state is shown by the dashed line, and the present state by the full line. Panel (*b*) shows two operations: a translation to the position indicated by the dotted line, followed by a rotation through an angle ω. Panel (*c*) shows an elongation of one of the line segments, and panel (*d*) shows a shear strain. The total shear strain is γ, with each line being rotated (in opposite directions) by half that amount. This figure may be compared with Figure 7.8 which shows a similar deformation, by plotting the movements of the two ends of a single line segment $P_0 P$.

7.4 Approach through Taylor's theorem

A more concise, algebraic approach to the approximation procedures used above may be given using Taylor's theorem. The familar one-dimensional statement of this theorem is as follows:

$$f(x) = f(x_0) + \frac{df}{dx}\delta x + \frac{d^2 f}{dx^2}\left(\frac{\delta x^2}{2}\right) + \cdots + \text{higher-order terms} \qquad (7.17)$$

In words, this states that in the vicinity of a point P_0 located at x_0, the function $f(x)$ may be represented by an infinite series composed of the product of the derivatives of $f(x)$ and powers of the small distance δx from x_0. Because δx is small, terms involving powers of δx can be neglected. In a linear approximation, all the terms are dropped, except for the first two. For two dimensions, Taylor's theorem becomes

$$f(x_1, x_2) = f((x_1)_0, (x_2)_0) + \frac{\partial f}{\partial x_1}\delta x_1 + \frac{\partial f}{\partial x_2}\delta x_2 + \cdots \tag{7.18}$$

where x_1, x_2 give the position of P and $(x_1)_0, (x_2)_0$ the position of P_0.

We now apply Equation (7.18) to our area element after the body has been deformed (Figure 7.2). We express the displacement $\mathbf{u}$ of point P to P' in terms of the displacement $\mathbf{u}_0$ of point P_0 to P_0'. The components u_1 and u_2 of $\mathbf{u}$ are functions of both x_1 and x_2; thus, applying Equation (7.18) we have

$$u_1 = (u_1)_0 + \frac{\partial u_1}{\partial x_1}\delta x_1 + \frac{\partial u_1}{\partial x_2}\delta x_2 + \cdots \tag{7.19}$$

$$u_2 = (u_2)_0 + \frac{\partial u_2}{\partial x_1}\delta x_1 + \frac{\partial u_2}{\partial x_2}\delta x_2 + \cdots \tag{7.20}$$

Here $(u_1)_0, (u_2)_0$ are the components of $\mathbf{u}_0$. Ignoring higher-order terms, and adding and subtracting $\frac{1}{2}(\partial u_2/\partial x_1)\delta x_2$ on the right-hand side of Equation (7.19), and $\frac{1}{2}(\partial u_1/\partial x_2)\delta x_1$ on the right-hand side of Equation (7.20), we obtain

$$u_1 = (u_1)_0 + \frac{1}{2}\left(\frac{\partial u_1}{\partial x_2} - \frac{\partial u_2}{\partial x_1}\right)\delta x_2 + \frac{\partial u_1}{\partial x_1}\delta x_1$$
$$+ \frac{1}{2}\left(\frac{\partial u_1}{\partial x_2} + \frac{\partial u_2}{\partial x_1}\right)\delta x_2 \tag{7.21}$$

$$u_2 = (u_2)_0 + \frac{1}{2}\left(\frac{\partial u_2}{\partial x_1} - \frac{\partial u_1}{\partial x_2}\right)\delta x_1 + \frac{\partial u_2}{\partial x_2}\delta x_2$$
$$+ \frac{1}{2}\left(\frac{\partial u_2}{\partial x_1} + \frac{\partial u_1}{\partial x_2}\right)\delta x_1 \tag{7.22}$$

From our earlier discussion of plane strain, we can identify the first pair of terms $((u_1)_0, (u_2)_0)$ as the two components of translation, the second pair as the two components of the displacement of P_0P due to rotation, the third pair of terms as the two components of the displacment due to dilatation, and the final pair of terms as the two components of the displacement due to shear strain.

Figure 7.8 gives a graphic representation; see also Figure 7.2. The end P_0 of the line moves by translation only and the other end P moves by

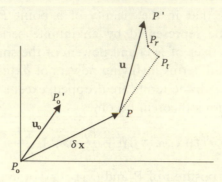

Fig. 7.8. The deformation of Figure 7.2 as it affects the line P_0P. First, a translation takes P_0P to P'_0P_t. Then P_t is taken to P_r by a rotation about P'_0. Finally a strain takes P_r to P'.

translation, rotation, and strain (elongation and shear). First P moves by a translation to P_t, then by a rotation to P_r (the rotation is around P'_0, through half the total angle $P_tP'_0P'$), and finally by a strain to P'. The effect of the strain on the original line is a further rotation, plus an elongation from P_r to P'. For infinitesimal strains, the order is not important, but it would be if the strains were finite. In rocks that have undergone finite strains, only the final result is observed and it is not generally possible to work out the exact history of how that state was reached.

In vector notation we could write Equations (7.21) and (7.22) concisely as

$$\mathbf{u} = \mathbf{u}_0 + \boldsymbol{\omega} \times \delta\mathbf{x} + \boldsymbol{\varepsilon}\delta\mathbf{x} \tag{7.23}$$

where $\mathbf{u}_0$ is the vector of translation and $\boldsymbol{\omega}$ is the rotation vector,

$$\boldsymbol{\omega} = \frac{1}{2}\left(\frac{\partial u_2}{\partial x_1} - \frac{\partial u_1}{\partial x_2}\right)\mathbf{k} \tag{7.24}$$

i.e., it is the vector whose cross-product with the position vector $\delta\mathbf{x}$ indicates a rotation about the x_3-axis (see Chapter 2). $\boldsymbol{\varepsilon}$ is the strain tensor ε_{ij},

$$\varepsilon_{ij} = \begin{bmatrix} \varepsilon_{11} & \varepsilon_{12} \\ \varepsilon_{21} & \varepsilon_{22} \end{bmatrix} = \begin{bmatrix} \dfrac{\partial u_1}{\partial x_1} & \dfrac{1}{2}\left(\dfrac{\partial u_2}{\partial x_1} + \dfrac{\partial u_1}{\partial x_2}\right) \\ \dfrac{1}{2}\left(\dfrac{\partial u_2}{\partial x_1} + \dfrac{\partial u_1}{\partial x_2}\right) & \dfrac{\partial u_2}{\partial x_2} \end{bmatrix} \tag{7.25}$$

Note that we could also consider the operator $(\boldsymbol{\omega}\times)$ to be a tensor Ω_{ij} (with typical element ω_{ij}):

$$\Omega_{ij} = \begin{bmatrix} 0 & \dfrac{1}{2}\left(\dfrac{\partial u_1}{\partial x_2} - \dfrac{\partial u_2}{\partial x_1}\right) \\ \dfrac{1}{2}\left(\dfrac{\partial u_2}{\partial x_1} - \dfrac{\partial u_1}{\partial x_2}\right) & 0 \end{bmatrix} \tag{7.26}$$

This tensor is not symmetric, like the strain tensor, but *antisymmetric*:

$$\omega_{12} = -\omega_{21}$$

Adding the strain tensor and the rotation tensor yields a tensor which is not symmetric, and whose elements consist of the displacement gradients:

$$
\begin{bmatrix}
\frac{\partial u_1}{\partial x_1} & \frac{1}{2}\left(\frac{\partial u_2}{\partial x_1} + \frac{\partial u_1}{\partial x_2}\right) \\
\frac{1}{2}\left(\frac{\partial u_2}{\partial x_1} + \frac{\partial u_1}{\partial x_2}\right) & \frac{\partial u_2}{\partial x_2}
\end{bmatrix}
$$

$$
+ \begin{bmatrix}
0 & \frac{1}{2}\left(\frac{\partial u_1}{\partial x_2} - \frac{\partial u_2}{\partial x_1}\right) \\
\frac{1}{2}\left(\frac{\partial u_2}{\partial x_1} - \frac{\partial u_1}{\partial x_2}\right) & 0
\end{bmatrix}
= \begin{bmatrix}
\frac{\partial u_1}{\partial x_1} & \frac{\partial u_1}{\partial x_2} \\
\frac{\partial u_2}{\partial x_1} & \frac{\partial u_2}{\partial x_2}
\end{bmatrix}
= \frac{\partial u_i}{\partial x_j} \quad (7.27)
$$

We can now see why the tensor component of shear strain was defined as half the total shear strain. We can also see why deformation is not defined in terms of simple displacement gradients: to do so would be to include in the off-diagonal gradients (i.e., terms like $\partial u_2/\partial x_1$) *both* rotation and shear strain.

7.5 Generalization to three dimensions

We will do this concisely, following the approach given at the end of the last section, and using tensor notation.

We consider two points, P_0 located at $((x_1)_0, (x_2)_0, (x_3)_0)$ and P, located close by at (x_1, x_2, x_3). During deformation P_0 is displaced to P_0' and P to P' (as in Figure 7.8).

Applying Taylor's theorem in three dimensions:

$$u_i = (u_i)_0 + \frac{\partial u_i}{\partial x_j}\delta x_j + \cdots \quad (7.28)$$

Adding and subtracting $\frac{1}{2}(\partial u_j/\partial x_i)\delta x_j$ on the right-hand side and neglecting higher-order terms, we obtain

$$u_i = (u_i)_0 + \frac{1}{2}\left(\frac{\partial u_i}{\partial x_j} - \frac{\partial u_j}{\partial x_i}\right)\delta x_j + \frac{1}{2}\left(\frac{\partial u_i}{\partial x_j} + \frac{\partial u_j}{\partial x_i}\right)\delta x_j \quad (7.29)$$

which can be written

$$u_i = (u_i)_0 + \Omega_{ij}\delta x_j + \varepsilon_{ij}\delta x_j \quad (7.30)$$

As before, the first term on the right-hand side is a translation, the second term is a rotation, and the third term includes both dilatation and shear strain. Ω_{ij} is the rotation tensor, whose typical element is shown in the previous equation (see also Equation 7.26), and ε_{ij} is the strain tensor. This

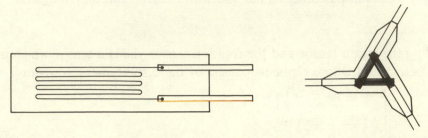

Fig. 7.9. Strain gauges. On the left is a single gauge, on the right a strain gauge rosette.

is the same equation as Equation (7.23), but expressed in tensor rather than vector notation.

7.6 Measurement of strain

Geologists measure strain by observing the shapes of objects whose original shape, before deformation, is either known (e.g., fossils) or can be hypothesized (e.g., spherical ooids). The topic is discussed thoroughly in texts on structural geology (e.g., Ramsay and Huber, 1983).

In experiments, very small strains are generally measured by using strain gauges. The use of these and other techniques is described by Dally and Riley (1991) and Hendry (1977).

A typical electrical strain gauge is shown on the left in Figure 7.9. It consists of a length of resistance wire, about 0.025 mm in diameter, wound as shown, and cemented to a paper backing, which in turn may be cemented to the test material. A small extension in the north–south direction does not change the resistance of the wire, but extension in the east–west direction can only take place by stretching the wire, and therefore changing its electrical resistance. The change in electrical resistance is directly proportional to the component of strain in that direction. Generally, a "rosette" of three strain gauges, oriented at 60 degrees to each other (see the right of Figure 7.9), is mounted and permits determination of the three components of strain in the plane of the surface (see below).

7.7 Strain ellipse, Mohr circle of strain

Strain gauges provide information only about elongation or contraction, not about shear strain. We will show how measurements of elongation, made in

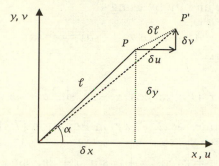

Fig. 7.10. Elongation of a line segment that makes an angle α with the x-axis.

three different directions on three different planes, may be used to determine the full strain tensor.

Figure 7.10 shows a line segment that has undergone a small strain in the xy-plane. The line segment makes an angle α with the x-axis. The original length of the line segment is ℓ and after strain it is approximately $\ell + \delta\ell$. We first derive an expression for the square of the length, after elongation, using the infinitesimal strain assumption (i.e., because $\delta\ell \ll \ell$, we can neglect squares of $\delta\ell$):

$$
\begin{aligned}
(\ell + \delta\ell)^2 &\approx \ell^2 + 2\ell\delta\ell \\
&\approx \ell^2 \left(1 + \frac{2\delta\ell}{\ell}\right)
\end{aligned}
\tag{7.31}
$$

Next we write the same relationship, but using components in the x- and y-directions:

$$
\begin{aligned}
(\ell + \delta\ell)^2 &= (\delta x + \delta u)^2 + (\delta y + \delta v)^2 \\
&\approx \delta x^2 + \delta y^2 + 2\delta u \delta x + 2\delta v \delta y
\end{aligned}
\tag{7.32}
$$

We make use of the following relations for displacements,

$$
\delta u = \frac{\partial u}{\partial x}\delta x + \frac{\partial u}{\partial y}\delta y \qquad \delta v = \frac{\partial v}{\partial x}\delta x + \frac{\partial v}{\partial y}\delta y
$$

and for elongations, from Equations (7.10) and (7.11), and for shear strain, from Equation (7.15),

$$
\varepsilon_x = \frac{\partial u}{\partial x}, \qquad \varepsilon_y = \frac{\partial v}{\partial y}, \qquad \varepsilon_{xy} = \frac{1}{2}\left(\frac{\partial v}{\partial x} + \frac{\partial u}{\partial y}\right)
$$

to rewrite the last two terms in Equation (7.32) giving:

$$
(\ell + \delta\ell)^2 \approx \delta x^2 + \delta y^2 + 2\varepsilon_x \delta x^2 + 2\varepsilon_y \delta y^2 + 4\varepsilon_{xy}\delta x \delta y
$$

We now eliminate δx and δy by using

$$\delta x = \ell \cos \alpha, \qquad \delta y = \ell \sin \alpha$$

and obtain

$$(\ell + \delta \ell)^2 \approx \ell^2 (1 + 2\varepsilon_x \cos^2 \alpha + 2\varepsilon_y \sin^2 \alpha + 4\varepsilon_{xy} \sin \alpha \cos \alpha)$$

We use this to substitute for $(\ell + \delta \ell)^2$ in Equation (7.31) and obtain

$$\varepsilon = \frac{\delta \ell}{\ell} = \varepsilon_x \cos^2 \alpha + \varepsilon_y \sin^2 \alpha + 2\varepsilon_{xy} \sin \alpha \cos \alpha \qquad (7.33)$$

which can also be written as

$$\varepsilon = \varepsilon_x \cos^2 \alpha + \varepsilon_y \sin^2 \alpha + \varepsilon_{xy} \sin 2\alpha \qquad (7.34)$$

If we have measured the elongation in three different directions in the xy-plane, then application of Equation (7.34) to each of the directions gives three equations in the three unknowns $\varepsilon_x, \varepsilon_y, \varepsilon_{xy}$ and therefore these three components of the stress tensor can be determined. Similar sets of measurements in the xz- and yz-planes enable determination of the complete strain tensor.

Note the exact analogy between Equation (7.34) for the elongation and Equation (4.10) in Section 4.9 for the normal stress acting on a plane whose normal makes an angle α with the x-axis. Similarly, we could derive an equation for the shear strain analogous to Equation (4.11) for the shear stress. This shows that there is a complete analogy between stress and infinitesimal strain. Both are second-order tensors, and both can be reduced to diagonal forms which give the principal stresses or strains on surfaces that have zero shear stresses or strains. Just as the state of stress can be represented by a stress ellipsoid, so the state of infinitesimal strain can be represented by a strain ellipsoid.

As we did in developing the Mohr circle of stress in Chapter 4, we now consider the case where the x- and y-axes are coincident with the directions of principal strain. In this case, $\varepsilon_{xy} = 0$, and Equation (7.34) reduces to

$$\varepsilon = \varepsilon_x \cos^2 \alpha + \varepsilon_y \sin^2 \alpha$$

where α is the angle between a line and the x-axis. This equation is exactly analogous to Equation (4.18). Using the same trigonometric identities, we can change the equation above into the analogue of Equation (4.20):

$$\varepsilon = \left(\frac{\varepsilon_x + \varepsilon_y}{2} \right) + \left(\frac{\varepsilon_x - \varepsilon_y}{2} \right) \cos 2\alpha \qquad (7.35)$$

The corresponding relation for the shear strain (cf. Equation 4.21) is

$$\frac{\gamma}{2} = \left(\frac{\varepsilon_x - \varepsilon_y}{2}\right) \sin 2\alpha \tag{7.36}$$

The result (as with stress) is that the expressions for the normal strain (elongation) and shear strain of a line making an angle α with the axis of largest principal strain (ε_x) define a circle. This is the Mohr circle of infinitesimal strain. Most authors use $\gamma/2$ to represent the shear strain on the Mohr circles of strain, and we must remember the factor of one-half (cf. Equation 7.14), because it is the tensor component of shear strain that is analogous to the shear stress, not the total (or engineering) component. Using the analogy with the stress Mohr circle, we note that the lines with the largest shear strain are oriented at an angle of 45 degrees to ε_x and the magnitude of the shear strain is equal to the difference $\varepsilon_x - \varepsilon_y$. The Mohr circle for infinitesimal strain is developed more fully in Means (1976) and in many structural geology texts.

7.8 Review problems

For examples of strain problems related to movement on faults see Turcotte and Schubert (1982, Chapter 2).

(1) **Problem:** For plane strains (Section 7.3) the term *simple shear* is used to describe a strain in which the displacement is confined to one direction (say x_1) and the displacements in the other direction are zero. An example is the shearing of a deck or cards, lying flat on a table. Show that simple shear has, in fact, both a shear and a rotational component.

Answer: From the definition of simple shear, $u_2 = 0$ for all x, and therefore $\partial u_2/\partial x_1 = 0$. The rotation, given by Equation (7.6), is therefore

$$\omega = -\frac{1}{2}\frac{\partial u_1}{\partial x_2}$$

The shear, given by Equation (7.15), is

$$\varepsilon_{12} = \frac{1}{2}\frac{\partial u_1}{\partial x_2}$$

Note two things about the rotational component: it does not involve internal distortion, and it is not necessary to have displacements along a curved path in order to produce "solid body" rotation.

(2) Problem:

(a) Show that if B_{ij} is a symmetrical tensor and C_{ij} is an antisymmetrical tensor and if

$$A_{ij} = B_{ij} + C_{ij} \qquad (7.37)$$

then

$$A_{ij} + A_{ji} = 2B_{ij}$$

(b) Use this result to decompose the tensor

$$A_{ij} = \begin{bmatrix} 5 & 3 & 1 \\ 3 & 7 & 2 \\ 3 & 4 & 3 \end{bmatrix}$$

into the sum of a symmetric and antisymmetric tensor. If A_{ij} were a displacement gradient tensor, then the symmetric part would be the strain tensor, and the antisymmetric part the rotation tensor.

Answer:

(a) If B_{ij} is symmetric and C_{ij} is antisymmetric then

$$B_{ij} = B_{ji}, \quad C_{ij} = -C_{ji}$$

so that

$$\begin{aligned} A_{ji} &= B_{ji} + C_{ji} \\ &= B_{ij} - C_{ij} \end{aligned} \qquad (7.38)$$

Adding Equations (7.37) and (7.38) gives

$$A_{ij} + A_{ji} = 2B_{ij}$$

(b) Adding the matrix to its transpose, we have

$$\begin{bmatrix} 5 & 3 & 1 \\ 3 & 7 & 2 \\ 3 & 4 & 3 \end{bmatrix} + \begin{bmatrix} 5 & 3 & 3 \\ 3 & 7 & 4 \\ 1 & 2 & 3 \end{bmatrix} = \begin{bmatrix} 10 & 6 & 4 \\ 6 & 14 & 6 \\ 4 & 6 & 6 \end{bmatrix}$$

Dividing by 2 gives:

$$B_{ij} = \begin{bmatrix} 5 & 3 & 2 \\ 3 & 7 & 3 \\ 2 & 3 & 3 \end{bmatrix}$$

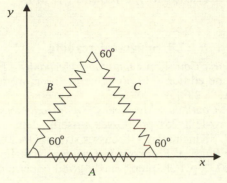

Fig. 7.11. Strain rosette.

Subtracting this from A_{ij} gives

$$C_{ij} = \begin{bmatrix} 0 & 0 & -1 \\ 0 & 0 & -1 \\ 1 & 1 & 0 \end{bmatrix}$$

which is antisymmetric.

(3) **Problem:** Show that for the strain rosette shown in Figure 7.11 the equation relating the observed strains $\varepsilon_A, \varepsilon_B, \varepsilon_C$ to the strains $\varepsilon_x, \varepsilon_y, \varepsilon_{xy}$ is

$$\begin{bmatrix} \varepsilon_A \\ \varepsilon_B \\ \varepsilon_C \end{bmatrix} = \begin{bmatrix} 1 & 0 & 0 \\ 0.25 & 0.75 & 0.866 \\ 0.25 & 0.75 & -0.866 \end{bmatrix} \begin{bmatrix} \varepsilon_x \\ \varepsilon_y \\ \varepsilon_{xy} \end{bmatrix}$$

Answer: We make use of Equation (7.34)

$$\varepsilon_A = \varepsilon_x \cos^2 \alpha + \varepsilon_{xy} \sin 2\alpha + \varepsilon_y \sin^2 \alpha$$

which relates the elongation of a line making an angle α to the x-direction to three components of strain in the plane. We change the notation to that used above and obtain

$$\varepsilon_A = [\cos^2 \alpha \ \sin^2 \alpha \ \sin 2\alpha] \begin{bmatrix} \varepsilon_x \\ \varepsilon_y \\ \varepsilon_{xy} \end{bmatrix}$$

So, a complete equation for all three measured strains $(\varepsilon_A, \varepsilon_B, \varepsilon_C)$ is

$$\begin{bmatrix} \varepsilon_A \\ \varepsilon_B \\ \varepsilon_C \end{bmatrix} = \begin{bmatrix} \cos^2 \alpha & \sin^2 \alpha & \sin 2\alpha \\ \cos^2(60° + \alpha) & \sin^2(60° + \alpha) & \sin 2(60° + \alpha) \\ \cos^2(120° + \alpha) & \sin^2(120° + \alpha) & \sin 2(120° + \alpha) \end{bmatrix} \begin{bmatrix} \varepsilon_x \\ \varepsilon_y \\ \varepsilon_{xy} \end{bmatrix}$$

Evaluating for $\alpha = 0$ leads to the required result.

7.9 Suggested reading

Fung, Y.-C., 1977. *A First Course in Continuum Mechanics.* Englewood Cliffs NJ,
    Prentice-Hall, second edition, 340 pp. (One of the more readable introductions
    to the subject. QA808.2.F85)

Hodge, P. G. Jr., 1970. *Continuum Mechanics. An Introductory Text for Engineers.*
    New York, McGraw-Hill, 251 pp. (Gives a useful elementary treatment of
    tensors. Chapter 7, "Kinematics", discusses strain. QA808.2.H6)

Johnson, A. M., 1970. *Physical Processes in Geology.* San Francisco, Freeman,
    Cooper and Co., 577 pp. (Chapter 5 has a good concise discussion of small
    strains. QE33.J58)

Long, R. R., 1961. *Mechanics of Solids and Fluids.* Englewood Cliffs NJ,
    Prentice-Hall, 156 pp. (Gives a useful concise, treatment of stress and strain,
    using tensor notation throughout. QA931.L84)

Means, W. D., 1976. *Stress and Strain. Basic Concepts of Continuum Mechanics for
    Geologists.* New York, Springer-Verlag, 339 pp. (QE604.M4)

Turcotte, D. L. and G. Schubert, 1982. *Geodynamics, Applications of Continuum
    Physics to Geological Problems.* New York, John Wiley and Sons, 450 pp.
    (Chapter 2 gives a good discussion of how strains are measured by observing
    relative displacements across faults. QE501.T83)

Verhoogen, J. *et al.*, 1970. *The Earth. An Introduction to Physical Geology.* New
    York, Holt, Rinehart and Winston, 748 pp. (An interesting attempt by the
    Berkeley department to produce an advanced level introduction to geology.
    Chapter 9, "Deformation of Rocks", is an excellent brief introduction to stress
    and strain. QE501.V38)

8

Elasticity

8.1 Constitutive equations

In Chapter 4 we developed the concept of the stress tensor, and in Chapter 7 that of the strain tensor. However, we have not yet developed any understanding of how the two are related to each other.

The way that the strain of a material depends on the applied stress defines the type of material. Real materials may display very complex strain–stress relationships (see Chapter 10) but two theoretical models have been developed which are relatively simple, and which provide a close approximation to the behaviour of real materials, at least under some conditions. The two ideal materials are: (i) a Hookian elastic solid, from which we obtain the theory of linear elasticity, and (ii) a Newtonian viscous fluid, which we will cover in Chapter 9.

The equations which relate the strain tensor to the stress tensor for any type of material are called *constitutive equations*. They cannot in practice be deduced from the general laws of mechanics (e.g., from Newton's laws) but depend upon experience, experiment, and rational analysis. Even today, there is generally no way in which particular constitutive relations can be deduced from fundamental physics (e.g., atomic theory, quantum theory), nor can the actual values of the elastic constants or viscosity of particular materials be predicted from such theory. They must be measured by observing the behaviour of the material in the laboratory or in the field. The most successful application of theory, which might be considered an exception to these comments, is the prediction from statistical mechanics of the viscosity of an ideal gas.

In this chapter we will develop the theory of linear elasticity, which is generally applicable only to small elastic deformations of materials. This limits the number of possible problems one may consider, but some of those that remain have considerable importance. A classic problem is the bending of beams and plates: the theory cannot be applied to folding, where the strains are large, but it can and has been applied to flexure of lithospheric plates under loads, such as those produced by volcanoes or sedimentary basins. Elastic waves (i.e., seismic waves) in rocks generally produce only infinitesimal deformations, and are another major field of application. As we have already seen in earlier chapters, there also many problems in rock mechanics where elasticity theory can be applied.

It turns out that there are interesting mathematical analogies between the constitutive equations for linear elastic materials and viscous materials. Both theories are linear: in elasticity theory strains are related linearly to stresses, and in viscous flow theory rates of strain are related linearly to stresses. In later chapters we will develop the theory of such common fluids as water and air. In many cases of interest, the theory cannot be applied directly because of the phenomenon of turbulence; however, it can be applied to the slow flow of rocks under conditions of high temperature and pressure, e.g., to the flow of the mantle of the earth.

8.2 Hooke's law

Robert Hooke (1635–1703) was a contemporary of Newton (1642–1727) and a protégé of Robert Boyle (1627–91). Hooke probably played an important role in provoking Newton to develop his theory of gravitation. He was an early member of the Royal Society of London and was appointed to the position of "Curator of Experiments" – so he was supposed to organize several experiments for display at each of the society's weekly meetings. He made contributions to many fields of science: microscopy, meteorology (he invented many instruments), combustion (even before phlogiston theory), astronomy, etc. He was also regarded as a quarrelsome eccentric.

In 1678 he published his law as an anagram: "ceiiinossssttuu" standing for "ut tensio sic uis (vis)" or "the power of any spring is in the same proportion with the tension thereof". Hooke did not clearly distinguish force from stress and applied his law in a rather general way to springs, but we can extend it to elastic extension more generally and write

$$\sigma = E\varepsilon \qquad\qquad (8.1)$$

where σ is the normal stress (positive for tension) and ε is the elongation or normal strain (positive for elongation, negative for compression).

The proportionality coefficient, E is called *Young's modulus*. Thomas Young (1773–1829) introduced the modulus in 1807, but what he suggested

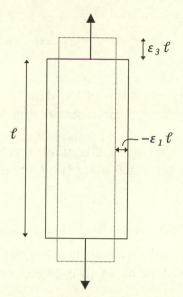

Fig. 8.1. Extension of a bar.

was not what is now defined as his modulus, nor was his suggestion original. The honour really belongs to Leonhard Euler (1707–83) who in 1727 introduced a "modulus of extension" in a work on bells. For more details about this and many other aspects of the history of mechanics, see Truesdell (1968, pp. 124–5).

Young's modulus is not sufficient to define the relationship between stress and strain, even in the simplest case (Figure 8.1). The extension of the bar obeys Hooke's Law, but the bar not only extends in the vertical direction; it also contracts in the horizontal direction. The ratio between the extension ε_3 and the contraction $-\varepsilon_1$ is named for the French mathematician Siméon Denis Poisson (1781–1840), *Poisson's ratio*, v (Greek nu):

$$v = -\varepsilon_1/\varepsilon_3 \tag{8.2}$$

If $\varepsilon_3 = \sigma_3/E$ and the contraction takes place equally in the two horizontal directions (x_2 and x_1) it follows that:

$$\varepsilon_2 = \varepsilon_1 = -\frac{v}{E}\sigma_3 \tag{8.3}$$

If we assume the material is incompressible, so that volume is conserved, then the dilatation, Equation (7.13), is zero, i.e.,

$$\theta = \varepsilon_1 + \varepsilon_2 + \varepsilon_3 = 0$$

so that

$$\frac{1}{E}(1 - 2v)\sigma_3 = 0 \tag{8.4}$$

and it follows that $v = 0.5$.

Most rocks show Poisson's ratios of less than 0.5, generally about 0.2–0.3. Negative values are found in some cases, as well as values larger than 0.5 (see Appendix B).

Values greater than 0.5 indicate dilatation (increase in volume) resulting from shearing of grains past each other (see Chapter 4, Subsection 4.2.2).

8.3 Generalized Hooke's law

In the general case, we must relate elements of the stress tensor to those of the strain tensor. This is done by an array of proportionality coefficients c_{ijkl} called *stiffnesses*:

$$\sigma_{ij} = c_{ijkl}\varepsilon_{kl} \tag{8.5}$$

Here σ_{ij} refers to a stress component in the j-direction acting on a plane whose pole is the the i-direction (Section 4.7). The ε_{kk} refer to normal strains and the ε_{kl}, for $k \neq l$, to the shear strain of the kl-plane (Section 7.3). The Einstein summation convention is used (see Section 2.16).

Note that in writing Equation (8.5), we assume that we can choose any initial state of stress as zero: the equation really relates the *increment* of strain to the *increment* of stress. The justification for this is that it is supported by experiment.

Alternatively the equation can be written in inverse form:

$$\varepsilon_{ij} = s_{ijkl}\sigma_{kl} \tag{8.6}$$

where the s_{ijkl} are called *compliances*. Note that, just to be difficult, c conventionally stands for stiffness, and s for compliance.

These tensor equations represent nine separate equations, one for each independent tensor component (i.e., each permutation of i and j). Each of the nine equations has nine terms (each permutation of k and l). There are at the outset 9×9, or 81 stiffnesses or compliances, (so the c_{ijkl} or s_{ijkl} are fourth-order tensors). Luckily, as we will now show, the number can be reduced to two!

First we remember that $\sigma_{ij} = \sigma_{ji}$ and $\varepsilon_{ij} = \varepsilon_{ji}$, so that $s_{ijkl} = s_{ijlk}$. Making use of these symmetries, we can change the notation. For strains, we renumber the components by writing

$$\varepsilon_{ij} = \begin{bmatrix} \varepsilon_{11} & \varepsilon_{12} & \varepsilon_{13} \\ \varepsilon_{21} & \varepsilon_{22} & \varepsilon_{23} \\ \varepsilon_{31} & \varepsilon_{32} & \varepsilon_{33} \end{bmatrix} \text{ as } \begin{bmatrix} \varepsilon_{1} & \varepsilon_{6} & \varepsilon_{5} \\ \varepsilon_{6} & \varepsilon_{2} & \varepsilon_{4} \\ \varepsilon_{5} & \varepsilon_{4} & \varepsilon_{3} \end{bmatrix} = \varepsilon_{r} \tag{8.7}$$

and similarly for the stresses σ_{ij} (which reduce to σ_s). Then we can write the generalized equation in a more compact form:

$$\varepsilon_r = s_{rs}\sigma_s \tag{8.8}$$

where it is understood that r and s range from one to six. Now there are six equations, each with six terms, and only 36 compliances to worry about.

A further great simplification is produced by assuming that the elastic material is *isotropic*, that is, the elastic moduli do not vary in different directions. We have already made such an assumption in a different context when we considered the coefficient of permeability in Section 6.5. Just as we expect that the permeability of real rocks is not truly isotropic, so we expect that the elasticity of rocks is not truly isotropic. Certainly, single crystals generally have markedly nonisotropic elasticity, and many rocks show a fabric with a strongly preferred orientation of the constitutent mineral grains. The assumption of isotropic elasticity is nevertheless generally made because many rocks are not *strongly* anisotropic and because the assumption makes it possible to reduce the number of compliances from 36 to two.

We do not give the derivation here (it is outlined briefly in Section 8.6), but if we assume isotropic elasticity the compliance matrix can be expressed entirely in terms of only two elastic constants. Using Young's modulus and Poisson's ratio, we obtain

$$s_{rs} = \begin{bmatrix} \frac{1}{E} & -\frac{v}{E} & -\frac{v}{E} & 0 & 0 & 0 \\ -\frac{v}{E} & \frac{1}{E} & -\frac{v}{E} & 0 & 0 & 0 \\ -\frac{v}{E} & -\frac{v}{E} & \frac{1}{E} & 0 & 0 & 0 \\ 0 & 0 & 0 & \frac{(1+v)}{E} & 0 & 0 \\ 0 & 0 & 0 & 0 & \frac{(1+v)}{E} & 0 \\ 0 & 0 & 0 & 0 & 0 & \frac{(1+v)}{E} \end{bmatrix} \tag{8.9}$$

Note that the three shear terms s_{44}, s_{55}, and s_{66} in Equation (8.9) are sometimes written as $2(1 + v)/E$. In this case, the corresponding shear terms in the strain tensor are total, or engineering, shear strains, which are twice the tensor components of shear strain $\varepsilon_{ij}(i \neq j)$ used here (Equation 7.16). It is preferable to use the tensor components in order to maintain consistency with the general form of Hooke's law given in Equation (8.8), which holds for both anisotropic and isotropic elastic materials.

We can write out the relations of Equation (8.9) for normal strains (elongations) in tabular form:

Stress		Strain components due to normal stress		
		in x_1-direction	in x_2-direction	in x_3-direction
ε_{ij}	(ε_r)	σ_{11} $(s=1)$	σ_{22} $(s=2)$	σ_{33} $(s=3)$
ε_{11}	(ε_1)	$\frac{1}{E}\sigma_{11}$	$-\frac{v}{E}\sigma_{22}$	$-\frac{v}{E}\sigma_{33}$
ε_{22}	(ε_2)	$-\frac{v}{E}\sigma_{11}$	$\frac{1}{E}\sigma_{22}$	$-\frac{v}{E}\sigma_{33}$
ε_{33}	(ε_3)	$-\frac{v}{E}\sigma_{11}$	$-\frac{v}{E}\sigma_{22}$	$\frac{1}{E}\sigma_{33}$

The table implies that

$$\varepsilon_1 = \varepsilon_{11} = \frac{1}{E}\left[\sigma_{11} - v(\sigma_{22} + \sigma_{33})\right]$$

$$\varepsilon_2 = \varepsilon_{22} = \frac{1}{E}\left[\sigma_{22} - v(\sigma_{11} + \sigma_{33})\right] \tag{8.10}$$

$$\varepsilon_3 = \varepsilon_{33} = \frac{1}{E}\left[\sigma_{33} - v(\sigma_{11} + \sigma_{22})\right]$$

Note that columns for $s = 4, 5, 6$ are not needed in the table because all corresponding compliances equal zero. Also, if only one normal stress differs from zero, that is, if $\sigma_{22} = \sigma_{33} = 0$, then $\varepsilon_{11} = \sigma_{11}/E$ and $\varepsilon_{22} = \varepsilon_{33} = -(v/E)\sigma_{11}$. So we recover the simple form of Hooke's law and Poisson's ratio given by Equations (8.1) and (8.2). Similarly, the expressions for the shear strains are

$$\varepsilon_4 = \varepsilon_{23} = \frac{(1+v)}{E}\sigma_{23} \tag{8.11}$$

$$\varepsilon_5 = \varepsilon_{13} = \frac{(1+v)}{E}\sigma_{13} \tag{8.12}$$

$$\varepsilon_6 = \varepsilon_{12} = \frac{(1+v)}{E}\sigma_{12} \tag{8.13}$$

The pattern of these equations is exactly what we would expect from the assumption of isotropic elasticity. The expression for ε_{22} is the same as the expression for ε_{11} except that the subscripts 22 and 11 switch places in the two expressions. If only normal stresses are applied, only elongations or contractions are produced. A shear stress σ_{12} applied in the x_2-direction to the plane normal to x_1 produces only a shear strain in the corresponding direction: it does not produce elongation or shear in other directions (as it might, if the material were not isotropic).

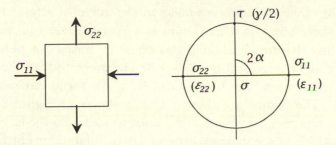

Fig. 8.2. Mohr circle of stress (or strain) for block with $\sigma_{11} = -\sigma_{22}$.

The complete set of equations can be written compactly in index notation:

$$\varepsilon_{ij} = \frac{(1+v)}{E}\sigma_{ij} - \frac{v}{E}\delta_{ij}\sigma_{kk} \qquad (8.14)$$

where $i, j, k = 1, 2, 3$. Note the use of Kronecker's delta and of the dummy subscript k.

Only two elastic moduli are needed to describe an isotropic solid. So far, we have used Young's modulus and Poisson's ratio. Several others are in common use for historical reasons, and because they are useful for particular applications. We introduce them in the next three sections.

8.4 Rigidity modulus

The rigidity modulus is particularly useful to describe shear. Consider the simple shear γ produced by a shear stress τ (Chapter 7, Problem 1). Then the *rigidity modulus G* is defined by:

$$\tau = G\gamma \qquad (8.15)$$

To understand how this is related to Young's modulus and Poisson's ratio it is easiest to consider a particularly simple type of deformation. Suppose that we apply normal stresses to two sides of a block, such that the stress, σ_{11}, on one pair of sides is a compression (considered positive), the stress on the second pair of sides, $\sigma_{22} = -\sigma_{11}$, is a tension, and the third pair of sides is unstressed, i.e., $\sigma_{33} = 0$ (Figure 8.2). We can represent the stresses and strains on a single Mohr diagram, shown on the right-hand side of the figure.

We showed in Chapter 4 that the shear stress τ is a maximum for the plane inclined at 45 degrees to the direction of the major principal stress, and in the present example, the normal stress σ on this plane is equal to zero, shown by the Mohr circle of stress shown in Figure 8.2.

The Mohr circle of *strain* is similar to the circle of stress: the plane on which the shear stress is a maximum is a plane of zero normal stress, so if there is any deformation along this plane it must be a pure shear, and cannot have any component of dilatation. The magnitude of the shear strain on the 45 degree plane can be obtained from the Mohr circle of strain (see Chapter 7). The strains are shown in parentheses in Figure 8.2. The two principal strains are ε_{11} (a contraction, corresponding to the largest principal stress σ_{11}, which is a compression) and ε_{22} (a dilatation, corresponding to the smallest principal stress, σ_{22}, which is a tension). The magnitude of the maximum shear can be derived from Equation (7.36). It is as follows:

$$\frac{\gamma}{2} = \varepsilon_{11} = -\varepsilon_{22} \tag{8.16}$$

Therefore $\gamma = 2\varepsilon_{11} = -2\varepsilon_{22}$. But for $\sigma_{33} = 0$ we know from Equations (8.10) that

$$\varepsilon_{22} = \frac{1}{E}(\sigma_{22} - v\sigma_{11}) \tag{8.17}$$

Thus

$$\gamma = -\frac{2}{E}(\sigma_{22} - v\sigma_{11}) \tag{8.18}$$

But here $\sigma_{11} = -\sigma_{22} = \tau$, so we can write

$$\gamma = \frac{2}{E}\tau(1+v) \quad \text{or} \quad \tau = \frac{E}{2(1+v)}\gamma \tag{8.19}$$

It follows that the rigidity modulus is not independent of Young's modulus and Poisson's ratio, but is related to them by

$$G = \frac{E}{2(1+v)} \tag{8.20}$$

8.5 Bulk modulus

Let the mean normal stress $\bar{\sigma}$ be given by

$$\bar{\sigma} = (\sigma_{11} + \sigma_{22} + \sigma_{33})/3$$

This is one-third of the first stress invariant (Chapter 4, Subsection 4.17.8) and in a fluid its magnitude is equal to that of the pressure (Chapter 5, Section 5.1). We can define the *bulk modulus of elasticity k* by

$$\bar{\sigma} = k\theta \tag{8.21}$$

where θ is the dilatation, defined by Equation (7.13):

$$\theta = \varepsilon_{11} + \varepsilon_{22} + \varepsilon_{33} \tag{8.22}$$

Equation (8.21) states that the dilatation of an elastic solid is directly proportional to the mean stress. From Equations (8.10),

$$\theta = \frac{1}{E}\left[\sigma_{11} - v(\sigma_{22} + \sigma_{33})\right] + \frac{1}{E}\left[\sigma_{22} - v(\sigma_{11} + \sigma_{33})\right]$$

$$+ \frac{1}{E}\left[\sigma_{33} - v(\sigma_{11} + \sigma_{22})\right]$$

$$= \frac{1}{E}(\sigma_{11} + \sigma_{22} + \sigma_{33}) - \frac{2v}{E}(\sigma_{11} + \sigma_{22} + \sigma_{33})$$

$$= \frac{1 - 2v}{E}(\sigma_{11} + \sigma_{22} + \sigma_{33})$$

$$= \frac{3(1 - 2v)}{E}\bar{\sigma} \tag{8.23}$$

Therefore

$$k = \frac{E}{3(1 - 2v)} \tag{8.24}$$

and the bulk modulus too is related directly to Young's modulus and Poisson's ratio.

8.6 Lamé's constants

In Section 8.3 we presented, though we did not derive, a complete set of constitutive relations for an elastic solid (Equation 8.14). They are in the form $\varepsilon_{ij} = f(\sigma_{ij})$, that is, the strains are functions of the stresses. For the purpose of developing the equations of motion, however, it is more convenient to express the stresses as a function of the strains, i.e., to write the constitutive equation in the form $\sigma_{ij} = g(\varepsilon_{ij})$. Developing the equations in this form leads to the definition of two different elastic constants, called Lamé's constants. Another advantage of these constants is that they are the ones that arise most naturally in the derivation of the theory of elastic solids. This derivation uses arguments about symmetry and isotropy. We will not develop the derivation fully, but will sketch its outline and show how Lamé's constants are related to the other elastic constants.

First we consider the case of a deformation such that the axes of the principal stresses coincide in direction with those of the principal strains. If the material is isotropic, then after such a deformation a rectangular block must remain rectangular: there are no shear strains. So we can write linear equations relating each of the three principal stresses σ_i to the three principal strains ε_i. For example, in the x_1-direction,

$$\sigma_1 = a\varepsilon_1 + b\varepsilon_2 + c\varepsilon_3 \tag{8.25}$$

It also follows from isotropy that $b = c$ (an elongation in one direction must give rise to equal contractions in the two normal directions). It is convenient to rewrite this equation in the form

$$\sigma_1 = (a - b)\varepsilon_1 + b(\varepsilon_1 + \varepsilon_2 + \varepsilon_3)$$
$$= 2\mu\varepsilon_1 + \lambda(\varepsilon_1 + \varepsilon_2 + \varepsilon_3) \tag{8.26}$$

where we have simply replaced a and b with two new coefficients μ and λ, called *Lamé's constants*, after the French engineer Gabriel Lamé (1795–1870: Lamé wrote the first book devoted to the mathematical theory of elasticity). The general form of Equation (8.26) is

$$\sigma_i = 2\mu\varepsilon_i + \lambda(\varepsilon_1 + \varepsilon_2 + \varepsilon_3) \tag{8.27}$$

Now consider the general transformation rules for obtaining the complete set of strains ε_{ij} from the principal strains ε_i. The transformation laws are of the type introduced in Chapters 2 and 4, namely

$$\varepsilon'_{ij} = \ell_{ip}\ell_{jq}\varepsilon_{pq} \tag{8.28}$$

where the ℓ_{ij} are sets of direction cosines that produce rotation from one set of orthogonal reference axes to another. (The dummy indices p and q are used to indicate summation.) Next we use the known properties of such sets of direction cosines to determine the strains with reference to axes x and y, where these two directions do *not* coincide with the axes of principal strains or stresses. This leads to the following equations:

$$\varepsilon_{xx} = \ell_{11}^2\varepsilon_1 + \ell_{12}^2\varepsilon_2 + \ell_{13}^2\varepsilon_3 \tag{8.29}$$
$$\varepsilon_{xy} = \ell_{11}\ell_{21}\varepsilon_1 + \ell_{12}\ell_{22}\varepsilon_2 + \ell_{13}\ell_{23}\varepsilon_3 \tag{8.30}$$
$$\sigma_{xx} = \ell_{11}^2\sigma_1 + \ell_{12}^2\sigma_2 + \ell_{13}^2\sigma_3 \tag{8.31}$$
$$\sigma_{xy} = \ell_{11}\ell_{21}\sigma_1 + \ell_{12}\ell_{22}\sigma_2 + \ell_{13}\ell_{23}\sigma_3 \tag{8.32}$$

Substituting for the principal stresses, using Equation (8.27) leads to

$$\sigma_{ij} = 2\mu\varepsilon_{ij} + \lambda\delta_{ij}\theta \tag{8.33}$$

where we have replaced subscripts x and y with i and j, θ is defined as before (as the first invariant of the strain tensor, which is equal to the sum of the principal strains) and δ_{ij} is Kronecker's delta.

Details of the derivation are given by Jaeger (1971, pp. 54–8) and Sommerfeld (1964, pp. 62–3). Lamé's constants arise naturally in this derivation and it can be shown that they are related to the moduli already discussed as follows:

$$\mu = \frac{E}{2(1+v)} = G \tag{8.34}$$

$$\lambda = \frac{vE}{(1+v)(1-2v)} \tag{8.35}$$

$$k = \lambda + \frac{2}{3}\mu \tag{8.36}$$

Only one of Lamé's constants, μ, has a simple physical interpretation (it is the shear modulus G). The other moduli have a more direct physical interpretation than λ, which is why they are preferred for most practical applications.

8.7 Measurement of stress in rocks

We considered simple models of stress in the earth in Chapter 4, but at that time we could not discuss how the stress in rocks is measured, because most of the techniques make use of elasticity theory. Generally, it is the strain that is actually measured, using strain gauges (Section 7.6) or surveying techniques. Then, by measuring, or assuming, the elastic constants of the rocks, it is possible to compute the stress. There is a problem common to most of these techniques (and common in other branches of science): to measure something, we must introduce a measuring device, and this may disturb the original value of what we are trying to measure.

Stress in rocks below the surface can only be measured directly from a borehole or mine passage: but excavating a mine passage, and even drilling a small borehole, will change the stress field in the surrounding rock. In general, the compressive stresses will be relaxed as the rock expands elastically into the cavity. In fact, one reason why it is important to know the distribution of rock stress in mines is in order to construct mine passages in a way that minimizes the probability of their collapse due to inherent rock stresses. The walls of poorly designed passages may even "burst" violently inwards. This happens where the maximum compressive stress parallel to the mine wall is large enough to cause the wall to buckle and break into the passage. Texts on rock mechanics (e.g., Jaeger and Cook, 1979; Goodman, 1980) describe the techniques that have been developed. What follows is a sketch of the principles involved. The techniques mostly make use of strain gauges attached to the rock surface to measure the way the rock expands or contracts in one or more directions. Stress is measured in two different ways.

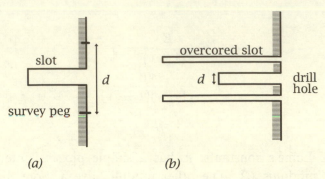

Fig. 8.3. Measurement of stress: (*a*) by flatjack method; (*b*) by overcoring.

Flatjack Method. The gauges are attached to a rock surface before the stresses on it are relaxed significantly by drilling. A slot is then drilled which relaxes the stress normal to the plane of the slot. The relaxation can be measured using the gauges, or by survey. A strong hydraulic jack is then placed in the slot, and pressure is built up until the gauges or surveys show that the induced elastic strain has been removed. The measured pressure produced by the jack is then equal to the stress normal to the surface of the slot before it was drilled (Figure 8.3(*a*)).

Overcoring Method. A narrow hole is drilled into the rock, and at the far end of the hole, strain gauges are attached to the end surface (normal to the axis of the hole). Then a cylindrical slot is drilled into the rock around the original hole and the surface with the gauge. This relaxes the stresses parallel to that surface, and the resulting strain is measured by the gauges. The measured strain is related to stress by using the theory of elasticity (Figure 8.3(*b*)).

Both these techniques make the assumption that the stress parallel to the measurement surface has not been significantly altered from its natural state. This is somewhat more unlikely in the first method, where the measurement surface is the wall of a mine passage, than in the second case, where it is at the end of a small-diameter borehole. The first method, however, has the advantage that the restoring stress is measured directly, and not deduced from elasticity theory. In the second case, the strain measurement must be combined with measurement of the elastic properties (moduli) of the rock.

To obtain a knowledge of the stress in three dimensions, measurement must be made on more than one surface, at different orientations: we described in Chapter 4 how such measurements can be used to calculate the principal stresses.

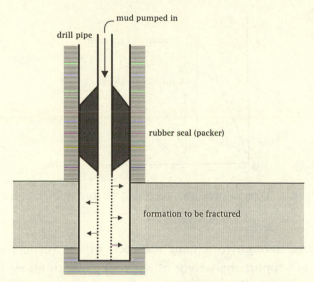

Fig. 8.4. Technique for fracturing a reservoir rock.

8.8 Hydraulic fracturing of petroleum reservoir rocks

A common problem in producing oil or gas from reservoirs is that, although the porosity and thickness of the reservoir saturated with hydrocarbons may be adequate, the permeability is so low that oil or gas flows towards the borehole too slowly for commercial production. In this case, petroleum engineers try to modify the reservoir around the borehole, to increase the permeability. One way is to use acid to enlarge the pores, and another is to pump fluids (water, oil or acid, and some sand) down the borehole at such high pressures that the reservoir rock is actually fractured. Sand entering the fractures prevents them from completely closing again after the pressure is released.

The actual technique is to close off the borehole above and below the reservoir with expandable rubber "packers" (Figure 8.4). Then fluid is pumped down at high pressure, until there is a sharp drop in pressure, indicating that the formation has fractured (*A* in Figure 8.5). Pumping is then stopped, the valves are opened to allow fluid to flow back out of the formation, and the pressure falls to a lower value. If the valves are then closed, the pressure rises slowly to a nearly constant value (the *shut-in pressure* p_s; *B* in Figure 8.5).

The interpretation is as follows: pressure in the borehole increases until it is large enough to expand the borehole against the rock pressures (i.e., until the tangential stresses around the borehole become tensile) or until the

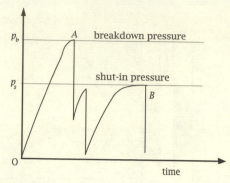

Fig. 8.5. Typical pressures recorded during well fracturing.

pressure is large enough to extend the rock in the vertical direction (i.e., large enough to support the weight of the rock column above the reservoir). In the first case vertically oriented fractures are formed, and in the second case horizontal fractures are formed.

In either case, we expect that the fractures will be formed most easily in the direction normal to the lowest principal stress σ_3.

For the first fractures to form, the deviatoric tensile stress produced must exceed the lowest tangential stress developed around the borehole (see below), plus the tensile strength of the formation. After the fracture has been formed, however, it can be kept open by a pressure which is equal to the lowest principal stress.

The orientation of the fractures can be determined by surveying the borehole after fracturing has taken place, and the magnitude of the lowest principal stress is determined fairly readily from the shut-in pressure. By making various other measurements and assumptions, and using the theory for stresses developed around a cylindrical cavity in an elastic solid, it is also possible to determine the magnitude of σ_2 (assuming σ_1 is vertical as is generally the case). It is claimed by some workers that the presence of the packers tends to suppress the formation of horizontally oriented fractures close to the borehole. They claim that vertically oriented fractures, produced close to the hole, turn into horizontal fractures at some distance away. Thus, the interpretation remains controversial in detail. It does seem, however, that hydraulic fracturing records can give reliable information about the magnitude of the lowest principal stress, and the orientation of the lowest stress in the horizontal plane.

The orientation, though not the magnitude, of the maximum and minimum horizontal stresses, may be determined in another, much simpler way. The

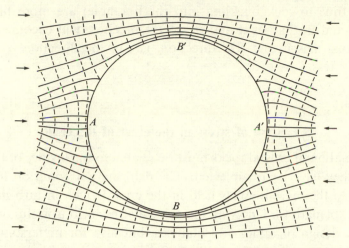

Fig. 8.6. Distribution of stress around a borehole. The full and broken lines are the trajectories of the maximum and minimum principal stresses, respectively. See the text for discussion.

shapes of boreholes drilled for oil are frequently surveyed by petroleum engineers, before casing is cemented into the hole to produce oil. It was discovered more or less by accident (cf. Gough and Gough, 1987) that many boreholes have roughly elliptical cross-sections, because of the "breakout" of sedimentary rock preferentially in one direction. Moreover, the breakout direction is remarkably consistent from one well to another, over large areas within sedimentary basins. It has been established that the breakouts tend to be normal to the direction of maximum horizontal stress.

To understand why consider Figure 8.6, which shows the stress field (Section 4.14) near the wall of a borehole of circular cross-section as calculated from elasticity theory. The maximum horizontal stress σ_2 acts along the line AA', and the minimum horizontal stress σ_3 acts along BB'. Because of the presence of the borehole, elasticity theory indicates that at the wall of the borehole the tangential stresses have a maximum value of $3\sigma_2 - \sigma_3$ at B and B', and a minimum value of $3\sigma_3 - \sigma_2$ at A and A'. The mud exerts a uniform pressure p on the wall of the borehole. If the pressure in the borehole is larger than the minimum horizontal pressure (as happens during borehole fracturing) then we expect fractures to open up normal to its direction, i.e., in the AA' direction. But if the pressure is not as large as this, as is normally the case, then there will be a larger difference between the pressure in the borehole and the tangential stresses at B and B', than between the pressure in the borehole and the tangential stresses at A and A'. This means that breakouts (due to rock buckling into the borehole) are more like to take place at B and B' than at A and A', so the bolehole will become ellipsoidal with the long axis in the BB' direction, i.e., in the direction of minimum horizontal stress.

8.9 State of stress in the crust of the earth

The deformation of crustal rocks must be related to the history of the stresses to which they have been subjected. The deformation going on today must be related to the present stress field in the crust (which in turn may have a historical explanation). Knowing the present stress field is useful not only for the correct design of mines and quarries, but also for understanding what deformations are taking place now and will take place in the future – so, for example, in understanding earthquakes and corresponding movements on large faults. Therefore, a large number of measurements have been made of the state of stress in the crust. The results have been summarized by McGarr and Gay (1978), Gough and Gough (1987), Hickman (1991), and Zoback (1992).

The early results were surprising. It might be expected that the maximum principal stress would be produced by gravity, so that it would be vertical. The other two principal stresses might be expected to be less. Their difference and orientation (in the horizontal plane) would be determined by tectonics. The Swiss geologist, Albert Heim (1849–1937), suggested long ago that because rocks tend to creep (deform slowly) under long-continued stresses, the stress field at depth would tend to be hydrostatic (i.e., all three principal stresses would be almost equal to the stress produced by the overlying load of rocks).

Actual measurements show that at the deepest levels that can be reached in mines, neither of these predictions is generally true (though it is still thought that they will be true at deeper levels in the lithosphere). In most cases, one of the principal stresses is almost vertical, though measurements in some shield-area mines show this is not always the case. The vertical stress is generally closely approximated by the load of overlying rock. In South Africa the vertical stress is the largest, and the horizontal stresses are lower than but proportional to it. In the Canadian shield, however, the vertical stress is generally not the largest principal stress: the reasons for this anomaly are not well understood. The measurements have established that there is frequently a difference between the stress relationships observed at shallower and greater depth in mines, so that it cannot be assumed that measurements made near the surface are typical of even the upper part of the crust.

McGarr (1980) found that generally there was a linear increase in the maximum shear stresses directly measured in the crust with depth of measurement (down to depths of about 5 km): in "soft rocks" the gradient in shear stress was 3.8 MPa km^{-1}, and in "hard rocks" it was 6.6 MPa km^{-1}. If extrapolated to 15 km, these trends would imply maximum shear stresses of about 100 MPa or more. These values are roughly what are predicted from theoretical considerations (see Chapter 4, Section 4.13).

Though the early measurements in mines were surprising, more recent regional compilations of stress directions determined in boreholes, and indirectly from earthquake focal mechanisms and from geological evidence such as the orientation of recent volcanic dykes and faults, show that in many areas there is a regional consistency of stress orientation that is clearly related to the boundaries and motions of tectonic plates. Intraplate regions have compressive stresses, and extensional stresses are found mainly in regions that are anomalously high and hot (thermal domes, probably related to heat or plastic rock plumes rising from the mantle). The directions of maximum horizontal stress are generally parallel and proportional to the directions of plate movement, which strongly suggests that whatever forces produce intraplate stresses also produce plate movements. Nevertheless, local structures, particularly major faults along plate boundaries, like the San Andreas fault, may rotate and modify stress fields in those areas.

8.10 Flexure of plates

8.10.1 Introduction

One important application of linear elasticity theory is to the bending of beams or plates. This is a classical problem in mechanical engineering but it has applications in plate tectonics. According to plate tectonic theory, the lithosphere of the earth may be considered to be composed of rigid plates, which overlie a relatively soft part of the mantle, called the *asthenosphere*. To a first approximation, the lithosphere may be regarded as an elastic plate, which may bend slightly or break along fractures, but does not deform plastically, and the asthenosphere may be regarded as a very viscous fluid. If we are concerned mainly with geological time scales of the order of a million years, it seems likely that we may ignore both very long-term ductile behaviour of the lithosphere (on time scales of tens to hundreds of millions of years) and the relatively short-term viscous behaviour of the asthenosphere (on time scales of tens of thousands of years). In this case, the response of

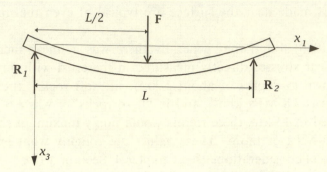

Fig. 8.7. Bending of a thin plate under an applied load.

the lithosphere to an imposed load, such as the growth of a large volcano (e.g., Hawaii) is essentially the bending of an elastic plate. In this section we will outline the theory briefly (for further details, see Forsyth, 1979; Turcotte, 1979; Turcotte and Schubert, 1982, Chapter 3). We will return in Chapter 10 to the question of whether or not a purely elastic model is appropriate for the long-term behaviour of the lithosphere.

The basic problem to be solved is shown in Figure 8.7. We consider the bending of a thin elastic plate, supported at the ends by two line supports, and loaded by a single line load, F. The plate is "thin" only in comparison with its length L and we will consider only "cylindrical" bending, that is bending in the two dimensions x_1 (horizontal) and x_3 (vertical, with the positive direction down). We ignore the weight of the plate itself because, for tectonic plates, that weight is supported by the buoyant force exerted by the asthenosphere. To determine the equilibrium we balance the forces and moments acting on the plate. Down is considered the positive direction so the force balance is $F-(R_1+R_2) = 0$, where R_1 and R_2 are the magnitudes of the forces exerted by the two supports. The moment balance (about $x_1 = 0$) is $F(L/2)-R_2L = 0$. Note that because x_3 is positive down and x_1 is positive to the right, this means that the moment exerted by F, which tends to bend the plate so that it is convex-up, is positive, while the moment exerted by the right-hand support force R_2, which tends to bend the plate concave-up, is negative. (Of course, the sign of the moment depends on the choice of origin – but in this section we will calculate equilibria about the left end of the plate.) We could go on to draw diagrams showing how the shear forces and moments acting on any part of the plate vary along the length of the plate (for an example, see Johnson, 1970, pp. 46–50). In our example, this would show that the moments are negative everywhere between the supports, with a minimum value (= largest absolute value) in the centre.

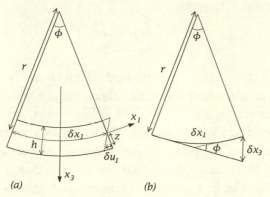

Fig. 8.8. Bending of a thin plate: a section of length δx_1 is shown. (*a*) Extension and compression parallel to x_1. (*b*) The displacement due to bending is $\delta x_3 = \phi \delta x_1$.

Thus far, however, we have simply treated the plate as a rigid body, acted on by exterior forces. As there is static equlibrium, the exterior forces and moments are required to balance. In problems concerning bending, however, we are concerned with a different type of balance, that between the exterior forces and moments that produce bending, and the elastic forces and moments that resist bending. Bending (and resisting) moments come in pairs, and it is useful to have a sign convention for such pairs, just as we have a sign convention for the pairs of forces that produce shear: the one we adopt here (which is by no means universal) is that moments that tend to bend the plate convex-up are positive bending moments.

8.10.2 *The small-strain assumption*

Linear elasticity theory assumes small strains. Bending a plate is possible only if we compress the concave side of the plate and stretch the convex side (Figure 8.8). The strains can be kept small only if the local radius of curvature r, is large compared with the thickness of the plate, h. Along the central plane we may assume that bending produces no strain in the direction parallel to the plate. If r is large, lengths along the plate are almost equal to their components in the x_1-direction, and we set the origin of x_3 at the centre of the section.

If the flexure is small, and uniform in the dimension of x_2, the strain is essentially *plane strain*. We may neglect the x_2-components of stress and strain, and Equations 8.10 reduce to:

$$\varepsilon_{11} = \tfrac{1}{E}(\sigma_{11} - v\sigma_{33})$$
$$\varepsilon_{33} = \tfrac{1}{E}(\sigma_{33} - v\sigma_{11}) \tag{8.37}$$

In the x_1-direction the plate is compressed as much at the concave surface as it is extended at the convex surface; therefore we may deduce that the strain in the x_3-direction, ε_{33}, is zero. In other words, we assume that for small angles of bending, we can neglect all strains except those taking place parallel to the centre of the plate. In particular we neglect all shear stresses. This is equivalent to an assumption that the plate is composed of a bundle of elastic fibers, that can be extended or compressed only in the x_1-direction. We make use of this assumption to eliminate σ_{33} from the equations above and reduce the equations of linear elasticity to the special form

$$\sigma_{11} = \frac{E}{(1 - v^2)}\varepsilon_{11} \tag{8.38}$$

There are also some important geometrical consequences of the small-strain assumption. Figure 8.8(*b*) shows the displacement δx_3 produced by bending. We have

$$\phi \approx \delta x_1/r \approx \delta x_3/\delta x_1 \tag{8.39}$$

so that

$$1/r \approx \delta x_3/\delta x_1\delta x_1$$

As δx_1 shrinks to zero,

$$\frac{1}{r} = \lim_{\delta x_1 \to 0}\frac{\delta x_3}{\delta x_1\delta x_1} = \frac{d^2 x_3}{dx_1^2} \tag{8.40}$$

Figure 8.8(*a*) shows that ϕ is also given by the shortening or elongation δu_1 of a plate "fiber" divided by its distance z from the centre line of the plate:

$$\phi = \delta u_1/z$$

Using the first part of Equation (8.39), we obtain

$$\delta u_1/\delta x_1 = z/r$$

But the left-hand side is equal to the infinitesimal strain in the x_1-direction, ε_{11}, at a distance z from the central plane; therefore, from Equation (8.40),

$$\varepsilon_{11} = \frac{z}{r} = z\left(\frac{d^2 x_3}{dx_1^2}\right) \tag{8.41}$$

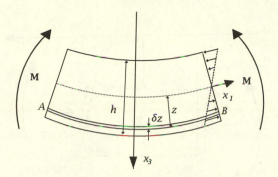

Fig. 8.9. Fiber stresses resulting from moments acting on the section of plate shown in Figure 8.8.

8.10.3 *Moments in a thin plate*

As we have seen, compressing the concave side of the plate and stretching the convex side produces *fiber stresses* parallel to the plate, as shown in Figure 8.9. For small angles of bending the stresses (which are actually parallel to the plate surface) may be considered to be compressions or extensions in the x_1-direction, and are therefore denoted σ_{11}. The asymmetrical distribution of stresses about the centre plane produces a moment that resists the applied *bending moment M*. For an element of thickness δz located a distance z from the origin, the elastic moment is

$$m = \sigma_{11} z \delta z$$

The moment is positive because it resists the bending moment (which is negative because it tends to bend the element concave-up). The total bending moment is obtained by summing the elastic moments across the plate, and reversing the sign:

$$M = -\int_{-h/2}^{h/2} \sigma_{11} z \, dz \qquad (8.42)$$

Now we use Equation (8.38) to express the bending moment in terms of the horizontal strains rather than the horizontal stresses:

$$M = \frac{-E}{(1-v^2)} \int_{-h/2}^{h/2} \varepsilon_{11} z \, dz \qquad (8.43)$$

To evaluate this integral we make use of the expression for ε_{11} that we derived in the previous subsection (Equation 8.41). Substituting in Equation (8.43)

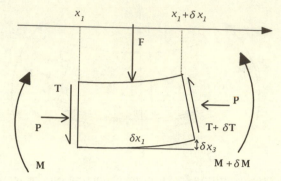

Fig. 8.10. Forces and moments on a section of a bending plate.

and integrating yields

$$M = -\left[\frac{Eh^3}{12(1 - v^2)}\right]\frac{d^2x_3}{dx_1^2} \tag{8.44}$$

The quantity in large brackets on the right-hand side of the equation is called the *flexural rigidity*, D of the plate. Note that it is not a property *only* of the elastic moduli (E and v) of the plate, but depends very strongly on its thickness, h.

8.10.4 Balance of forces and moments

The final step in our analysis is to balance forces and moments on a small section of the plate. We are once again concerned only with forces and moments acting on the section from outside, not with elastic forces inside the element. In Figure 8.10 we again show a section with width δx_1, at a distance x_1 along the plate. We have shown a single force, of magnitude F, acting on this part of the plate. More generally, however, it could be a distribution of line loads along the plate $F(x_1)$, so the total force would be $F(x_1)\delta x_1$. There is no direct resisting force acting at x_1, but the two supports give rise to forces T and $T + \delta T$ which act on the two ends of the element as shown in the figure, so the total force (acting in approximately the negative x_3-direction) is $-\delta T$. There may also be applied horizontal forces, P, but we assume they are independent of x_1 and essentially cancel out in that direction. From the balance of vertical components of force acting on the element we obtain, after taking limits as $\delta x_1 \to 0$:

$$F(x_1)dx_1 - dT = 0 \quad \text{or} \quad dT/dx_1 = F(x_1) \tag{8.45}$$

To balance moments we consider the moments (about the origin) produced by the forces acting on the two ends of the element. The moment due to the shear forces is $Tx_1 - (T + \delta T)(x_1 + \delta x_1) \approx -T\delta x_1$. It is negative because $T + \delta T$ acts in the negative x_3-direction. The horizontal forces acting on the two ends, though equal and opposite, act at different levels, so the net moment is $Px_3 - P(x_3 - \delta x_3) = P\delta x_3$. We do not need to consider the moments produced by $F(x_1)$ separately because these are included in the total bending moment $-M$. The portion of $-M$ acting on the element being considered is $-\delta M$. Summing and taking limits as $\delta x_1 \to 0$ gives

$$-dM - T\,dx_1 + P\,dx_3 = 0$$

or

$$\frac{dM}{dx_1} = -T + P\frac{dx_3}{dx_1} \tag{8.46}$$

Differentiating this equation with respect to x_1, and using Equation (8.45) gives

$$\frac{d^2M}{dx_1^2} = -F(x_1) + P\frac{d^2x_3}{dx_1^2} \tag{8.47}$$

We now substitute the expression for M derived in the previous subsection (Equation 8.44) and obtain

$$D\frac{d^4x_3}{dx_1^4} = F(x_1) - P\frac{d^2x_3}{dx_1^2} \tag{8.48}$$

where D is the flexural rigidity defined by Equation (8.44). This is the basic differential equation for two-dimensional flexure of a thin plate.

8.10.5 Application to loading of the lithosphere

The differential equation of flexure has many applications in the earth sciences, and particularly to bending of the lithosphere. We will consider one particularly simple, but useful application. For others see Turcotte and Schubert (1982, Chapter 3).

The lithosphere has been defined as the relatively rigid outer part of the earth, overlying the ductile asthenosphere. The plates of plate tectonic theory are pieces of the lithosphere. The high rigidity of the plates is due to a combination of factors but mainly to their low temperature and ultramafic petrology. One of the major observations supporting this general model is isostatic depression or rise of the earth's crust due to loading or unloading. The main loads imposed by nature on the crust are sea water (during eustatic rises of sea level), large lakes, ice sheets, volcanoes, thrust

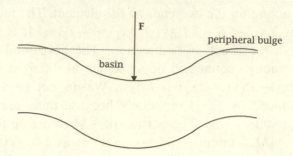

Fig. 8.11. Basin and peripheral bulges created by a line load.

sheets, and sediment in sedimentary basins. It is well demonstrated, either directly or by indirect inference, that the crust responds by sinking until a static balance is restored, and the observed amount of sinking suggests that the main response is depression of the lithosphere into the underlying asthenosphere (Section 5.5).

Isostatic depression is possible only under two conditions: (i) the asthenosphere must be able to flow away from depressed areas, and (ii) the lithosphere must be able to move vertically along fractures, or to bend elastically beneath the load. Geological observations indicate that the usual response to load is that the lithosphere bends rather than breaks. How it bends provides data from which we can estimate the flexural rigidity and therefore the thickness and elastic properties of the lithosphere as a whole. Bending also creates basins, which generally fill with sediment, thus further loading the crust and producing more bending. Figure 8.11 shows a simple two-dimensional model of such a basin. The load is idealized as a single line load. Before bending, the weight of the lithosphere is supported everywhere by the asthenosphere, but flexure raises some parts and depresses others, so we must add to the vertical component of forces a term that results from the departure from local isostatic compensation.

Before dismissing such a model as completely unrealistic one should know, first, that the model gives quite good results for the observed depression of the crust around the Hawaiian Islands (Watts and Cochran, 1974; Turcotte and Schubert, 1982, pp. 125–6) and, second, that it can easily be generalized to distributed (rather than line) loads.

An important property of this model is that the horizontal applied stress in the lithosphere can be neglected ($P = 0$). The load, F, may be assumed to be due to a column of rock at $x_1 = 0$. Elsewhere the local departure from

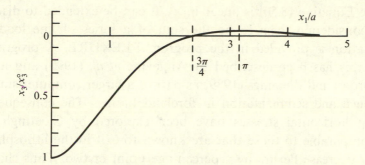

Fig. 8.12. Solution of flexure equation, for a line load. Only the right half of the profile is shown (after Turcotte and Schubert, 1982, p. 126).

isostatic compensation is easily calculated to be

$$F(x_1) = x_3 \delta \gamma \qquad (8.49)$$

x_3 is not only the vertical elevation of the upper surface of the lithosphere, but also the vertical displacement of the base of the lithosphere, due to flexure, and $\delta \gamma$ is the difference in unit weight between the asthenosphere and the material filling any depression created by the flexure. The Hawaiian islands are surrounded by a topographic depression, which is filled with water, so in that case the material is water: in other basins, the filling material may be only partly water. If the bulges rise above sea level, then the material would be air (of negligible density). With this modification, Equation (8.48) becomes

$$D \frac{d^4 x_3}{dx_1^4} + \delta \gamma \, x_3 = F(x_1) \qquad (8.50)$$

$F(x_1)$ is zero everywhere, except at $x_1 = 0$.

This equation has an exact solution, which is shown in dimensionless form in Figure 8.12 for positive values of x_1. The vertical deflection is scaled by x_3^0, the maximum deflection, which is found at $x_1 = 0$ and has the value

$$x_3^0 = \frac{Fa^3}{8D} \qquad (8.51)$$

and the horizontal distance is scaled by a, which is given by

$$a = (4D/\delta \gamma)^{1/4} \qquad (8.52)$$

The maximum height of the bulge is found a distance π from the central load (in these scaled units) and has a height of $-x_3^0 e^{-\pi}$.

The theory is given in more detail in Turcotte and Schubert (1982, Chapter

3). Because Equation (8.50) is linear in x_3, it can be extended to distributed loads by considering them to be the sum of a series of line loads. One implementation is provided in the program FLEXURE. A program very similar to this has been described by Angevine *et al.* (1990) and has been used by Jordan and Flemings (1989) as part of a larger computer simulation of subsidence and sedimentation in foreland basins. The consequences of considering horizontal stresses have been explored by Cloetingh (1988). Stresses comparable to those that are known to exit in the lithosphere can increase or decrease flexure by a percentage point or two. Thus changes in stress patterns, perhaps resulting from plate reorganization, might produce vertical movements of the order of tens of metres.

Flexural rigidities obtained by fitting calculated to observed deflections yield an estimate for the thickness of the lithosphere, when combined with reasonable estimates of Young's modulus and Poisson's ratio (say $E = 70$ GPa, $v = 0.25$) derived from seismic observations. The thicknesses are generally a few tens of kilometres, which is lower than those obtained by some other methods. They must be interpreted not as the distance from the earth's surface to the top of the asthenosphere, but as the thickness of that part of the lithosphere that has elastic rigidity. Part of the lithosphere does not contribute much to its elasticity, either because it is too fractured (near surface), or because it is too hot and ductile (see Section 4.13).

8.11 Equations of motion: Navier's equation

We now have all the tools needed to discuss the motion and deformation of an elastic material. In Chapter 4 we introduced the description of stress in three dimensions. In Chapter 7 we covered the description of strain, and in this chapter we have presented the constitutive equations relating stress and strain for a linearly elastic material. Now, we may write the equations of motion for an element in a continuous elastic material (which will include stress terms), and use the elastic constitutive relations to replace the stress terms with the corresponding elastic strains. The results are governing equations that describe the motion and deformation of elastic materials in response to applied forces.

To begin, we consider the balance of forces acting on a small cube of material (Figure 8.13). From Newton's second law

$$\text{mass} \times \text{acceleration} = \text{body forces} + \text{surface forces}$$

The left-hand side of this equation (equal in magnitude to the "inertia

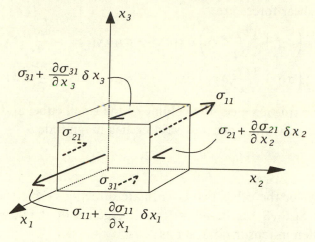

Fig. 8.13. Forces acting on a small cube, in the x_1-direction.

forces"), per unit volume of solid, may be written:

$$\text{density} \times \text{acceleration} = \rho \left(\frac{\partial^2 x_j}{\partial t^2} \right)$$

Because the displacements are considered to be small, we have only to worry about the *local* acceleration, which is defined to be the acceleration at a given position. We can ignore any acceleration that might result from the change in position (if this remark seems obscure, defer further enlightenment until Section 9.3).

The *body force* per unit volume is proportional to the mass density. It may be written in a purely formal way as:

$$\text{body force} = \text{density} \times \text{force per unit mass}$$
$$= \rho X_j \qquad (8.53)$$

If the body force is gravity, then $X_3 = g$ and $X_1 = X_2 = 0$.

The *surface forces* are the stresses acting on the sides of the small cube times the surface area of the face, δA. Figure 8.13 shows only the stresses acting in the x_1-direction. There are three components: a net normal force acting on the face normal to x_1 and net shear forces acting on the faces normal to x_2 and x_3. The net normal force is

$$\left(\sigma_{11} + \frac{\partial \sigma_{11}}{\partial x_1} \delta x_1 \right) \delta x_2 \delta x_3 - \sigma_{11} \delta x_2 \delta x_3 = \frac{\partial \sigma_{11}}{\partial x_1} \delta V \qquad (8.54)$$

and the net shear forces are

$$\left(\sigma_{21} + \frac{\partial \sigma_{21}}{\partial x_2} \delta x_2\right) \delta x_1 \delta x_3 - \sigma_{21} \delta x_1 \delta x_3 = \frac{\partial \sigma_{21}}{\partial x_2} \delta V \tag{8.55}$$

$$\left(\sigma_{31} + \frac{\partial \sigma_{31}}{\partial x_3} \delta x_3\right) \delta x_1 \delta x_2 - \sigma_{31} \delta x_1 \delta x_2 = \frac{\partial \sigma_{31}}{\partial x_3} \delta V \tag{8.56}$$

where $\delta V = \delta x_1 \delta x_2 \delta x_3$, i.e., the volume of the small cube. So the net surface force in the x_1-direction per unit volume has magnitude

$$\frac{\partial \sigma_{11}}{\partial x_1} + \frac{\partial \sigma_{21}}{\partial x_2} + \frac{\partial \sigma_{31}}{\partial x_3}$$

and similarly for the other two coordinate directions. The total force (with components S_1, S_2, S_3) is therefore the divergence of the stress vector, and may be written in tensor notation as

$$S_j = \frac{\partial \sigma_{ij}}{\partial x_i}$$

The complete set of equations of motion may now be written in terms of stresses as

$$\rho \frac{\partial^2 x_j}{\partial t^2} = \rho X_j + \frac{\partial \sigma_{ij}}{\partial x_i} \tag{8.57}$$

At this point, the equations of motion are quite general and can be applied to any continuum that is subject to infinitesimal displacements. We now use the theory of elasticity to replace the stresses by strains. The basic relationship is that given by Equation (8.33):

$$\sigma_{ij} = 2\mu\varepsilon_{ij} + \lambda\delta_{ij}\theta$$

This same equation can be written in terms of displacement gradients as

$$\sigma_{ij} = \mu\left(\frac{\partial u_i}{\partial x_j} + \frac{\partial u_j}{\partial x_i}\right) + \lambda\delta_{ij}\frac{\partial u_k}{\partial x_k} \tag{8.58}$$

Note that we have used the dummy subscript k instead of i where summation is implied. By grouping together the two terms with repeated subscripts (and changing the notation) we obtain the divergence of the stress:

$$\frac{\partial \sigma_{ij}}{\partial x_i} = (\mu + \lambda)\frac{\partial^2 u_k}{\partial x_j \partial x_k} + \mu\frac{\partial^2 u_j}{\partial x_k \partial x_k} \tag{8.59}$$

Using this we can now write the equations of motion in terms of displacement gradients:

$$\rho\frac{\partial^2 x_j}{\partial t^2} = \rho X_j + (\mu + \lambda)\frac{\partial^2 u_k}{\partial x_j \partial x_k} + \mu\frac{\partial^2 u_j}{\partial x_k \partial x_k} \tag{8.60}$$

This set of equations is known as *Navier's equations of motion* after the French engineer Claude Navier (1785–1836) who derived an incomplete form of the equations in 1821. The complete form was given by Cauchy in the following year. Note that we can write the set as a single tensor or vector equation (see below) and refer to it as Navier's equation (singular), or write out the equations for each of the components separately, and refer to the set of equations as Navier's equations (plural). Both usages are common (and similar remarks apply to the Navier–Stokes equation, to be introduced in the next chapter).

To explore the physical meaning of Equation (8.60), and at the same time gain a little experience in switching notation, we expand a two-dimensional form in the $x, y; u, v$ notation. For the x-direction, setting $j = 1$ and summing for $k = 1, 2$ gives

$$\rho \frac{\partial^2 x}{\partial t^2} = \rho X_x + (\mu + \lambda) \left(\frac{\partial^2 u}{\partial x^2} + \frac{\partial^2 v}{\partial x \partial y} \right) + \mu \left(\frac{\partial^2 u}{\partial x^2} + \frac{\partial^2 u}{\partial y^2} \right) \tag{8.61}$$

Each term has the dimensions of a force per unit volume. The left-hand side is the magnitude of the inertia force, and the first term on the right-hand side is the body force. We can see the significance of the remaining terms on the right-hand side (the surface forces) if we rewrite them as

$$2\mu \frac{\partial}{\partial x} \left(\frac{\partial u}{\partial x} \right) + \mu \frac{\partial}{\partial y} \left(\frac{\partial v}{\partial x} + \frac{\partial u}{\partial y} \right) + \lambda \frac{\partial}{\partial x} \left(\frac{\partial u}{\partial x} + \frac{\partial v}{\partial y} \right)$$

The meaning of each term becomes clear when we refer back to Chapter 7 (Equation 7.21). The first term is the component due to changing elongation in the x-direction, the second term is the component due to changing pure shear, and the third term is the component due to changing dilatation. Thus, the changing shear in the y-direction produces a single component in the x-direction, arising from μ, which we have seen is the same as the rigidity modulus G. The changing elongation or dilatation produces two components, one arising from each of the Lamé moduli. We might have expected this because the bulk modulus k is related to both Lamé constants (see Equation 8.36). In solids, both moduli are different from zero. By definition, fluids have no rigidity (though they resist compression or dilatation) so $\mu = 0$ and the surface forces consist only of the third term.

We can also write Navier's equations in vector notation:

$$\rho \frac{\partial^2 \mathbf{u}}{\partial t^2} = \rho \mathbf{X} + (\mu + \lambda) \nabla (\nabla \cdot \mathbf{u}) + \mu \nabla^2 \mathbf{u} \tag{8.62}$$

Other forms can be obtained, for example, by taking the gradient of all the terms, or by taking the curl of all the terms, and these may be useful in some applications.

8.12 Application to seismic waves

We are all familiar with the oscillatory motion that results from extending and then releasing an elastic solid (e.g., the expansion and contraction of a spring) or from bending and then releasing an elastic beam. In the earth, slow deformation of the crust builds up stress until the crust fractures (see Chapter 4), and there is rapid sliding of crustal blocks, which is followed by a brief period of oscillatory motion. This is what we know as an earthquake, and the oscillatory motions set up by the 'quake are seismic waves. They are of great importance not only for the damage that they do at the earth's surface but also for the information they have provided about the nature of the rocks that they travel through.

We are therefore interested not only in how rocks deform elastically when there is an equilibrium of applied forces and moments, but also in how elastic solids accelerate when they are subject to applied forces.

8.12.1 The wave equation

Before discussing the application of Navier's equation to seismic waves, let us consider the following partial differential equation, generally called the one-dimensional *wave equation*:

$$\frac{\partial^2 \psi}{\partial t^2} = C^2 \frac{\partial^2 \psi}{\partial x^2} \tag{8.63}$$

To demonstrate that this is the equation of a waveform that is translated without change in shape and indicate the significance of the terms in it, we show that, if f is a differentiable function of x, then $f(x - Ct)$ satisfies the wave equation (and therefore so does $f(x + Ct)$).

Let $u = x - Ct$, and let our differentiable function be $f(u)$. Now we will use the chain rule to differentiate with respect to x and t. We use the prime notation to indicate first and second derivatives with respect to (wrt) x. Note that f is a function *only* of u, so df/du is not a partial derivative, even though u is a function of two variables x and t.

- Differentiate wrt x:

$$\frac{\partial f(u)}{\partial x} = f' = \frac{df}{du}\frac{\partial u}{\partial x}$$

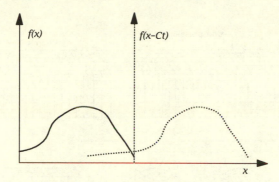

Fig. 8.14. A one-dimensional wave.

$$= \frac{df}{du}\frac{\partial(x - Ct)}{\partial x}$$

$$= \frac{df}{du} \tag{8.64}$$

Differentiating again wrt x:

$$\frac{\partial^2 f}{\partial x^2} = f'' = \frac{\partial^2 f}{\partial u^2} \tag{8.65}$$

• Differentiate wrt t:

$$\frac{\partial f}{\partial t} = \frac{df}{du}\frac{\partial u}{\partial t}$$

$$= \frac{df}{du}\frac{\partial(x - Ct)}{\partial t}$$

$$= -C\frac{df}{du} = -Cf' \tag{8.66}$$

where we have used Equation (8.64) to make the last substitution.
Differentiating again wrt t:

$$\frac{\partial^2 f}{\partial t^2} = -C\frac{df'}{du}\frac{\partial u}{\partial t}$$

$$= (-C)^2\frac{\partial^2 f}{\partial u^2}$$

$$= C^2\frac{\partial^2 f}{\partial x^2} \tag{8.67}$$

which is the wave equation.

The nature of the function $f(x - Ct)$ is shown in Figure 8.14. Translation of the f-axis by a distance Ct transforms the function $f(x)$ into $f(x - Ct)$: the nature of the function is otherwise unchanged. In other words, $f(x - Ct)$ is

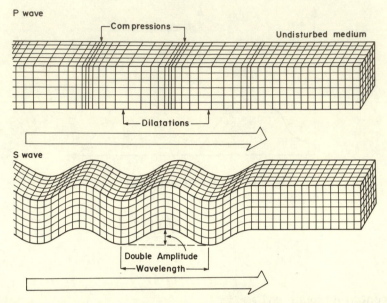

Fig. 8.15. Motion in P and S waves. After Phillips (1968, pp. 100f).

the equation of a function which is moving in the positive x-direction at a speed C. This is what we mean by a wave, and the speed at which the wave form moves is called the *celerity*. $f(x + Ct)$ would be the equation of a wave form moving in the negative x-direction.

Note that we do not need to specify the nature of the function f: it need not be a sinusoidal function or even periodic (e.g., it could be a single disturbance – which would produce what is known as a *solitary wave*). In considering waves that produce small strains, however, we must generally limit the form of f to a low-amplitude wave, in which case the exact form of the wave is not likely to be important and for simplicity we might as well assume a sine or cosine function.

8.12.2 Dilatation (primary) waves

We now return to Navier's equation and develop the theory for waves traveling through the interior of a continuous elastic medium. We assume that we are not interested in waves traveling along a free surface (or along an interior density discontinuity), and that we can neglect gravity, the only body force (in a fluid, it is balanced by buoyancy). We consider two types of waves in this and the next subsections.

For the first type of wave, assume that the displacement u_1 in the x_1-

direction is a function only of x_1 and t, i.e., u_1 results only from an initial compression, σ_{11}, and propagates in the x_1-direction with time. This occurs when a rod is struck on one end, so that internal displacements travel down (but not across) the rod (Figure 8.15, upper part). Then Navier's equation simplifies to

$$\rho\frac{\partial^2 u_1}{\partial t^2} = (\mu + \lambda)\frac{\partial^2 u_1}{\partial x_1^2} + \mu\frac{\partial^2 u_1}{\partial x_1^2} \tag{8.68}$$

Omitting the subscript, this becomes

$$\rho\frac{\partial^2 u}{\partial t^2} = (2\mu + \lambda)\frac{\partial^2 u}{\partial x^2} \tag{8.69}$$

and by comparing this equation with the one-dimensional wave equation (Equation 8.63) we see immediately that this is the equation of a wave with celerity C_p, where

$$C_p = \sqrt{\frac{2\mu + \lambda}{\rho}} \tag{8.70}$$

One solution to this equation is

$$u = \alpha \cos(2\pi/L)(x - C_p t) \tag{8.71}$$

where α is an arbitrary constant. This equation indicates that the displacement u propagates as a cosine wave, with wavelength L, travelling in the x-direction with celerity C_p. Such waves are called P or *primary waves* because they travel faster than other kinds of waves and are therefore the first to arrive at a seismometer.

8.12.3 Distortion (secondary, or shear) waves

The second type of wave we consider has displacements only in the x_2-direction, not in the x_1-direction (Figure 8.15). We set $u_1 = 0$, $u_2 = f(x_1, t)$, and $u_3 = 0$. Then Navier's equation simplifies to

$$\rho\frac{\partial^2 u_2}{\partial t^2} = \mu\frac{\partial^2 u_2}{\partial x_1^2} \tag{8.72}$$

Changing the notation ($u_2 = v$, $x_1 = x$), we see that Equation (8.72) too is a form of the one-dimensional wave equation (8.63), with celerity C_s where

$$C_s = \sqrt{\mu/\rho} \tag{8.73}$$

One solution is:

$$v = \beta \cos(2\pi/L)(x - C_s t) \tag{8.74}$$

where β is an arbitrary constant. This equation indicates that the displacement in the y-direction (that is, normal to the direction of wave propagation) moves as a cosine wave with wavelength L and celerity C_s. These are called *secondary waves*, also shear or S waves, by seismologists.

It can be shown (e.g., Sommerfeld, 1964, pp. 330–3) that there are also waves at the free surface of an elastic body, for example, at the surface of the solid earth, whose celerity is given by

$$C_{\text{surf}} = \sqrt{\mu/2\rho} \tag{8.75}$$

So these waves have somewhat lower celerities than the S waves. For further discussion of seismic waves, see Bullen and Bolt (1985).

The S and surface waves depend only on μ, the rigidity modulus. As μ is zero for a fluid, these waves can be transmitted only by solids (the surface waves discussed here are not at all the same kind of wave as the gravity waves observed on the surface of a sea or lake). By contrast, P waves can be transmitted through fluids because their celerity depends on λ which is also a function of the bulk modulus ($\lambda = k - \frac{2}{3}\mu$). The bulk modulus is finite for all real continua, including fluids.

Seismic P waves are the same kind of wave as sound waves, which are, of course, transmitted through air and water. The fact that S waves are not transmitted through the outer core of the earth is the main evidence that the outer core is fluid.

8.12.4 Irrotational and solenoidal displacements

If we neglect body forces, and make use of the vector identity

$$\mathbf{A} \times \mathbf{B} \times \mathbf{C} = \mathbf{B}(\mathbf{A} \cdot \mathbf{C}) - \mathbf{C}(\mathbf{A} \cdot \mathbf{B})$$

then the vector form of the Navier equations, Equation (8.62), can be rewritten as

$$\rho \frac{\partial^2 \mathbf{u}}{\partial t^2} = (2\mu + \lambda)\nabla(\nabla \cdot \mathbf{u}) - \mu\nabla \times \nabla \times \mathbf{u} \tag{8.76}$$

If $\mathbf{u}$ is *irrotational*, then by definition curl $\mathbf{u} = 0$ or $\nabla \times \mathbf{u} = 0$ and Equation (8.76) reduces to

$$\rho \frac{\partial^2 \mathbf{u}}{\partial t^2} = (2\mu + \lambda)\nabla^2 \mathbf{u} \tag{8.77}$$

which is the equation of a P wave.

If **u** is *solenoidal*, then by definition grad $\mathbf{u} = \nabla \cdot \mathbf{u} = 0$, and it follows that the medium is incompressible. In this case, Equation (8.76) reduces to

$$\rho \frac{\partial^2 \mathbf{u}}{\partial t^2} = \mu \nabla^2 \mathbf{u} \tag{8.78}$$

which is the equation of an S wave. This is not to say, of course, that S waves can *only* be transmitted through incompressible media, but it tells us that the compressibility is irrelevant to the transmission of these waves.

8.13 Review problems

(1) **Problem:** Suppose that the axes x_2 and x_3 define a horizontal plane at the earth's surface, and assume that the near-surface materials are at rest and behave like an elastic solid. Use Equation (8.10) to show that the coefficient of earth pressure k, defined by Equation (4.26) as

$$\sigma_{22} = \sigma_{33} = k\sigma_{11}$$

is given in this case by

$$k = v/(1 - v)$$

where v is Poisson's ratio.

Answer: Consider

$$\varepsilon_{22} = \frac{1}{E}[\sigma_{22} - v(\sigma_{11} + \sigma_{33})]$$

By hypothesis $\sigma_{22} = \sigma_{33} = k\sigma_{11}$. ε_{22} is the elastic strain in one of the horizontal directions, and we can assume that it is equal to zero, because confinement makes it impossible for the near surface materials to expand horizontally while they are at rest. Therefore the equation reduces to

$$0 = k\sigma_{11} - v(\sigma_{11} + k\sigma_{11})$$

or

$$k = v/(1 - v)$$

(2) **Problem:** The construction and filling of a reservoir imposes a substantial load on the lithosphere. The resulting subsidence can be measured by careful leveling, and is due to two different responses: (i) elastic compression of the crust, and (ii) isostatic sinking.

  (a) How could you distinguish the two responses, based on resurveying the reservoir over a period of many years?

(b) Assuming that the elastic part of the crust beneath the reservoir is composed of granite with $E = 50$ GPa and $v = 0.25$, what would be the elastic subsidence after filling a reservoir to a depth of 100 metres? Use the result obtained in Problem 1, assume that the largest principal stress remains vertical, and that subsidence is due to elastic compression of a thickness of crust about equal to the size of the reservoir (say 1 km). Note that we do not expect the effect of the load to extend much deeper than that, because it spreads out laterally (see the answer for a more sophisticated approach to the problem).

Answer:

(a) The elastic response should take place almost instantaneously, therefore it should be measurable as soon as the reservoir is filled. The isostatic response depends on flow in the asthenosphere, which takes thousands of years for completion. Precise leveling, however, should be able after about ten years to detect some continuing subsidence.

(b) The extra load imposed is a vertical normal stress $\sigma_{11} = 100\gamma_w = 980$ kPa. The vertical strain is given by

$$\varepsilon_{11} = \frac{1}{E}[\sigma_{11} - v(\sigma_{22} + \sigma_{33})]$$

If we use the result obtained in Problem 1:

$$\sigma_{22} = \sigma_{33} = k\sigma_{11} = \frac{v}{1-v}\sigma_{11}$$

then we obtain

$$\varepsilon_{11} = \frac{1}{E}\sigma_{11}\left(1 - \frac{2v^2}{1-v}\right)$$

$$= 1.63 \times 10^{-5}$$

This is a strain, so to get the subsidence we must multiply by the thickness of the compressed crust to obtain the total elastic subsidence of 16.3 mm. Note that this is only a very rough calculation, because incremental load concentrated in a relatively small area (like the water in a reservoir) produces principal stress trajectories that are not strictly vertical, but diverge at depth. Goodman (1980, p. 179) gives the equation for the mean displacement h beneath a flexible plate of radius r loaded with a pressure p, resting on an extensive elastic medium (technically known as an "infinite half space of elastic isotropic material"):

$$h = 1.7\frac{p(1 - v^2)r}{E}$$

Applying this equation to the problem yields the more realistic (but not very different) estimate of 15.7 mm.

Elastic subsidence beneath reservoirs and large structures is a real phenomenon of concern to engineers and environmental scientists. Subsidence is only one effect: the deformation produced frequently sets off small earthquakes, some of which are due to a decrease in rock strength produced by an increase in pore pressure transmitted from the reservoir to the rocks beneath it (see Section 5.12). Gupta and Rastogi (1976) describe many examples, and a more recent review is given by Simpson (1986).

(3) **Problem:** A fluid may have elastic as well as viscous properties. For a *fluid*,

(a) What are Lamé's constants λ and μ?

(b) Show that the Navier equations of elasticity reduce to

$$\rho \frac{\partial^2 \theta}{\partial t^2} = k \nabla^2 \theta$$

where θ is the dilatation.

Answer:

(a) For a fluid the rigidity modulus is zero (by definition). So $\mu = G = 0$. Now note that $k = \lambda + 2\mu/3$, so for a fluid $\lambda = k$, the bulk modulus.

(b) The Navier equations are

$$\rho \frac{\partial^2 x_j}{\partial t^2} = (\mu + \lambda) \frac{\partial^2 u_k}{\partial x_j \partial x_k} + \mu \frac{\partial^2 u_j}{\partial x_i \partial x_i}$$

If $\mu = 0$ and $\lambda = k$ these reduce to:

$$\rho \frac{\partial^2 x_j}{\partial t^2} = k \frac{\partial^2 u_k}{\partial x_j \partial x_k}$$

The right-hand side is $k(\partial \theta / \partial x_j)$. Applying the operator $\partial / \partial x_j$ to both sides:

• Left-hand side:

$$\rho \frac{\partial}{\partial x_j} \left(\frac{\partial^2 x_j}{\partial t^2} \right) = \rho \frac{\partial^2}{\partial t^2} \left(\frac{\partial x_j}{\partial x_j} \right) = \rho \frac{\partial^2}{\partial t^2} \left(\frac{\partial u_j}{\partial x_j} \right) = \rho \frac{\partial^2 \theta}{\partial t^2}$$

• Right-hand side:

$$k \frac{\partial}{\partial x_j} \left(\frac{\partial \theta}{\partial x_j} \right) = k \frac{\partial^2 \theta}{\partial x_j \partial x_j} = k \nabla^2 \theta$$

(4) **Problem:**

    (a) Show that the rate of change of density with pressure in a compressible fluid is

$$\frac{d\rho}{dp} = \frac{\rho}{k}$$

    where k is the bulk modulus.

    (b) What is the rate of change of pressure with depth?

    (c) Near the surface, seawater has a density of 1025 kg m^{-3} and transmits sound at a speed of 1450 m s^{-1}. What is its bulk modulus k?

    (d) What is the density of the surface water, if it is carried to a depth of 5000 m (without change in temperature)?

Answer:

    (a) Conservation of mass for an element of fluid requires

$$\delta(\rho V) = 0$$

Using the product rule gives

$$\rho \delta V + V \delta \rho = 0$$

or

$$-\frac{\delta V}{V} = \frac{\delta \rho}{\rho}$$

The left-hand side is the dilatation θ, which is related to the change in pressure δp and the bulk modulus k by

$$\delta p = k\theta$$

So we can write

$$\frac{\delta p}{k} = \frac{\delta \rho}{\rho}$$

or

$$\frac{\delta \rho}{\delta p} = \frac{\rho}{k}$$

In the limit as $\delta p \to 0$,

$$\frac{d\rho}{dp} = \frac{\rho}{k}$$

    (b) For a small increment of depth δz,

$$\delta p \approx \rho g \delta z$$

ρ varies with depth, but as $\delta z \to 0$ it tends to a particular value ρ, so

$$\frac{dp}{dz} = \rho g$$

(c) For a fluid $k = C_p^2 \rho$, which yields a value of about 2.2 GPa for seawater.

(d) From the chain rule:

$$\frac{d\rho}{dz} = \frac{d\rho}{dp}\frac{dp}{dz}$$

so

$$\frac{d\rho}{dz} = \frac{\rho}{k}\rho g$$

or

$$\frac{d\rho}{\rho^2} = \frac{g}{k} dz$$

Integrating:

$$-\frac{1}{\rho} = \frac{g}{k} z + C$$

At $z = 0$, $\rho = \rho_0$ so $C = -1/\rho_0$, and

$$-\frac{1}{\rho} = \frac{g}{k} z - \frac{1}{\rho_0}$$

and solving for ρ gives

$$\rho = \frac{\rho_0 k}{k - \rho_0 g z}$$

For a depth of 5000 m in seawater, the density is 1049 kg m^{-3}.

(5) **Problem:**

(a) Derive a *dimensionless* form of the x-component of the two-dimensional Navier equation of elasticity for a substance of zero rigidity ($\mu = 0$),

$$\rho \frac{\partial^2 x}{\partial t^2} = \lambda \left(\frac{\partial^2 u}{\partial x^2} + \frac{\partial^2 v}{\partial x \partial y} \right)$$

(b) When you do this, you will obtain a dimensionless coefficient,

$$M^2 = \lambda / \rho c_0^2$$

where $c_0 = \ell_0 / t_0$ is a representative *velocity*, which arises from the ratio of the representative length ℓ_0 and the representative time t_0 used to make the equations dimensionless.

M is generally called the *Mach number* and is regarded as the ratio between the speed of sound C, in the substance and the representative velocity. Explain this interpretation. The Mach number can also be regarded as a ratio of two different kinds of forces: what kinds of forces are they?

(c) We might obtain the number M^2 more directly from simple dimensional analysis, by assuming that the important variables in the problem are a velocity, the elastic modulus λ, and the density ρ. Explain why it is better to obtain M by making a dimensionless form of the governing equation.

Answer:

(a) First divide through by the density, then make all the x, y, u, and v terms dimensionless by dividing by a characteristic length ℓ_0, and all t terms dimensionless by dividing by a characteristic time t_0. Denote the dimensionless terms by capitals (e.g., $X = x/\ell_0$). This gives

$$\frac{t_0^2}{\ell_0}\left(\frac{\partial^2 X}{\partial T^2}\right) = \ell_0 \frac{\lambda}{\rho}\left(\frac{\partial^2 U}{\partial X^2} + \frac{\partial^2 V}{\partial X \partial Y}\right)$$

Dividing through by t_0^2/ℓ_0 and writing $c_0^2 = \ell_0^2/t_0^2$ gives

$$\frac{\partial^2 X}{\partial T^2} = \frac{1}{c_0^2}\frac{\lambda}{\rho}\left(\frac{\partial^2 U}{\partial X^2} + \frac{\partial^2 V}{\partial X \partial Y}\right)$$

(b) The speed of sound in a fluid is simply the speed of P waves. For $\mu = 0$, we know from Equation (8.70) that this is

$$C_p = \sqrt{\frac{\lambda}{\rho}}$$

so the interpretation of the Mach number follows.

(c) It is better to make use of the governing equation to derive the Mach number because we know from the derivation that the governing equations express, in the most general way possible, how elastic waves travel through fluids. Navier's equations have now been tested by a century of use, so we can have great confidence in them. The dimensionless form tells us that there will always be dynamic similarity (see Section 3.4) provided that the Mach number is the same. If we use simple dimensional analysis, we must worry that perhaps we have not included all the relevant variables, and so we should test the result obtained theoretically by carrying out experiments.

8.14 Suggested reading

Angevine, C. L., P. L. Heller and C. Paola, 1990. *Quantitative Sedimentary Basin Modeling.* Amer. Assoc. Petroleum Geol. Continuing Education Course Notes, No. 32, 132 pp. (see pp. 36–58. QE472.A455)

Bolt, B. A., 1988. *Earthquakes.* New York, W. H. Freeman, 282 pp. (A clear, simple introduction to seismology. QE534.2.B64)

Bullen, K. E. and B. A. Bolt, 1985. *An Introduction to the Theory of Seismology.* Cambridge University Press, fourth edition, 499 pp. (The second chapter is a concise introduction to elasticity, and the third is a general introduction to wave theory. Chapters 4 and 5 treat seismic body and surface waves respectively. QE534.2.B85)

Jaeger, J. C., 1971. *Elasticity, Fracture and Flow: with Engineering and Geological Applications.* New York, John Wiley and Sons, third edition, 268 pp. (Chapter 2 has a concise treatment of elasticity. QA931.J22)

Johnson, A. M., 1970. *Physical Processes in Geology.* San Francisco, Freeman, Cooper, and Co. 577 pp. (Chapters 2 to 5 have a good discussion of beam theory and folding of rocks. QE33.J58)

Sommerfeld, A., 1964. *Mechanics of Deformable Bodies.* New York, Academic Press, 396 pp. (Volume II of his *Lectures on Theoretical Physics.* A very lucid and concise treatment of stress, strain, elasticity and flow of viscous fluids by a well-known theoretical physicist. QC20.S69 v.2)

Timoshenko, S. P., 1953. *History of Strength of Materials.* New York, McGraw-Hill, 452 pp. (Includes short biographies of all the major contributors to the theory of elasticity. TA405.T58H)

Truesdell, C., 1968. *Essays in the History of Mechanics.* New York, Springer-Verlag, 384 pp. (Readable essays by a leading modern student of continuum mechanics and the history of science. QC122.T7)

Turcotte, D. L. and G. Schubert, 1982. *Geodynamics: Applications of Continuum Physics to Geological Problems.* New York, John Wiley and Sons, 450 pp. (see Chapter 3. QE501.T83)

The state of stress in the crust is currently an active field of research. For reviews see:

Gough, D. I. and W. I. Gough, 1987. Stress near the surface of the earth. Ann. Rev. Earth Planetary Sci., v.15, pp. 545–66.

Hickman, S. H., 1991. Stress in the lithosphere and the strength of active faults. Rev. Geophysics, Supplement, April 1991, pp. 759–75.

McGarr, A. and N. C. Gay, 1978. State of stress in the earth's crust. Ann. Rev. Earth Planetary Sci., v.6, pp. 405–36.

Zoback, M. L., 1992. First- and second-order patterns of stress in the lithosphere: the World Stress Map project. Jour. Geophys. Res., v.97, pp. 11 703–28 (and see other papers in No. B8).

9
Viscous fluids

9.1 Constitutive equations

A fluid is defined as a substance that deforms continuously under the action of a shear stress, no matter how small that stress is. (This definition, though it is certainly good enough for our purposes, is replaced by a different one in more advanced treatments of continuum mechanics: for a concise statement, see Truesdell, 1966.) Commonly occurring fluids include gases (e.g., air) and liquids (e.g., water, lava, magma). Also, solids such as ice and most crystalline rocks, under conditions of relatively high temperature and confining pressure, can flow at slow rates of deformation as though they were viscous fluids. Under the right conditions, therefore, they can be regarded as very viscous fluids.

For such materials, it is clear that the constitutive equations must relate stress to rate of strain, rather than to strain itself, since a small stress may, over a long period of time, produce an indeterminately large strain.

As we remarked in Section 8.1, description in terms of infinitesimal strains, used in elasticity theory as an approximation valid only for small strains, can be applied as an exact description of instantaneous rates of strain in a fluid. The reason why the theory becomes exact, as applied to fluids, is that increments of the displacements u, v and w may be made infinitesimally small simply by considering infinitesimally small increments of time. This does not necessarily mean that the theory of a viscous fluid is a linear theory, because we have still to write constitutive equations relating instantaneous stresses to instantaneous rates of strain. The constitutive equations may be nonlinear; but in the commonly used theory the equations are linear. Fluids which obey such linear equations are described as *Newtonian fluids* (and fluids which do not are called *non-Newtonian*).

As a logical (but potentially confusing) extension to the analogy between

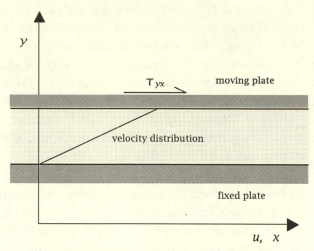

Fig. 9.1. Plane strain of a viscous fluid.

infinitesimal strain in solids and instantaneous strain rate in fluids, the same symbols $u, v,$ and w are used, with different meanings, in both cases. In the theory of elasticity, $u, v,$ and w are used to represent components of displacement in the x-, y-, and z-directions. In viscous flow theory, the same symbols describe the components of velocity (rate of displacement). The use of these symbols is so common for both cases that we have chosen to maintain the common usage, and note which meaning is used if there is the possibility of ambiguity. In the remainder of this chapter (and most of the rest of the book), $u, v,$ and w will refer to velocity components.

In the simplest case, we consider an incompressible fluid which is subject to a shear stress in the x-direction, on the xz-plane, and which is responding by a plane strain, as shown in Figure 9.1. Then all velocity gradients are zero except $\partial u/\partial y$, and all components of shear stress are zero except τ_{yx}. It may be that normal stresses are not zero, but we can assume they do not affect the flow, because the fluid is incompressible. Then, we assume a linear constitutive equation which can be written as

$$\tau_{yx} = \mu \, (\partial u/\partial y) \tag{9.1}$$

The constant of proportionality μ (a modulus, analogous to the rigidity modulus for elastic solids) is called the *coefficient of dynamic viscosity*.

That many common fluids respond in this way to plane shear is easily verified by experiment. For example, plane shear can be approximated by shearing a fluid in the small space (annulus) between two cylinders that differ slightly in diameter. We can rotate one of the cylinders (generally the outer one) at a constant rate (to produce a constant rate of shear) and measure the tangential stress by measuring the torque on the other cylinder. Carrying out the experiment for several different rates of shear allows a test of whether the relationship between stress and rate of shear is in fact linear, as Equation (9.1) would predict. An instrument of this type is also commonly used to measure the viscosity of fluids, and is called a *Couette viscometer* after Maurice Couette (1858–1943; see Donnelly, 1991, for the history of studies of flow between rotating cylinders, and an illustration of Couette's original apparatus).

In deriving the constitutive equations for an elastic solid, we made use of the isotropy assumption to reduce the required number of moduli to two (see Section 8.6, in which Lamé's constants are derived). For crystalline solids, true isotropy is an implausible assumption. We expect that isotropy is much more likely to be true of fluids, simply because they lack any long-range molecular structure. Though we are concerned with relating stress to rate of deformation, rather than to deformation, we can develop an argument for coefficients of viscosity which follows exactly the same lines as the one we sketched for Lamé's constants – and so we expect that *two* coefficients of viscosity will be required. By the same argument, we expect that the full constitutive equation has a form analogous to Equation (8.33):

$$\tau_{ij} = 2\mu\varepsilon_{ij} + \lambda\delta_{ij}\phi \tag{9.2}$$

where ε_{ij} is the rate of strain tensor (sometimes written $\dot{\varepsilon}_{ij}$ to distinguish it from the strain tensor),

$$\varepsilon_{ij} = \frac{1}{2}\left(\frac{\partial u_i}{\partial x_j} + \frac{\partial u_j}{\partial x_i}\right) \tag{9.3}$$

ϕ is the divergence of the velocity,

$$\phi = \frac{\partial u_1}{\partial x_1} + \frac{\partial u_2}{\partial x_2} + \frac{\partial u_3}{\partial x_3} \tag{9.4}$$

and δ_{ij} is Kronecker's delta. We write τ_{ij} for stress, instead of σ_{ij}, for a reason that will soon be apparent.

Equations (9.2), (9.3) and (9.4) are strictly analogous to the similar equations for elastic solids in Chapter 8, except that we are now considering *rates* of strain (or *velocity* gradients) rather than strains (or displacement gradients). μ is the coefficient of dynamic viscosity, and λ is called the *second coefficient of viscosity*.

We now make another distinction relative to the elastic constitutive relations. For a fluid *at rest*, $\varepsilon_{ij} = 0$ and $\phi = 0$, so $\tau_{11} = \tau_{22} = \tau_{33} = 0$. For these conditions, we have already defined the normal stress to be equal to the hydrostatic pressure $-p$ (negative because positive pressures correspond to negative dilatations and vice versa). It is useful and convenient to keep the hydrostatic pressure in our constitutive relation for viscous fluids because p is clearly definable and useful in many practical problems and also because it is isotropic and does not give rise to any velocities or velocity gradients. For the latter reason, p is not related to any deformations of the fluid and we write the constitutive equation using a *total stress tensor* σ_{ij}:

$$\sigma_{ij} = \tau_{ij} - p\delta_{ij}$$
$$= -p\delta_{ij} + 2\mu\varepsilon_{ij} + \lambda\delta_{ij}\phi \tag{9.5}$$

The stress tensor τ_{ij} that is related to the rate of strain by Equation (9.2) may be called the *viscous stress tensor*. Thus the total stress is equal to the pressure plus the viscous stress.

Applying the definition of ϕ and ε, we can write the equation in the form

$$\sigma_{ij} = -p\delta_{ij} + \lambda\delta_{ij}\frac{\partial u_k}{\partial x_k} + \mu\left(\frac{\partial u_i}{\partial x_j} + \frac{\partial u_j}{\partial x_i}\right) \tag{9.6}$$

Writing out the first term gives

$$\sigma_{11} = -p + \lambda\left(\frac{\partial u_1}{\partial x_1} + \frac{\partial u_2}{\partial x_2} + \frac{\partial u_3}{\partial x_3}\right) + 2\mu\frac{\partial u_1}{\partial x_1}$$

If the fluid is at rest, there are no velocity gradients and terms of the form $\partial u/\partial x$ are equal to zero. It follows that $\sigma_{11} = -p$. Of course, the same applies to the total normal stress in other directions. We can continue to make the distinction between hydrostatic pressure and stresses related to deformation even when the fluid is not at rest, but in this case there may be additional normal stresses (sometimes called "dynamic pressures") generated by the flow. Therefore we need a more complete definition of fluid pressure.

We define the "mechanical mean pressure", $-\bar{p}$, as

$$-\bar{p} = \bar{\sigma} = \frac{\sigma_{11} + \sigma_{22} + \sigma_{33}}{3} \tag{9.7}$$

Then the stress tensor $\tau'_{ij} = \sigma_{ij} - \bar{\sigma}\delta_{ij}$ is called the *deviatoric stress tensor* : it is the part of the total stress that differs from the mean normal stress. But how is the mechanical mean pressure related to hydrostatic pressure?

In the general case, it can be shown from Equation (9.2) (Aris, 1962, pp. 110–11) that

$$-\bar{p} = -p + \left(\lambda + \frac{2\mu}{3}\right)\frac{\partial u_k}{\partial x_k} \tag{9.8}$$

This suggests an alternative way of defining the second viscosity, that is, using a "bulk viscosity" $k = \lambda + (2\mu/3)$. Then, we can write the difference between the hydrostatic pressure (the pressure used in thermodynamics) and the mechanical mean presure as

$$p - \bar{p} = k \frac{\partial u_k}{\partial x_k}$$

This difference is zero if we assume an incompressible fluid.

It can also be shown theoretically that if the viscous fluid is a monatomic gas, the mechanical and thermodynamic pressures must be the same. So in this case $k = 0$, and $\lambda = -2\mu/3$. k has been determined experimentally for some other types of fluid and has been found to be very small. This means that the mechanical mean pressure is almost the same as the hydrostatic pressure, and the viscous stress tensor is almost the same as the deviatoric stress tensor.

We may ask if the second, or bulk, viscosity is ever of any practical significance. The answer is that it is, but only in a very limited class of problems. For example, fluids transmit P waves (shock waves) and the rate of decay of the wave can be related to k. Generally, we can assume that $\lambda = -2\mu/3$, and therefore we need to consider only one coefficient of viscosity, even if the fluid is not incompressible.

For many fluids, neither λ nor k is of any significance, because the fluid dilatation ϕ is very small. ϕ is exactly zero for incompressible fluids (from mass conservation, see Chapter 6). For such fluids, the constitutive relation (Equation 9.6) reduces to

$$\sigma_{ij} = -p\delta_{ij} + \mu \left(\frac{\partial u_i}{\partial x_j} + \frac{\partial u_j}{\partial x_i} \right) \tag{9.9}$$

Let us verify that Equation (9.2) does give Equation (9.1) for the simple case illustrated in Figure 9.1. From Equation (9.2),

$$\tau_{yx} = \tau_{21} = 2\mu\varepsilon_{21} = 2\mu \left[\frac{1}{2} \left(\frac{\partial u_2}{\partial x_1} + \frac{\partial u_1}{\partial x_2} \right) \right]$$

But, by hypothesis, $\partial u_2/\partial x_1 = 0$, so that $\tau_{21} = \mu(\partial u_1/\partial x_2)$ or, in the x, y notation, $\tau_{yx} = \mu(\partial u/\partial y)$.

For a more complete discussion of the constitutive equations for a viscous fluid see Aris (1962, Chapter 5), Long (1961, Chapter 3), Whitaker (1968, Chapter 5), or a text on continuum mechanics such as Fung (1977) or Malvern (1969).

9.2 Equation of motion

The equation of motion for a viscous fluid is analogous to that for an elastic solid. For a unit volume of fluid:

$$\begin{matrix} \text{mass} \times \\ \text{acceleration} \end{matrix} = \begin{matrix} \text{body} \\ \text{forces} \end{matrix} + \begin{matrix} \text{pressure} \\ \text{forces} \end{matrix} + \begin{matrix} \text{viscous} \\ \text{forces} \end{matrix}$$

The left-hand side is equal in magnitude to the "inertia forces". The differences between this equation and the equation of motion for elasticity, given at the start of Section 8.11, are:

(1) In considering the inertia forces, it is no longer possible to neglect the changes in velocity that result from changes in position, as well as changes in time. Thus, not only may a fluid particle accelerate because the entire fluid body is being accelerated (this is the only acceleration we considered in Chapter 8), but it may also accelerate because the particle is moving toward a region with a different velocity. For example, flow in a converging channel must accelerate as the channel narrows, even though there is no change with time in the amount of water flowing through the channel when the flow is steady.

(2) As we have seen, it is convenient to separate the surface forces into two parts, a normal component equal to the hydrostatic pressure (even though the fluid is actually in motion) and a part due to the deviatoric stress. For an incompressible fluid, the deviatoric stress is entirely due to viscous shear stresses.

In the following section we concentrate on the second difference. We leave the first for detailed consideration in Section 9.3.

9.2.1 Body forces

As in Chapter 8, the body forces can be written conventionally as

$$\text{body forces} = \rho X_j$$

In the flow of fluids, generally the only body force is gravity: it can then be written

$$\text{body force} = \rho X_j = -\rho g_j \tag{9.10}$$

where g_j is the component of g parallel to x_j. If we choose the x_3-coordinate as vertical, the x_1- and x_2-components of the body force are then zero, and the x_3 component is $-\rho g$.

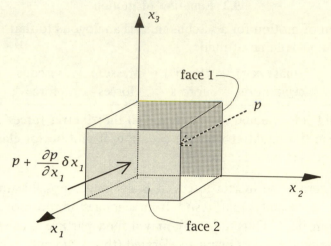

Fig. 9.2. Pressure acting on a fluid element.

9.2.2 *Pressure forces*

The pressure is defined as the mean pressure (see discussion in previous section). Consider the pressure in the x_1-direction, exerted on a small fluid element (see Figure 9.2). For fluids, we reverse the usual convention, and consider compression to be positive. Then the pressure forces acting on the two faces normal to the x_1-axis are:

$$\text{force on face } 1 = p\delta x_2\,\delta x_3$$
$$\text{force on face } 2 = \left(p + \frac{\partial p}{\partial x_1}\delta x_1\right)\delta x_2\delta x_3$$

So the difference, or net force on the element acting in the x_1-direction is

$$\begin{matrix}\text{force on} \\ \text{face 1}\end{matrix} - \begin{matrix}\text{force on} \\ \text{face 2}\end{matrix} = -\frac{\partial p}{\partial x_1}\delta x_1\delta x_2\delta x_3 = -\frac{\partial p}{\partial x_1}\delta V \qquad (9.11)$$

where δV is the volume of the element. Thus, the magnitude of the pressure force per unit volume is just the pressure gradient, $\partial p/\partial x_1$. Remember that a positive pressure gradient produces a force acting in the negative x_1-direction:

$$\text{pressure force} = p_j = -\frac{\partial p}{\partial x_j} \qquad (9.12)$$

that is, the force is directed from higher to lower pressures.

9.2.3 Viscous forces

By a similar type of argument (see Section 8.11) we may show that forces due to the viscosity are given by summing the stress gradients:

$$\text{viscous forces} = \frac{\partial \tau_{ij}}{\partial x_i} = \frac{\partial \tau_{1j}}{\partial x_1} + \frac{\partial \tau_{2j}}{\partial x_2} + \frac{\partial \tau_{3j}}{\partial x_3} \tag{9.13}$$

Now we incorporate the constitutive relation for a Newtonian fluid,

$$\tau_{ij} = 2\mu\varepsilon_{ij}$$

and use the expression for ε_{ij} given in Equation (9.3) to obtain:

$$\tau_{ij} = 2\mu\left[\frac{1}{2}\left(\frac{\partial u_j}{\partial x_i} + \frac{\partial u_i}{\partial x_j}\right)\right]$$

$$= \mu\left(\frac{\partial u_j}{\partial x_i} + \frac{\partial u_i}{\partial x_j}\right) \tag{9.14}$$

The viscous forces are given by the divergence of τ_{ij},

$$\frac{\partial \tau_{ij}}{\partial x_i} = \mu\left(\frac{\partial^2 u_j}{\partial x_i \partial x_i} + \frac{\partial^2 u_i}{\partial x_i \partial x_j}\right)$$

but the second term on the right-hand side is simply

$$\frac{\partial}{\partial x_j}\left(\frac{\partial u_i}{\partial x_i}\right)$$

For an incompressible fluid,

$$\phi = \frac{\partial u_i}{\partial x_i} = 0$$

and so the viscous forces reduce to

$$\frac{\partial \tau_{ij}}{\partial x_i} = \mu\frac{\partial^2 u_j}{\partial x_i \partial x_i} \tag{9.15}$$

In x, y, z and u, v, w notation, the x-component of this equation can be written

$$\frac{\partial \tau_x}{\partial x} = \mu\left(\frac{\partial^2 u}{\partial x^2} + \frac{\partial^2 v}{\partial x^2} + \frac{\partial^2 w}{\partial x^2}\right)$$

This derivation has been restricted to incompressible fluids. Real fluids are compressible, although liquids are more nearly incompressible than gases. Compressibility is most important when it produces shock waves, a phenomenon confined to large Mach numbers (Chapter 8, Problem 5). Most natural flows have low Mach numbers, so compressibility has little effect, even if the fluid is a relatively easily compressible gas. Compressibility is not the only property commonly ignored in fluid mechanics. Viscosity is a type of internal friction, which therefore generates heat, and viscosity itself is temperature dependent. Yet because the heat capacity of most fluids is high, we consider viscosity to be a constant unaffected by the flow. In both cases, it is possible to develop forms of the governing equations that take account of these complications.

We have now determined the terms on the right-hand side of the equation of motion given at the beginning of the section, and will proceed in the next section to a consideration of the left-hand side.

9.3 Kinematics of flow and material acceleration

Consider a particle of fluid moving within a flow. We can describe the motion in at least two ways:

(1) with reference to a fixed set of coordinates: so a given particle of fluid occupies a given position only for an instant of time. This method describes the way the flow is varying at a particular place, as time passes, so that the velocity can be written as a function of the coordinates of that position and time. It is sometimes called the *Eulerian* method.
(2) with reference to a particular fluid particle: as time passes, the coordinates of the particle change. This method describes the changing position of a particle, so that the velocity of the particle can be written as a function of the coordinates of the particle and of time. It is sometimes called the *Lagrangian* method (neither "Eulerian" nor "Lagrangian" correctly describes the techniques used by Euler and Lagrange: but the terms are commonly used despite their historical inaccuracy). The changing coordinates of a moving particle of fluid, or *material particle*, are described as the *material coordinates* (Malvern, 1969, pp. 138–41, gives a more complete discussion).

To show how we deal with the large displacements of a fluid particle that usually occur, we consider first how a scalar quantity T, such as temperature, varies as a function of the material coordinates x_1, x_2, x_3 and of time t. For a small change δT we can write:

$$\delta T = \frac{\partial T}{\partial t}\delta t + \frac{\partial T}{\partial x_1}\delta x_1 + \frac{\partial T}{\partial x_2}\delta x_2 + \frac{\partial T}{\partial x_3}\delta x_3$$

If we now say that δt is the time taken for the fluid particle to undergo the displacement $(\delta x_1, \delta x_2, \delta x_3)$, then $\delta x_1/\delta t, \delta x_2/\delta t, \delta x_3/\delta t$ are the components of the velocity **u** of the fluid particle. The meaning of δT is then the change in temperature between the two points linked by this displacement that occurs in the time taken for the fluid particle to move between them. The limit of $\delta T/\delta t$ as δt tends to zero is written as DT/Dt and is known as the *material derivative* (though some authors write "total derivative" or "substantive derivative").

$$\frac{DT}{Dt} = \frac{\partial T}{\partial t} + \frac{\partial T}{\partial x_1}\frac{dx_1}{dt} + \frac{\partial T}{\partial x_2}\frac{dx_2}{dt} + \frac{\partial T}{\partial x_3}\frac{dx_3}{dt}$$

$$= \frac{\partial T}{\partial t} + u_1\frac{\partial T}{\partial x_1} + u_2\frac{\partial T}{\partial x_2} + u_3\frac{\partial T}{\partial x_3}$$

$$= \frac{\partial T}{\partial t} + \mathbf{u}\cdot\nabla T \tag{9.16}$$

Following the above reasoning, the material derivative is also known as *differentiation following the motion of the fluid.* The right-hand side of Equation (9.16) is the sum of two terms.

- $\partial T/\partial t$, the *local* term, measures the rate at which the temperature of the fluid particle would change if it remained fixed at a point.
- $\mathbf{u}\cdot\nabla T$, the *convective* or *advective* term, measures the rate at which the temperature of the moving fluid particle would change if the pattern of temperature gradients in space were constant in time.

Thus, we may re-express Equation (9.16) as

total rate of change
  of T with time

$$= \quad\begin{array}{c}\text{local rate of change}\\\text{of } T \text{ with time}\end{array} + \begin{array}{c}\text{convective rate of change}\\\text{of } T \text{ with time}\end{array} \tag{9.17}$$

If we apply this to the weather, observed at a single station, the "local" change is the overall warming during the day or cooling at night (due to changes in the amount of heat coming from the sun), but the "convective" change is due to the movement of warmer or colder air masses into the area. We note that in fluid mechanics "convective" means "arising from the general motion of the fluid" and is not restricted to the transfer of heat.

Applying Equation (9.16) to the case where T is a vector, say the velocity **u**, gives the material derivative of velocity i.e., the *material acceleration* of a fluid particle:

$$\frac{D\mathbf{u}}{Dt} = \frac{\partial\mathbf{u}}{\partial t} + \mathbf{u}\cdot\nabla\mathbf{u} \tag{9.18}$$

For the x_1-coordinate this can be written

$$\frac{Du_1}{Dt} = \frac{\partial u_1}{\partial t} + u_1\frac{\partial u_1}{\partial x_1} + u_2\frac{\partial u_1}{\partial x_2} + u_3\frac{\partial u_1}{\partial x_3} \tag{9.19}$$

The material acceleration can also be written in tensor notation:

$$\frac{Du_j}{Dt} = \frac{\partial u_j}{\partial t} + u_i\frac{\partial u_j}{\partial x_i} \tag{9.20}$$

As discussed above, we may write Equation (9.20) as

$$\begin{array}{ccc} \text{material} & = & \text{local} & + & \text{convective} \\ \text{acceleration} & & \text{acceleration} & & \text{acceleration} \end{array}$$

To obtain the forces per unit volume on the fluid particle, we simply multiply the accelerations by the mass per unit volume, which is the mass density, ρ. We assume the fluid is incompressible, so the density is a constant. The *local* part of the acceleration is due to a general change in time in the forces at each point and the *convective* part is due to movement of the particle into a region of different velocity, such as a constriction, as discussed at the beginning of Section 9.2.

If the flow is *steady*, i.e., it does not change with time, then the local accelerations are zero:

$$\frac{\partial u_j}{\partial t} = 0$$

However, there may still be convective accelerations.

If the flow is *uniform*, i.e., it does not change in the downflow direction, and if we define x_1 as the flow direction, then:

(1) $u_2 = 0$ and $u_3 = 0$, and their derivatives are also equal to zero;
(2) the velocity gradient $\partial u_1/\partial x_1$ is equal to zero.

In this case, it follows that the convective accelerations are zero, though there may still be local accelerations. If the flow is both steady and uniform there are no accelerations at all.

In a steady flow, the velocity at any point does not vary with time. So, lines that are everywhere tangent to flow velocities (*streamlines*) also do not change with time. Flow cannot cross such streamlines, and fluid particles within an envelope of such lines (i.e., within a *streamtube*) are confined within that tube; in uniform flow, streamlines are parallel, straight lines. The path travelled by a fluid particle is a *pathline*, and the line formed by the fluid particles that have passed a particular point in space is a *streakline*. In steady flows, streamlines, pathlines, and streaklines are all the same; in unsteady flows, they will in general be different lines. Tritton (1988, Chapter 6) shows computed examples and Olfe (1987) provides a computer demonstration. FLINES is a modified version of Olfe's program.

9.4 Navier–Stokes equation

9.4.1 General form

In Sections 9.2 and 9.3 we derived all the components necessary to put together the complete equations of motion for a viscous, incompressible fluid subject to a gravity force. Equation (9.20), when multiplied by the density, gives the acceleration terms on the left-hand side, and Equations (9.10), (9.12), and (9.15), respectively give the gravity, pressure and viscous forces on the right-hand side:

$$\rho \left(\frac{\partial u_j}{\partial t} + u_i \frac{\partial u_j}{\partial x_i} \right) = -\rho g_j - \frac{\partial p}{\partial x_j} + \mu \frac{\partial^2 u_j}{\partial x_i \partial x_i} \tag{9.21}$$

This is called the *Navier–Stokes equation* after the French engineer Claude Navier (1785–1836) and the English mathematician George Stokes (1819–1903) who derived it in the first half of the nineteenth century. Some important parts of the equation had been derived earlier: Euler equated the material accelerations to the pressure and gravity terms (so the equation for a fluid with zero viscosity is called Euler's equation, see Section 9.5). Cauchy gave the first demonstration that the surface forces were given by the divergence of the stress tensor. His equation equating the material accelerations to the body and surface forces is called *Cauchy's first law* and is actually an equation with broader application than the Navier–Stokes equation, since it is the basis of all theories of continua. But it could not be applied to viscous fluids until Navier and Stokes developed the equation for viscous forces. Navier arrived at an incomplete form of the equation based on an erroneous molecular model of a fluid: Stokes gave a more rigorous mathematical derivation.

The Navier–Stokes equation has been written above in the compact tensor notation. Below, the x- and z-components are written out for the two-dimensional case in the x, z and u, w notation (z is chosen to be vertical):

$$\rho \left(\frac{\partial u}{\partial t} + u \frac{\partial u}{\partial x} + w \frac{\partial u}{\partial z} \right) = -\frac{\partial p}{\partial x} + \mu \left(\frac{\partial^2 u}{\partial x^2} + \frac{\partial^2 u}{\partial z^2} \right) \tag{9.22}$$

$$\rho \left(\frac{\partial w}{\partial t} + u \frac{\partial w}{\partial x} + w \frac{\partial w}{\partial z} \right) = -\rho g - \frac{\partial p}{\partial z} + \mu \left(\frac{\partial^2 w}{\partial x^2} + \frac{\partial^2 w}{\partial z^2} \right) \tag{9.23}$$

The equations have been written so that each term is a force per unit volume.

If we divide through by ρ, each term becomes an acceleration (i.e., force per unit mass), and the coefficient of the viscous term becomes μ/ρ. This ratio of the viscosity to the density recurs frequently in fluid mechanics: it is called the *kinematic viscosity*, v.

The Navier–Stokes equation is very important because it is the fundamental equation for a very large class of fluid motions. However, analytical solutions can be obtained easily for only four classes of problems (Whitaker, 1968, Chapter 5):

(1) fluids at rest, or moving with constant velocity, or moving at velocities so slow that accelerations may be neglected ("slow viscous", "creeping", or "Stokes" flows);

(2) uniformly accelerated flows, i.e., flows in which every fluid particle undergoes the same acceleration;

(3) one-dimensional flows, e.g., Couette flow, or flow through a straight pipe of circular cross-section;

(4) irrotational flows, i.e., flows for which the vorticity is zero and there exists a velocity potential (see Chapter 6, Section 6.5). As we have seen, it is relatively easy to obtain solutions to the equations of motion for such flows, using either analytical or numerical methods. The flow of "ideal" fluids (with zero viscosity) is always irrotational (unless vorticity is imposed by the initial conditions, see Section 9.6).

A further limitation is that the Navier–Stokes equation may be applied directly to laminar but not to turbulent flow. The Navier–Stokes equation certainly does apply to turbulent flows, but because of the chaotic nature of such a flow, resulting from dynamic instability, the equation cannot be used to provide practical information about the average characteristics of the flow, e.g., the mean velocity. Instead, recourse must be made to empirically determined "laws" of turbulent flow (see Chapter 11). In some cases, these empirical equations are equivalent to an assumption either that the Newtonian fluid has a much higher viscosity than its true viscosity, or that it behaves in bulk like a non-Newtonian fluid.

Because of these limitations most practical problems are solved either by resorting to numerical solutions (which is not easy for the full Navier–Stokes equation), or by resorting to the simplifications listed above, or by avoiding the Navier–Stokes equation completely. To do this, engineers define a control volume that encompasses the problem at hand (e.g., a section of pipe, a tank, a fixed volume in space) and then conserve mass and momentum only for the volume as a whole, thereby avoiding the often complex details of the flow within the volume. This approach is often useful, but it provides no insight regarding the internal fluid dynamics.

The fact that the Navier–Stokes equation can be solved exactly for only a limited number of cases does not limit its usefulness. Applied mathematicians are constantly extending the range of cases that can be treated analytically; and almost all cases can be analyzed with numerical methods. Also, much can be learned about the nature of flows from examination of the equations. For example, an exact solution for a simple, ideal, flow geometry may also provide a perfectly good approximate solution to a real flow problem with a more complex, but approximately similar, geometry. In order to know when such an approximation is valid, we must consider each term in the full governing equations that we wish to simplify or neglect. In the the next subsection, we will introduce a different form of the Navier–Stokes equation that, among other advantages, can be used to determine when such approximations can be made.

9.4.2 Nondimensional form

To do this, we write a dimensionless version of the Navier–Stokes equation. The same thing was done with the Navier equation of elasticity in Problem 3 of Chapter 8. We start by selecting representative values of length and velocity to nondimensionalize each term in the full equations. For example, we might pick the mean depth and mean velocity as the representative values for flow in an open channel. In the case of a sphere settling in fluid, we might choose the sphere diameter and the terminal settling velocity to nondimensionalize the equations.

If we designate the representative length and velocity as l_o and v_o respectively, we then make all lengths and velocities in the problem dimensionless by dividing by l_o and v_o (e.g., the dimensionless velocity in the x-direction is $U = u/v_o$). We nondimensionalize time using l_o/v_o and we nondimensionalize pressure using ρv_o^2, which arises from Bernoulli's equation (Chapter 6, and the next section of this chapter) and represents the fluid pressure due to its motion. To illustrate, we nondimensionalize the vertical component of the Navier–Stokes equation (Equation 9.23). We choose the z-direction because an important dimensionless parameter arises from nondimensionalizing the gravity term. First, we make the substitutions that nondimensionalize each term. These are as follows:

$$Xl_o = x, \quad Zl_o = z, \quad Uv_o = u, \quad Wv_o = w, \quad Tl_o/v_o = t, \quad P = p/\rho v_o^2$$

Noting that l_o and v_o are constants and may be moved outside the derivatives, we collect representative terms and obtain

$$\frac{\rho v_o^2}{l_o}\left(\frac{\partial W}{\partial T}+U\frac{\partial W}{\partial X}+W\frac{\partial W}{\partial Z}\right)=-\rho g-\frac{\rho v_o^2}{l_o}\frac{\partial P}{\partial Z}+\frac{\mu v_o}{l_o^2}\left(\frac{\partial^2 W}{\partial X^2}+\frac{\partial^2 W}{\partial Z^2}\right)$$

Dividing through by $\rho v_o^2/l_o$ gives

$$\frac{\partial W}{\partial T}+U\frac{\partial W}{\partial X}+W\frac{\partial W}{\partial Z}=-\frac{gl_o}{v_o^2}-\frac{\partial P}{\partial Z}+\frac{\mu}{\rho l_o v_o}\left(\frac{\partial^2 W}{\partial X^2}+\frac{\partial^2 W}{\partial Z^2}\right) \tag{9.24}$$

All the quantities in capitals (upper case) are dimensionless. The lower case quantities have dimensions but are in two groups, each of which is dimensionless. Both of these groups have been introduced previously and they are the two most important dimensionless parameters in fluid mechanics. The first term on the right-hand side is the reciprocal of the square of the Froude number Fr, where

$$Fr=\frac{v_o}{(gl_o)^{1/2}}$$

and the coefficient in the third term is the reciprocal of the Reynolds number Re, where

$$Re=\frac{\rho v_o l_o}{\mu}$$

The process of nondimensionalizing the Navier–Stokes equation demonstrates the physical meaning of these two numbers. Recalling that we have divided by $\rho v_o^2/l_o$ (the combination of representative values that resulted from nondimensionalizing the inertia terms), we see that the Froude number is the ratio of inertia to gravity forces in the flow and that the Reynolds number is the ratio of inertia to viscous forces.

Equation (9.24) contains all of the information in the original dimensional form of the corresponding Navier–Stokes equation. Because the equation is dimensionless, it is directly useful for modeling. If the Reynolds number and Froude number are identical for two viscous flows, the two flows must be dynamically similar, even though the actual length scale and velocity of the flows, and the density and viscosity of the fluids, may not be the same. Unfortunately, the limited range of density and viscosity of modeling fluids make it almost impossible to achieve identical Froude and Reynolds numbers for length ratios between model and reality that differ much from unity (see Chapter 3, Problem 3).

This scaling problem can be resolved in two important situations. First, the Froude number is only important in flow problems with a free surface (because it is equal to the ratio between the flow velocity and the speed of long waves that develop on the surface, or on the interface between the two fluids). It is of no significance in problems where the density is constant and there is no free surface, so that gravity forces are balanced by buoyancy. Second, for many natural flows the Reynolds number is very large and the flow is turbulent. In these cases, it is found that dynamic similarity does not depend strongly on the Reynolds number and a reasonable approximation to dynamic similarity can be achieved by ensuring that the Froude number is identical in both flows.

This latter point is evident by inspection of Equation (9.24). The coefficient included in the viscous term in the equation is the inverse of the Reynolds number. If the Reynolds number is very large, the dimensionless viscous term will be much smaller than the other terms in the equation and may be safely ignored. This points out a second important use of the dimensionless Navier–Stokes equation: it provides a means of evaluating the relative importance of the viscous effects. This is particularly important because the viscous term introduces much of the mathematical difficulty in the Navier–Stokes equation. Such scaling arguments are not just applied to the viscous term. The different components of the velocity or its derivatives can often be assumed to be negligibly small or exactly equal to zero. This allows the equation to be simplified before being solved. We will do this frequently in Sections 9.8–9.10. The correct choice of terms to include in an analysis can make the difference between arriving at an approximately correct solution and no solution at all.

An important point is that the representative lengths and velocities chosen to nondimensionalize the governing equations need not exactly equal particular values for the flow at any time or place; the point is to represent the general scale of the problem. Once the equations are nondimensionalized, we are concerned with the relative order of magnitude of the different terms. A term that is only a factor of two or three smaller than the others could not be safely neglected. A term that is many orders of magnitude smaller can safely be neglected (e.g., values of the Reynolds number can exceed 10^6).

If we need to consider the flow not just of a single layer of fluid, but of two layers having a small density difference $\Delta\rho$, and if we make the assumption that the density difference is important only for buoyancy, not for inertia (the *Boussinesq approximation*), then the definition of the Reynolds number of the lower layer remains essentially unchanged, but the Froude number is replaced by the *densiometric Froude number*, $Fr' = u/\sqrt{g\Delta\rho l_o/\rho}$. The densiometric Froude number proves to be the significant parameter that determines the stability of the interface between the two layers, and the degree of mixing that takes place. It is thus very important in stratified flows, such as the flow of oil over water, or of a more saline layer beneath a less saline layer, or the flow of turbidity currents (where the density increment is due to the presence of suspended sediment).

9.5 Inviscid flows: the Euler and Bernoulli equations

The simplest way to make the Navier–Stokes equation more tractable is to neglect the viscous stress term, which is mathematically equivalent to setting the viscosity equal to zero, giving rise to the term *inviscid flow*. This may be done exactly for irrotational flow (see Section 9.6). It may also be done approximately if the Reynolds number of the flow is very large. The result of neglecting the viscous term is the *Euler equation*,

$$\rho \left(\frac{\partial u_j}{\partial t} + u_i \frac{\partial u_j}{\partial x_i} \right) = -\rho g_j - \frac{\partial p}{\partial x_j} \tag{9.25}$$

This equation may be directly integrated to yield an important equation of fluid mechanics, the Bernoulli equation, named for Daniel Bernoulli (1700–82). We have already derived this equation from energy considerations in the context of flow through porous media (Section 6.2). We derive it again here as an application of the momentum equation (the Navier–Stokes equation). We can make the problem one-dimensional by integrating the Euler equation along a streamtube with an infinitesimally small cross-section. This might seem to be unnecessarily complex if one does not know the exact location of the streamtube, but it turns out that the problem is much simplified because (for a suitably small streamtube) there is negligible cross-tube variation in the density, velocity, and pressure, so we can neglect cross-tube derivatives of these properties. Besides, in many cases, we need not be concerned with the detailed location of all parts of the streamtube.

Let the direction along the axis of the streamtube be defined as the *s*-direction, and let the velocity within the tube be *v*. Of course, the *s*-direction changes as the streamtube curves in space, and the magnitude of *v* may vary along the tube (but, by definition, the direction of *v* is the *s*-direction). For this simple, one-dimensional case we may write the equation of motion as

$$\rho \left(\frac{\partial v}{\partial t} + v \frac{\partial v}{\partial s} \right) = -\frac{\partial p}{\partial s} - \rho g_s \tag{9.26}$$

where g_s is the component of gravity acting in the *s*-direction. If the streamtube is oriented at a vertical angle α to the *x*-direction, then $g_s = g \sin \alpha$. But $\sin \alpha = dz/ds$. Using these results, and assuming steady flow, the

Euler equation becomes

$$\rho v \frac{\partial v}{\partial s} = -\frac{\partial p}{\partial s} - \rho g \frac{dz}{ds}$$

which can also be written

$$\frac{\rho}{2} \frac{\partial (v^2)}{\partial s} = -\frac{\partial p}{\partial s} - \rho g \frac{dz}{ds}$$

Integrating between two positions 1 and 2 along the tube gives

$$\frac{\rho}{2}(v_2^2 - v_1^2) = -(p_2 - p_1) - \rho g(z_2 - z_1)$$

Rearranging in separate groups referring to the two positions gives

$$p_1 + \frac{\rho v_1^2}{2} + \rho g z_1 = p_2 + \frac{\rho v_2^2}{2} + \rho g z_2 = \text{constant} \tag{9.27}$$

This is the *Bernoulli equation*. Because the left-hand side contains terms relevant only to point 1 and the right-hand side contains terms relevant only to point 2, both sides must equal a constant. In Chapter 2 we derived the statement of the conservation of mechanical energy by integrating the work done by conservative forces. Here, the Bernoulli equation expresses the conservation of energy along a streamtube, and we have derived it by integrating Newton's second law along the tube. In Chapter 2, we ignored any nonconservative forces such as friction. Here too, we have ignored the nonconservative viscous forces, which are the source of energy loss in fluid flows.

It might appear that ignoring viscosity would seriously limit the application of the Bernoulli equation. But in fact the equation is a good approximation at large Reynolds numbers, provided that the velocity gradients along the streamtube are much larger than those across it (we expect strong cross-tube gradients mainly in regions close to a solid boundary). In the form shown, the various terms have the dimensions of stress and can be interpreted as pressures. We call terms such as $\rho v^2/2$ *dynamic pressures*.

Now consider a streamtube that encounters a solid boundary (Figure 9.3). At the solid boundary, the velocity and, therefore, the dynamic pressure drops to zero. According to the Bernoulli equation, the pressure at this point must be equal to the sum of the two terms, p_0 and $\rho v_0^2/2$, at some point far from the cylinder, provided the streamline is horizontal ($z = z_0$). That is, the pressure p at the cylinder is greater than the pressure p_0 by an amount equal to the dynamic pressure $\rho v_0^2/2$:

$$p = p_0 + \frac{\rho v_0^2}{2}$$

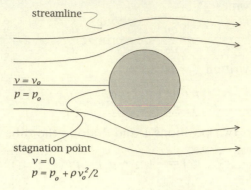

Fig. 9.3. Flow around a cylinder, showing the stagnation point. Far from the cylinder $v = v_0, p = p_0$; at the stagnation point the velocity falls to zero, and the pressure rises to a maximum, $p = p_0 + \rho v_0^2/2$.

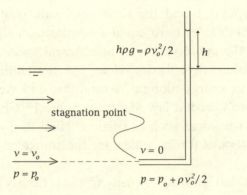

Fig. 9.4. The Pitot tube, used for measuring flow speed.

The pressure at the cylinder is the maximum that can be reached along that streamtube and is called the *stagnation pressure* (because it is achieved only when the velocity falls to zero, that is, the flow "stagnates"). Assuming the streamtube is horizontal, p_0 is the minimum fluid pressure along the streamtube, called the *static pressure*, which in this case is equal to the hydrostatic pressure.

These principles are the basis for the *Pitot tube*, a commonly used method of measuring velocity in fluids (Figure 9.4). The mouth of the Pitot tube is directed into the flow and, at the mouth, the velocity falls to zero and the stagnation pressure is developed. Because the pressure at p_0 is hydrostatic ($p_0 = \rho g z$, where z is the depth below the water surface), the corresponding pressure at the mouth of the Pitot tube causes water in the tube to rise just to the water surface. The additional pressure causing water in the tube to rise above the water surface is exactly equal to the pressure due to the fluid motion, so the value of v_0 may be determined from the height h to which the fluid in the tube rises above the water surface and $\rho g h = \rho v_0^2/2$.

Because this height cannot be measured very accurately over a moving water surface, a double tube (one inside the other) is actually used in practice. The inner tube functions just as the Pitot tube; the outer (annular) tube has small holes oriented normal to the flow direction and just downstream of the main tube opening. This is called a *Pitot static tube*. If the tube is small enough, there are essentially no energy losses in the flow from the tip of the tube to the small holes in the outer tube and the sum of the Bernoulli terms still equals a constant. Then, the pressure measured in the outer tube is equal to the static pressure and the pressure difference between the inner and outer tubes is equal to the dynamic pressure $\rho v_0^2/2$. The method works well at high Reynolds numbers, although the pressure differences are often quite small and a very sensitive pressure transducer is useful for making accurate measurements. For example, for a velocity of 0.05 m s^{-1}, the dynamic pressure for water is 1.25 Pa, which causes water to rise in a manometer tube by only 0.13 mm. Small velocities of this magnitude must often be measured in experimental studies.

9.6 Irrotational flows

One aspect of inviscid flow theory that was investigated thoroughly in the nineteenth century is vorticity. The *vorticity* is equal to the antisymmetric part of the rate of strain tensor (cf. the rotation tensor described in Section 7.7). In a two-dimensional flow, there is a single component given by

$$\omega_z = \frac{\partial v}{\partial x} - \frac{\partial u}{\partial y} \tag{9.28}$$

Applying this equation to the simple shear shown in Figure 9.1 reveals at once that this flow is not irrotational, because $\partial v/\partial x$ is zero and $\partial u/\partial y$ is not. As all viscous flows approximate simple shear in a small region very close to a solid boundary (Section 11.3), it is clear that an initially irrotational viscous flow soon generates vorticity by shearing along its solid boundary.

An inviscid fluid, however, is assumed to slip, not shear, past a solid boundary, so no vorticity is generated there. If we assume that the flow is not only inviscid but also has no initial vorticity then the flow must remain irrotational. In viscous flows, viscosity not only creates vorticity by shear at solid boundaries, but also destroys it by viscous "damping" within the flow.

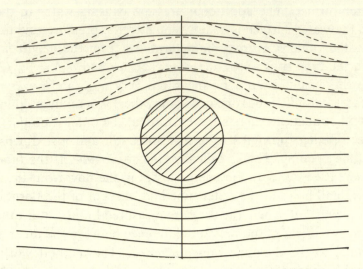

Fig. 9.5. Pattern of flow of an inviscid fluid past a circular cylinder normal to the flow. The solid lines show the inviscid flow solution; the broken lines in the upper part of the figure show the solution for slow viscous flow.

By taking the curl of all terms in Euler's equation of motion (Equation 9.25), it may be converted from an equation that relates velocities to one that relates vorticities. Examination of this equation leads to some interesting insights about the nature of vorticity (for an elegant summary, see Lighthill, 1986).

Many two-dimensional irrotational flows with simple boundary conditions can be represented by fluid potentials (see Chapter 6), and these are much easier to solve analytically than most viscous flows. Much of classic fluid dynamics was devoted to solving potential-flow problems, and a useful compendium of the solutions has been published by Kirchhoff (1985).

One example of such a flow is shown in Figure 9.5 for flow around a cylinder. It is compared with the solution for slow viscous flow (discussed in Section 9.11). Both solutions are symmetrical, but there is an important difference. Not only is the flow deflected more by the effects of viscosity, but the inviscid solution is obtained by assuming an unrealistic slip at the surface of the cylinder, and it predicts that equal pressures are developed on the front and rear sides – so that there is no net drag. This result, known as *d'Alembert's paradox*, is contrary to all experimental evidence and shows that viscous effects at a solid boundary can never be neglected, even at high Reynolds numbers.

Since real flows are viscous, it might be thought that potential and vorticity theory are interesting only to applied mathematicians, and of little use in understanding the real world. This is far from the truth, however, because these theories apply to many real flows, particularly large-scale geophysical flows in the atmosphere and oceans, wherever such flows are not strongly affected by solid boundaries. The analytical problem is how to express realistic boundary conditions for such flows.

9.7 Boundary layers

The possible applications of the Navier–Stokes equation were greatly increased when it was proposed by the German engineer Ludwig Prandtl (1875–1953), at the beginning of the twentieth century, that it is possible to divide many flows into two regions:

(1) the main flow, where velocity gradients and viscous stresses are small, and the flow is approximately irrotational;
(2) a thin region close to the boundary (the boundary layer) where velocity gradients and viscous stresses are very large, the flow is approximately two-dimensional, and considerable simplification of the Navier–Stokes equation is possible.

A boundary layer (BL) can be characterized by two lengths: L in the flow (x-) direction, and δ in the y-direction, normal to x. We assume that the boundary layer is thin ($\delta \ll L$) and that the curvature of the boundary is small enough that flow can be considered two-dimensional. For incompressible flow, continuity in the BL implies

$$\frac{\partial u}{\partial x} + \frac{\partial v}{\partial y} = 0$$

We will show that this implies $|u| \gg |v|$; then we will use this result to show that the x-component of the Navier–Stokes equation (without body forces) for steady flow in a viscous BL reduces to

$$\rho \left(u\frac{\partial u}{\partial x} + v\frac{\partial u}{\partial y} \right) = -\frac{\partial p}{\partial x} + \mu\frac{\partial^2 u}{\partial y^2} \tag{9.29}$$

As an approximation, we write for the BL

$$\frac{\partial u}{\partial x} \approx \frac{u}{L} \qquad \frac{\partial v}{\partial y} \approx \frac{v}{\delta}$$

From the continuity equation, these two quantities must be nearly equal in magnitude:

$$u/L \approx -v/\delta \quad \text{or} \quad \delta/L \approx -v/u$$

From the basic boundary layer assumption that $L \gg \delta$, it then follows that $|u| \gg |v|$. In general, this means that BL theory can be applied if the longitudinal length scale L is at least an order of magnitude greater than the transverse length scale δ.

Now write the equations for steady flow in two dimensions:

$$\rho \left(u\frac{\partial u}{\partial x} + v\frac{\partial u}{\partial y} \right) = -\frac{\partial p}{\partial x} + \mu \left(\frac{\partial^2 u}{\partial x^2} + \frac{\partial^2 u}{\partial y^2} \right) \tag{9.30}$$

$$\rho \left(u\frac{\partial v}{\partial x} + v\frac{\partial v}{\partial y} \right) = -\frac{\partial p}{\partial y} + \mu \left(\frac{\partial^2 v}{\partial x^2} + \frac{\partial^2 v}{\partial y^2} \right) \tag{9.31}$$

First we consider Equation (9.30). Because $L \gg \delta$, we expect velocity gradients across the BL to be much larger than those along it. So $\partial u/\partial y \gg \partial u/\partial x$, but the two products

$$u\frac{\partial u}{\partial x} \quad \text{and} \quad v\frac{\partial u}{\partial y}$$

should be of comparable magnitude because $v \ll u$. We can make this conclusion explicit by using our result from continuity to make the substitution $v \approx u(\delta/L)$, which shows that both terms are of order of magnitude u^2/L.

Again, because gradients in the x-direction are much less than those in the y-direction, we expect that the first viscous term in Equation (9.30) is at least two orders of magnitude smaller than the second, so we may neglect the first term. The other viscous term is of order of magnitude u/δ^2. Because this term is multiplied by μ, and the inertia terms are multiplied by ρ, we cannot neglect either without knowing specific values of ρ and μ. Now, we assume that the pressure gradient is at least the same order of magnitude as the other terms, because it is the only force available to drive the flow. So, we drop only one term from the x-component of the Navier–Stokes equation, giving

$$\rho \left(u\frac{\partial u}{\partial x} + v\frac{\partial u}{\partial y} \right) = -\frac{\partial p}{\partial x} + \mu\frac{\partial^2 u}{\partial y^2}$$

Next we consider the y-component and show that all terms (including the pressure term) must be very small compared with the corresponding terms for the x-direction. In Equation (9.31), the orders of magnitude of the two inertia terms are uv/L and v^2/δ. Using the substitution $v \approx u(\delta/L)$ again, we find that both of these terms are of order of magnitude $u^2(\delta/L^2)$

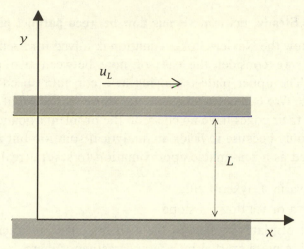

Fig. 9.6. Definition diagram for flow between parallel plates.

and are, therefore, a factor δ/L smaller than the corresponding terms in Equation (9.30). The first viscous term in Equation (9.31) is clearly very small and we neglect it. The order of magnitude of the second viscous term is $v/\delta^2 \approx u/\delta L$, which is, again, a factor of δ/L smaller than the corresponding term in Equation (9.30). There are no terms left except $\partial p/\partial y$, so it also must be much smaller than $\partial p/\partial x$:

$$\frac{\partial p}{\partial y} \ll \frac{\partial p}{\partial x}$$

This is a very important result in BL theory, because it states that the pressure within the BL varies only with x and is everywhere equal to the pressure just outside the BL; the latter pressure can be predicted from inviscid flow theory.

Even though viscous stresses are large in the BL, many BLs expand in thickness to the point where flow within the layer becomes turbulent. Therefore, it is appropriate to defer further discussion to Chapter 11, which deals with turbulent flows. Note that many real flows of interest to geologists, such as flow in a river must be considered to be "all" BL (with no adjacent "free" flow), because all parts of the flow are strongly affected by the boundary. BLs can only be recognized at relatively large Reynolds numbers: at very low Reynolds numbers, the effects of the boundary extend for very large distances into the flow.

9.8 Steady, uniform viscous flow between parallel plates

To illustrate how the Navier–Stokes equation is solved for a simple geometric configuration, we consider the case of flow between two parallel plates (Figure 9.6). The upper plate is moving at a constant speed u_L relative to the lower plate. We can consider the plates to be so wide that complications introduced at their edges have no effect on the problem at hand. We consider this case not only because it yields an analytical solution but also because it may be applied as a reasonable approximation to several real problems:

(1) flow of lava in a dyke or sill;
(2) flow of lava or ice down a slope;
(3) movement of plates by flow of the asthenosphere, or alternatively, drag on a moving plate exerted by a viscous asthenosphere.

For this steady, uniform flow, the Navier–Stokes equation can be greatly simplified:

(1) all acceleration terms (the "inertia" terms on the left-hand side of the equation) are zero;
(2) we need only consider components of flow in the x_1-direction, i.e., all terms involving u_2 and u_3 are zero.

In this section we are going to be considering two-dimensional problems, so we set the x-direction horizontal and the y-direction vertical (Figure 9.6). Gravity acts only in the y-direction. Taking all the simplifictions into account, the y- (vertical) component of the Navier–Stokes equation (i.e., Equation (9.23), allowing for our change in notation) reduces to

$$\frac{\partial p}{\partial y} = -\rho g$$

which we can directly integrate to obtain

$$p = -\rho g y + C$$

Evaluating the constant of integration at $y = L$, where $p = p_L$, we get, after rearranging,

$$p = p_L + \rho g(L - y) \tag{9.32}$$

which demonstrates that the pressure distribution is hydrostatic between the two plates. Note that the downstream pressure gradient (which, together with the motion of the upper plate, drives the flow) is given by $\partial p_L/\partial x$. The

x-component of the Navier–Stokes equation is

$$\frac{\partial p}{\partial x} = \mu \frac{\partial^2 u}{\partial y^2} \tag{9.33}$$

We set the boundary conditions as $u = 0$ at $y = 0$ and $u = u_L$ at $y = L$. Integrating Equation (9.33) gives

$$\frac{du}{dy} = \frac{1}{\mu} \int \frac{dp}{dx} dy = \frac{y}{\mu} \frac{dp}{dx} + C_1 \tag{9.34}$$

and integrating again gives

$$u = \int \left(\frac{y}{\mu} \frac{dp}{dx} + C_1 \right) dy = \frac{y^2}{2\mu} \frac{dp}{dx} + C_1 y + C_2 \tag{9.35}$$

From the first boundary condition, it follows that $C_2 = 0$. From the second boundary condition,

$$u_L = \frac{L^2}{2\mu} \frac{dp}{dx} + C_1 L$$

so

$$C_1 = \frac{u_L}{L} - \frac{L}{2\mu} \frac{dp}{dx}$$

So the complete solution may be written:

$$u = \frac{1}{2\mu} (y^2 - Ly) \frac{dp}{dx} + \frac{u_L y}{L} \tag{9.36}$$

To evaluate this solution, we must specify dp/dx. We show three forms of the velocity distribution in Figure 9.7. For the case where $dp/dx = 0$, Equation (9.36) simplifies to

$$u = \frac{u_L y}{L} \tag{9.37}$$

For a given value of u_L and L, therefore, the graph of u against y is a straight line.

This type of flow is called *Couette flow*: it is illustrated in Figure 9.1, and results from plane shear of the fluid by movement of the upper plate at a velocity of u_L relative to the lower plate.

Instead of producing a flow by moving the upper plate, we can hold both plates stationary and impose a negative pressure gradient, in order to drive the fluid through the space between them. For this case, $u_L = 0$ in Equation (9.36) and

$$u = \frac{1}{2\mu} \frac{dp}{dx} (y^2 - Ly) \tag{9.38}$$

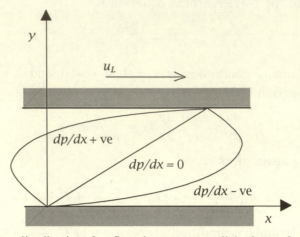

Fig. 9.7. Velocity distribution for flow between parallel plates, for three types of pressure gradient (positive, zero, and negative gradients).

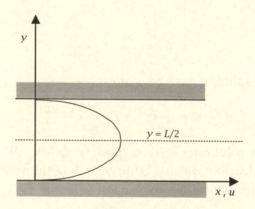

Fig. 9.8. Flow of a viscous fluid between two stationary plates.

which is the parabolic velocity distribution shown in Figure 9.8. Note that at the central plane $(y = L/2)$, $du/dy = 0$, which suggests that the lower half of this solution should apply to flow with a free surface: at the surface, the shear stress, and consequently also du/dy, are equal to zero. For the free-surface case, the (gauge) pressure is zero, that is, $p_L = 0$ and $\partial p/\partial x = 0$. Because no other forces are available to drive the flow, a free-surface flow must be driven by the gravitational force acting down a slope.

To develop the solution for this case, we write the equation of motion for the sloping plane in Figure 9.9. In this case, the component in the y-direction

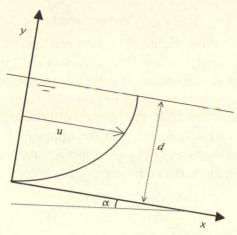

Fig. 9.9. Flow of a viscous fluid down an inclined plane.

is

$$\frac{\partial p}{\partial y} = -\rho g \cos \alpha$$

Upon integrating, and using the boundary condition that $p = 0$ at the free surface ($y = d$), we get

$$p = \rho g \cos \alpha (d - y)$$

which is again a hydrostatic pressure distribution. The x-component of the Navier–Stokes equation now includes a gravity component, but no downstream pressure gradient ($\partial p / \partial x = 0$):

$$\rho g \sin \alpha = \mu \partial^2 u / \partial y^2 \tag{9.39}$$

This equation is identical to Equation (9.33), except that the pressure gradient has now been replaced by a body-force term. Equation (9.39) can be integrated in a manner identical to that for Equation (9.33). The two boundary conditions are $u = 0$ at $y = 0$ and $\partial u / \partial y = 0$ at $y = d$. The first is identical to that used for flow between parallel plates and the second is consistent with the parallel plate solution for $y = L/2$, as noted above. The solution to Equation (9.39) is

$$u = -\frac{\rho g \sin \alpha}{2\mu}(y^2 - 2yd) \tag{9.40}$$

which is identical to Equation (9.38), except that dp/dx has been replaced by $-\rho g \sin \alpha$ and L has been replaced by $2d$.

9.9 Flow through a tube

The solution for two-dimensional flow between two stationary plates can easily be modified to apply to flow through a straight tube of circular cross-section. We convert to cylindrical coordinates, with y the radial distance measured from the centre of the tube, of radius r. Then the force driving the fluid through the tube is simply the (negative) pressure gradient times the cross-sectional area of the tube πr^2, and the opposing force at any distance y from the centre is the viscous drag on a surface of area $2\pi y$. Equating the two forces leads directly to the equation

$$\frac{du}{dy} = -\frac{1}{2\mu}\frac{dp}{dx}y$$

Integrating this equation from $y = 0$ to $y = r$, and applying the boundary condition that $u = 0$ at $y = r$, leads to

$$u = \frac{1}{4\mu}\frac{dp}{dx}(r^2 - y^2) \tag{9.41}$$

The equation can be integrated again, over the cross-section of the tube, to give the total discharge Q, or the average velocity $\bar{u}$,

$$Q = \pi r^2 \bar{u} = \frac{\pi r^4}{8\mu}\frac{dp}{dx} \tag{9.42}$$

which is generally known as *Poiseuille's equation* after the French physicist, J. L. M. Poiseuille (1799–1869). The cgs unit of viscosity (the poise) was also named for him, and although most scientists do not use a named SI unit of viscosity, in France it is called the poiseuille.

Poiseuille's equation has been used to construct "capilliary" models of porous media (see Chapter 6) and even to derive Darcy's law. Real pores are interconnected and highly variable in geometry, so these models fail to reproduce some of the most important aspects of flow through rock pores.

9.10 Geological applications

9.10.1 Feeder dykes for flood basalts

The pressure gradient available to force basalt magma into a vertical dyke must arise from the difference in density between the magma (about 2650 kg m^{-3} at 1200 °C) and the overlying rock column. Suppose this difference ($\delta\rho$) is about 300 kg m^{-3}, then the vertical pressure gradient is $\delta\rho g$ or about -3000 Pa m^{-1}.

The viscosity of basalt magma is about 100 Pa s (N m^{-2}s). Let the average rate of flow per unit breadth of dyke (i.e., for unit distance in the z-direction) be q. Then $q = UL$, which may be found by integrating the velocity given by Equation (9.38) between the limits $y = 0$ and $y = L$, where L is the width of the dyke in metres:

$$q = \int_{y=0}^{y=L} u\, dy = \frac{dp/dx}{2\mu} \left[\frac{y^3}{3} - \frac{L}{2} y^2 \right]_0^L$$

$$= -\frac{dp/dx}{2\mu} \left(\frac{L^3}{6} \right)$$

Using the values indicated above

$$q = \frac{3000}{12 \times 100} L^3 = 2.5 L^3 \, \mathrm{m^2\, s^{-1}}$$

This solution is based on the Navier–Stokes equation. Therefore, the solution applies to viscous laminar flow, but not to turbulent flow. As we shall see in Section 11.2, the flow will become turbulent when the Reynolds number exceeds some critical value, which for channels is of the order of 1000. Hence, our solution is good for values of q within the range

$$Re = \frac{\rho UL}{\mu} = \frac{\rho q}{\mu} < 1000$$

or

$$q < \frac{1000\mu}{\rho} = \frac{1000 \times 100}{2650} = 38$$

If we substitute $q = 2.5 L^3$ into this inequality, we get $L < 2.5$ m as an approximate limit for laminar flow in the feeder dyke.

Shaw and Swanson (1970) and Swanson *et al.* (1975) have applied calculations of this type to linear vent systems in the Columbia River basalts. Estimated rates of eruption were 1 km^3 per day per kilometre of vent, or about 12 m^{3}s^{-1} per metre. So the flow would have been laminar, and vents much wider than about 1.7 m would not have been required.

9.10.2 *Flow of lava*

It is hard to measure the viscosity of newly erupted lava directly. One way to estimate it is from field observations of the surface speed of lava of known thickness, flowing down known slopes.

The maximum velocity at the surface of the flow is given by Equation (9.40), with y equal to the flow depth d:

$$u_{\max} = \frac{\rho g \sin \alpha}{2\mu} d^2 \qquad (9.43)$$

$u_{\max}$, α, and d may be measured in the field, so the viscosity may be calculated. A similar technique may also be used to calculate the viscosity of glacier ice.

In both these examples, we assume that ice and basalt lava are Newtonian fluids: in reality, there are significant differences between these materials and Newtonian fluids, so the viscosity estimates are only approximate (see Chapter 10).

9.10.3 Plate motions

The evidence from isostasy (Section 5.5) and plate tectonics indicates that at least part of the mantle can flow like a fluid. Thus, although the evidence from seismology indicates that the whole mantle is solid in the sense that it can respond to short-period stresses as an elastic solid, we must suppose that it can flow in response to long-period stresses in the way that glacier ice can flow. As a first approximation we can treat the soft, upper part of the mantle (the asthenosphere) as a Newtonian fluid with a constant viscosity. This model is almost certainly incorrect in detail, but it served as a basis of much of the early theoretical work on plate tectonics (e.g., McKenzie, 1969). Even today, many theoretical papers treat the mantle as a Newtonian fluid, though with viscosity that may vary with depth (e.g., see reviews in Peltier, 1989).

The appropriate value for the viscosity of the mantle can be determined from studies of isostatic rebound resulting from the draining of large pluvial lakes or the melting of ice caps (Subsection 9.10.4). Estimates vary somewhat but are generally in the range 10^{19}–10^{20} Pa s (1 Pa s = 10 poises). These viscosities are about the same as those of cold glass, under atmospheric pressure.

We can use the theory of flow between two plates to make some calculations, based on some simple plate tectonic models. The models are not very realistic, but the calculations are nevertheless informative.

One real question is as follows: is flow of the mantle driven by movement of the plates or vice versa? Let us make some calculations:

(1) **Model 1. Plate driven by flow in asthenosphere.** We assume the necessary pressure gradient is provided by a small slope on the plate (Figure 9.10).

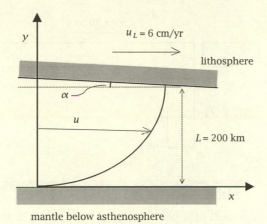

Fig. 9.10. Model of a plate driven by flow in an asthenosphere.

If the slope is small enough, the vertical components of velocity can be neglected and we can use the parallel plate model. If there is no other force acting in the plate, the shear stress on its base must be zero: if this were not so, the plate would accelerate. It follows that

$$\left(\frac{du}{dy}\right)_{y=L} = 0$$

If we differentiate Equation (9.36), and set $du/dy = 0$ at $y = L$ we obtain

$$0 = \frac{dp}{dx}\left(\frac{L}{\mu} - \frac{L}{2\mu}\right) + \frac{u_L}{L}$$

Solving for dp/dx using $u_L = 2 \times 10^{-9}$ m s^{-1} (6 cm yr^{-1}), $L = 2 \times 10^5$ m, and $\mu = 10^{20}$ Pa s, gives $-dp/dx = 10$ Pa m^{-1}. We now use Equation (9.32) to determine the pressure gradient. In this case $p_L = 0$ but L varies with x, so

$$\frac{dp}{dx} = \rho g \tan \alpha$$

Solving for the slope:

$$\tan \alpha = \frac{dp/dx}{\rho g} = 3 \times 10^{-4}$$

that is, there should be a change in elevation of about 0.6 km across a plate 2000 km wide.

Is there any support for this? Kaula (1972) reports satellite observations indicating positive isostatic gravity anomalies of about 20 milligals at spreading ridges (one milligal is 10^{-5} m s^{-2}). A positive anomaly

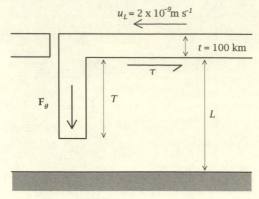

Fig. 9.11. Model of a plate pulled at one end by a sinking slab.

means there is more mass under the ridge than expected, indicating a higher elevation of the asthenosphere (perhaps held up by rising thermal convection currents?).

How much higher? Elder (1976, p. 18) gives an equation for the increment in gravity δg produced by a slab of rock of thickness δL and density ρ:

$$\delta g = 2\pi G \rho \delta L$$

where $G = 6.67 \times 10^{-11}$ N m^2 kg^{-2} is the universal gravitational constant. For $\rho = 3000$ kg m^{-3}, $\delta L = 600$ m, $\delta g = 6 \times 10^{-4}$ m s$^{-2} = 60$ milligals.

On our highly over-simplified model we would expect an isostatic anomaly three times larger than observed. So the model seems unlikely to be correct.

(2) **Model 2. A passive plate, pulled from one end.** The question here is: what pull is necessary to overcome the viscous drag of the mantle on the base of the plate?

We assume no pressure gradient in the mantle, so the velocity gradient is linear and $\tau = \mu u_L / L = 10^6$ Pa. This is the force exerted by the mantle on unit area of the plate, so for a strip of plate 1 m wide and 2000 km (2×10^6 m) long, the force needed to pull the strip is $F = 2 \times 10^{12}$ N.

We next inquire whether a force of this magnitude could be produced by the sinking of a dense slab of lithosphere. In Figure 9.11 we show the simplest possible model. We assume the plate sinks vertically at a subduction zone, and pulls the rest of the plate along as it sinks (in reality it sinks at an angle that varies from about 20 to 50 degrees.) Suppose the sinking slab is of length T, and that it has the same composition as,

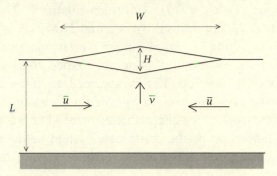

Fig. 9.12. Model for isostatic rebound.

but is on the average 100 °C colder than, the surrounding asthenosphere. If the coefficient of thermal expansion is that of olivine (4×10^{-5} °C^{-1}), and the density ρ is 3400 kg m^{-3}, the difference in density would be $\delta\rho = 4 \times 10^{-5} \times 100 \times \rho \approx 14$ kg m^{-3}. The gravity force acting on the sinking slab, per 1 m width, is related to its length T by $F_g \approx Tt \times 14g$. Equating F and F_g to find the length of slab necessary gives a length of about 100 km. We know that sinking slabs extend to depths of the order of 600 km beneath subduction zones (from the evidence of deep focus earthquakes). McKenzie (1969) gave a similar, but more sophisticated calculation of the gravity force, which takes account of the fact that the slab warms up as it sinks. The net conclusion is the same, however, namely that the pull exerted by a sinking slab may well be an important driving force in plate motions.

In this calculation, we have neglected several other forces that might oppose the motion of plates. The most significant seems to be the viscous drag that the mantle exerts on the sinking slab itself. Modern opinion favours subduction pull, as described in Model 2, as the main, but not the only, force driving plate motion. Its importance is confirmed by the observation that the fastest-moving plates are those that have a subducting boundary. For further discussion see Richter (1978), Forsyth and Uyeda (1975: summarized in Uyeda, 1978, pp. 197–201), or Cox and Hart (1986).

9.10.4 *Viscosity of the mantle*

The viscosity of the mantle cannot be measured directly, so estimates must be made from its response to surface forces such as the load of an ice sheet. One way to approach the problem of mantle viscosity is by using the theory

of flow on a slope (Elder, 1976). Consider Figure 9.12. An ice sheet of
maximum thickness H is removed by melting. How fast will the surface of
the earth rise after melting? In what follows we assume a very simple model:
the ice sheet is two-dimensional, of diamond-shaped cross-section and width
W. When it is fully developed it is assumed to have reached isostatic
equilibrium, i.e., the pressure exerted by this thickness of ice is balanced by
a buoyant pressure exerted by the asthenosphere. The ice pressure is given
by ρgh. Setting the origin of the coordinate system below the centre of the
ice sheet, the thickness h at a horizontal distance x from the origin is given
by

$$h = H \left(1 - \frac{x}{W/2} \right)$$

So the pressure gradient across the ice sheet is equal to

$$\frac{\partial p}{\partial x} = -\frac{\rho g H}{W/2}$$

When the ice is removed by melting, this same pressure gradient causes
the asthenosphere to flow up and inwards from the sides as shown in the
figure. We ignore the elasticity of the lithosphere between the ice and the
asthenosphere, and assume that the asthenosphere has a thickness L, and is
underlain by a mantle of much higher viscosity.

Let the rate of vertical movement be $\bar{v}$, i.e., v averaged over W. Let
the rate of horizontal flow of the asthenosphere be $\bar{u}$, i.e., u averaged over
L. From continuity, $W\bar{v} = 2L\bar{u}$ and so $\bar{v} = (2L/W)\bar{u}$. But $\bar{u}$ is simply
the average velocity between two (nearly) parallel plates, so we can use the
theory of Couette flow and write

$$\bar{u} = \frac{1}{L} \int_0^L u\,dy$$

Using Equation (9.38) for u and the expression for the pressure gradient
derived above, we obtain the average horizontal and vertical velocities:

$$\bar{u} = \frac{\rho g H L^2}{6\mu W}$$

$$\bar{v} = \frac{\rho g H L^3}{3\mu W^2}$$

The rate of isostatic rebound $\bar{v}$ can be estimated, for example, by dating
raised beaches; H and W can be estimated from geological evidence, and
by comparison with modern ice sheets. L is estimated from seismological
data. Thus we can determine the mantle viscosity. Setting $\bar{v} = 3 \times 10^{-9}$ m

s^{-1}, $H = 5$ km, $L = 200$ km, $W = 1000$ km, gives $\mu \approx 4 \times 10^{19}$ Pa s. In actual calculations, much more sophisticated models are used. For details see Officer (1974), Cathles (1975), Ranalli (1987), Lliboutry (1987), and Peltier (1989). These confirm simple estimates based on the theory given, or on dimensional analysis, and indicate that the asthenosphere must have viscosities of the order of 10^{19}–10^{20} Pa s.

In the simple plate models discussed in this section we have assumed, in order to simplify the calculations, that there is a low-viscosity asthenosphere, 200 km thick, overlying a much more viscous part of the mantle. This is almost certainly wrong: most recent studies suggest that there is, at most, about one order of magnitude difference in viscosity between the asthenosphere and the underlying mantle. But there is still debate about the thickness of the asthenosphere (estimates vary from 75 to 500 km) and whether or not there is any rapid change in viscosity from the asthenosphere to the lower mantle (e.g., Fjeldskaar and Cathles, 1991; Sabadini *et al.*, 1991).

9.11 Viscous flow past a sphere or cylinder

In Chapter 2, we discussed the drag force on a sphere and stated that a linear relation between the drag force and relative velocity existed for very slow flow around a small sphere. In Chapter 3, Subsection 3.3.3, we reached a similar conclusion based on a dimensional analysis and presented an empirical curve relating the dimensionless drag force (the drag coefficient) to the Reynolds number. For the particular case of slow "creeping" flow past a sphere, the Navier–Stokes equation may be solved exactly. This solution was originally obtained by Stokes.

The equation is simplified by setting the inertia terms (the left-hand side of the equation) equal to zero, ignoring body force terms, and making use of the spherical symmetry of the boundary conditions. The solution obtained gives the complete velocity and pressure field around the sphere, and it can therefore be used to calculate the total drag force acting on the sphere (*Stokes' law*):

$$F_D = 3\pi\mu Du \tag{9.44}$$

where D is the diameter of the sphere and u is the velocity of the fluid far upstream (i.e., the relative velocity – the equation applies equally well to the case where a sphere is moving at a velocity u through a fluid at rest). A similar solution, obtained for slow viscous flow around a cylinder, is shown by the broken lines in Figure 9.5. We might be tempted to conclude from the symmetry of the streamlines that the drag is entirely due to viscous shear

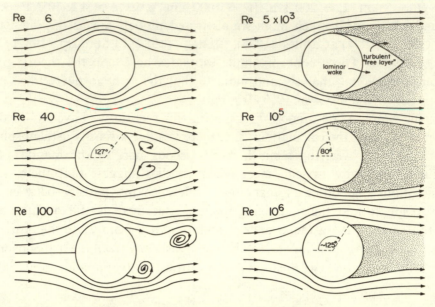

Fig. 9.13. Changing pattern of flow past a circular cylinder normal to the flow as *Re* increases (from Middleton and Southard, 1984).

stress, but this is not the case. The theory developed by Stokes tells us that one-third of the drag force is due to the pressure difference between the front and back of the sphere, and two-thirds is due to the viscous drag on the surface.

Stokes' law is valid only for low Reynolds numbers ($Re < 0.5$). But low Reynolds numbers can be produced either by small sizes or large viscosities. So in water, the equation applies only to the settling of very small particles ($D < 0.1$ mm), but in the asthenosphere it may be applied to slabs of cold lithosphere a hundred kilometres across. In both cases, drag estimates are only approximations because the solid bodies are not spherical.

Both the slow-viscous-flow solution and the inviscid-flow solution show a perfectly symmetrical pattern of streamlines around the sphere (Figure 9.5): neither solution predicts flow separation and wake formation, which are observed in all experiments at higher Reynolds numbers ($Re > 50$, Figure 9.13). But the inviscid-flow solutions do describe the flow well outside the boundary layer and wake region at high Reynolds numbers, so they are useful (for example, for computing the pressure force on the front of the cylinder).

There have been many careful experimental studies of flow around a sphere or a cylinder, generally using wind tunnels. Many of our ideas about flow instability and separation and wake development derive from these studies, but we do not have space to describe them here. We return to the subject briefly in Chapter 11. Details of the flow patterns shown in Figure 9.13 can be related directly to the experimentally measured drag coefficients (see Figure 3.2). A more complete description is given by Middleton and Southard (1984) and Tritton (1988).

9.12 Review problems

(1) **Problem:** Demonstrate that if a two-dimensional flow is irrotational and incompressible, then the sum of the viscous forces acting on any fluid element is equal to zero.

Answer: If the flow is irrotational,

$$\frac{\partial u}{\partial y} = \frac{\partial v}{\partial x}$$

so, differentiating again, with respect to (wrt) x and y:

$$\frac{\partial^2 u}{\partial x \partial y} = \frac{\partial^2 v}{\partial x^2} \quad \text{and} \quad \frac{\partial^2 u}{\partial y^2} = \frac{\partial^2 v}{\partial x \partial y}$$

Now continuity, for an incompressible fluid states that

$$\frac{\partial u}{\partial x} + \frac{\partial v}{\partial y} = 0$$

Differentiating again wrt to x gives

$$\frac{\partial^2 u}{\partial x^2} + \frac{\partial^2 v}{\partial x \partial y} = 0 \quad \text{so} \quad \frac{\partial^2 u}{\partial x^2} + \frac{\partial^2 u}{\partial y^2} = 0$$

And similarly, differentiating wrt y gives

$$\frac{\partial^2 v}{\partial x^2} + \frac{\partial^2 v}{\partial y^2} = 0$$

But these two expressions (multiplied by μ) are the viscosity terms in the two-dimensional Navier–Stokes equation. So the viscous forces are equal to zero.

It is possible to generalize this result to the three-dimensional case. We make use of the vector form of the Navier–Stokes equation, in which the viscosity term appears as $\mu \nabla^2 \mathbf{v}$. This is expanded using the vector identity

$$\nabla^2 \mathbf{v} = \nabla(\nabla \cdot \mathbf{v}) - \nabla \times (\nabla \times \mathbf{v})$$

For an incompressible fluid $\nabla \cdot \mathbf{v}$ is zero and if the vorticity is zero $\nabla \times \mathbf{v}$ is zero, so both of the viscosity terms are equal to zero.

(2) **Problem:** (after Turcotte and Schubert, 1982, p. 237) Assume a plate moving at a velocity of 70 mm yr^{-1}, and an asthenosphere with a thickness of 200 km, a viscosity of $\mu = 10^{19}$ Pa s, a base having zero velocity, and zero pressure gradient in the direction of movement.

(a) What is the shear stress on the base of the lithosphere?

(b) If this shear stress (0.111 MPa) acts on 6000 km of lithosphere with a thickness of 100 km, what tensional stress must be applied at a trench to overcome the basal drag?

Answer:

(a) 70 mm yr^{-1} is 0.07 m per 3.15 $\times 10^7$ seconds, or 2.222 $\times 10^{-9}$ m s^{-1}. There is no pressure gradient so the velocity gradient is linear and the shear stress is given by

$$\tau = \mu(du/dy)$$
$$= \mu(u/L)$$
$$= \frac{10^{19} \times 2.222 \times 10^{-9}}{2 \times 10^5}$$
$$= 0.111 \text{ MPa}$$

(b) The total force per unit width is 0.111 $\times 6 \times 10^6$ N, so the tension (a force per unit area) acting on a section of lithosphere 1 m wide and 100 km thick is

$$\sigma = \frac{0.111 \times 6 \times 10^6}{10^5} = 6.66 \text{ MPa}$$

(3) **Problem:** In Section 2.7 we demonstrated that we can use the maximum distance that large bombs are thrown out of a volcano to estimate the speed of ejection v_0. How can we use this speed to estimate the pressure in the magma chamber, and the total energy of the eruption?

Answer: We assume that the gaseous and solid material move at almost the same speed, have a bulk density ρ, and are propelled up the volcanic vent by the difference in pressure between the magma chamber p and the pressure at the mouth of the vent. The latter is close to atmospheric, thus essentially zero. The magma chamber is $-z$ m below the vent, which we use as the datum. The chamber is large, so the speed of moving magma, gases, etc. within it is close to zero. Applying Bernoulli's equation, we find that

$$p - \rho g z = \rho v_0^2/2$$

The ejection speed is very large, of the order of 100 m s^{-1}. We can neglect $\rho g z$ if the depth is very shallow ($p \approx \rho v_0^2/2$). It is, of course, hard to estimate ρ, but it may be assumed to be about half the density of the volcanic material.

The total energy of the eruption can be estimated if the total mass m_t of ejecta can be estimated. Then, since it must all have moved up the vent at about the same speed v_0, the total kinetic energy E_t must have been about $m_t v_0^2/2$.

Though this method has been used to estimate magma pressure in explosive eruptions, it is deeply flawed, because the velocities developed in volcanic vents are so large that they exceed the velocity of sound (i.e., the Mach number is greater than unity; see Chapter 8, Problem 5). Under these conditions, a theory of incompressible flow is no longer adequate. For a review of the application of compressible flow theory to this and similar problems see Wilson (1980) and Kieffer (1989).

9.13 Suggested reading

9.13.1 On fluid mechanics:

Unfortunately, there is no recent authoritative book on the history of fluid mechanics. The best source is probably the following:

Rouse, H. and S. Ince, 1957. *History of Hydraulics*. State University of Iowa, Iowa Institute of Hydraulic Research, 269 pp. (TC15.R86)

There are many texts on fluid mechanics, mostly written for engineering students. Tritton's book is the best of those that take a physical approach. Vogel's book provides a very readable, nonmathematical introduction, and Van Dyke has anthologized some of the finest photographs ever taken of flow phenomena.

Acheson, D. J., 1990. *Elementary Fluid Dynamics*. Oxford, Clarendon Press, 397 pp. (A clear exposition, with a new approach. More material on waves and vortex motion than found in most introductory texts. TA357.A276)

Aris, R., 1962. *Vectors, Tensors, and the Basic Equations of Fluid Mechanics*. New York, Dover, 287 pp. (A detailed discussion of the basic stress and rate of strain tensors, and the constitutive equations relating them to each other. QA911.A69)

Batchelor, G. K., 1970. *An Introduction to Fluid Dynamics*. Cambridge University Press, 615 pp. (A leading text, at a fairly advanced level. QA911.B33)

Daily, J. W. and D. R. F. Harleman, 1966. *Fluid Dynamics*. Reading MA, Addison-Wesley, 454 pp. (One of the best engineering texts: good discussion of Navier–Stokes equation. QC151.D22)

Hughes, W. F., 1979. *An Introduction to Viscous Flow*. New York, McGraw-Hill, 219 pp. (TA518.2.H83)

Lighthill, M. J., 1986. *An Informal Introduction to Theoretical Fluid Mechanics*. Oxford, Clarendon Press, 260 pp. (Simple yet elegant exposition. QA911.L46)

Olfe, D. B., 1987. *Fluid Mechanics Programs for the IBM PC.* New York, McGraw-Hill, 174 pp. + diskette. (A collection of computer programs to demonstrate phenomena in fluid mechanics: source code in BASIC. QA901.O49)

Tritton, D. J., 1988. *Physical Fluid Dynamics.* Oxford, Clarendon Press, second edition, 519 pp. (Written by a geophysicist, so covers all the interesting geophysical topics. Moderately advanced. QC151.T74)

Truesdell, C., 1966. *Six Lectures on Modern Natural Philosophy.* New York, Springer-Verlag, 117 pp. (These lectures are perhaps the most accessible introduction to the modern reformulation of continuum mechanics. QA802.T75)

Van Dyke, M. D., 1982. *An Album of Fluid Motion.* Stanford CA, The Parabolic Press, 176 pp. (A work of beauty as well as science. TA357.V35)

Vogel, S., 1981. *Life in Moving Fluids. The Physical Biology of Flow.* Princeton NJ, Princeton University Press, 352 pp. (A nonmathematical introduction to fluid mechanics with applications to aquatic organisms. QH505.V63)

White, F. M., 1986. *Fluid Mechanics.* New York, McGraw-Hill, 732 pp. (Good engineering text, with many problems. TA357.W48)

Whitaker, S., 1968. *Introduction to Fluid Mechanics.* Englewood Cliffs NJ, Prentice-Hall, 457 pp. (An engineering text, but introduces tensors and covers the Navier–Stokes equation more completely than most such texts. QA911.W38)

9.13.2 Geological applications

Cox, A. and R. B. Hart, 1986. *Plate Tectonics, How It Works.* Oxford, Blackwell, 392 pp. (An introduction to plate tectonics, with a chapter on mechanisms. QE511.4)

Elder, J., 1976. *The Bowels of the Earth.* Oxford University Press, 222 pp. (Perhaps somewhat cryptic, but stimulating reading. QE501.2E4)

Fjeldskaar, W. and L. Cathles, 1991. The present rate of uplift of Fennoscandia implies a low-viscosity asthenosphere. Terra Nova, v.3, pp. 393–400.

Forsyth, D. and S. Uyeda, 1975. On the Relative Importance of the Driving Forces of Plate Motion. Geophys. Jour. Roy. Astronomical Soc., v.43, pp. 163–200.

Lliboutry, L. A., 1987. *Very Slow Flows of Solids: Basics of Modeling in Geodynamics and Glaciology.* Boston, Martinus Nijhoff, 510 pp. (Advanced, but assumes little prior knowledge. QE501.3.L57)

Middleton, G. V. and J. B. Southard, 1984. *Mechanics of Sediment Movement.* Society of Economic Paleontologists and Mineralogists Short Course Notes 3, second edition, 401 pp. (Introduces basic fluid and sediment mechanics using very little mathematics. QE471.2.M53)

Ranalli, G., 1987. *Rheology of the Earth. Deformation and Flow Processes in Geophysics and Geodynamics.* Boston, Allen and Unwin, 366 pp. (Derives basic fluid equations and applies them to the mantle. QE501.3)

Turcotte, D. L. and G. Schubert, 1982. *Geodynamics: Applications of Continuum Physics to Geological Problems.* New York, John Wiley and Sons, 450 pp. (Excellent discussion of simple models of plate tectonics. QE501.T83)

Uyeda, S., 1978. *The New View of the Earth: Moving Continents and Moving Oceans.* San Francisco, W. H. Freeman, 217 pp. (A simple introduction to plate tectonics. QE511.5. U9313)

10

Flow of natural materials

10.1 Introduction

In Chapter 9 we referred to three factors which complicate the application of the theory of viscous fluids to real problems:

(1) the difficulty of solving the equations;
(2) differences between the Newtonian flow law and the flow laws actually followed by some natural materials;
(3) the development of flow instability (turbulence) at the high Reynolds numbers characteristic of many natural flows.

We consider the second factor in this chapter, and the third factor in Chapter 11.

The main geological materials to be considered are:

(1) **Ice:** this can be considered to be a very simple crystalline rock composed of a single mineral (belonging to a crystal class, hexagonal, of high symmetry). Ice at the earth's surface is close to its melting point, just as rocks are in some parts of the lower crust, or in the asthenosphere. The flow of ice, therefore, may be useful as a readily observable analogue of the processes acting in more deep-seated, inaccessible regions of the earth.
(2) **Mud**, or debris–water mixtures: water closely approximates a Newtonian fluid, but the addition of large amounts of sediment changes its mechanical properties considerably.
(3) **Lava:** studies of the morphology of lava flows have shown that lava, too, differs significantly from a Newtonian fluid. Presumably the same is also true of magma, though there are differences between the two (e.g., presence of bubbles and small crystals in lavas).

337

(4) **Rocks**, under conditions of high temperature and pressure: structures
     observed in high-grade metamorphic rocks clearly indicate that some
     have undergone extensive plastic deformation or flow. Isostasy and plate
     tectonics require flow of the asthenosphere, which is certainly not a simple
     fluid. Although we can often model the behaviour of these materials to a
     close enough approximation by assuming that they behave as Newtonian
     fluids with a high viscosity, this is certainly not the whole truth, any more
     than it is for ice.

Thus we often have to deal with materials that can flow like fluids, but
differ from simple Newtonian fluids in certain respects. To understand
these materials better we need to understand something about relationships
between stress and strain (or rate of strain), which are somewhat more
complex than those shown by simple elastic solids or viscous fluids. The
study of such relationships is called *rheology*.

If we bear in mind the saying of Heraclitus "$\pi\alpha\nu\tau\alpha\ \rho\epsilon\iota$" (panta rhei –
everything flows), a saying that is particularly true if we are considering long
periods of geological time, then we realize how important rheology is to the
earth sciences.

10.2 Mechanical models of natural materials

Real materials are infinite in their complexity. Science simplifies nature
by considering, as a start, only certain ideal materials, whose properties
can be described by simple stress–strain, or stress–rate-of-strain relations.
Many natural materials show combinations of properties represented by
the basic mechanical models considered so far: elastic, viscous, and plastic.
For example, many rocks show both strength (as in a brittle solid) and
viscosity (as in a fluid). If this is true then an approximation to the real
material behaviour may be obtained by combining (in series or in parallel)
the properties of two or more ideal materials. Such combinations are often
represented diagrammatically using the simple physical models that represent
the basic mechanical properties: an ideal elastic material is represented by a
spring, and an ideal viscous material by a dashpot (i.e., a simple hydraulic
shock absorber), and an ideal rigid solid (with a finite strength) by a block
that will not slide over a surface until some critical stress is exceeded. We
will consider the following examples (see Figure 10.1):

Ideal plastic (Saint Venant model). A material with a finite strength and
     zero viscosity. It is a combination of a rigid solid and an ideal (inviscid)
     fluid (Figure 10.1(*a*));

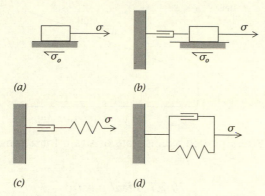

Fig. 10.1. Mechanical models of materials with complex rheology: (a) ideal plastic (σ_0 = strength); (b) Bingham plastic; (c) Maxwell body; (d) Kelvin body.

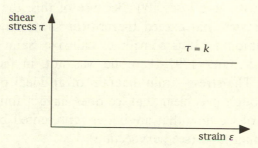

Fig. 10.2. Stress–strain diagram for an ideal plastic.

Bingham plastic. A material with a strength and viscosity that are both finite. It is a combination of a rigid solid and a Newtonian fluid (Figure 10.1(b));

Maxwell body. A material with both elasticity and viscosity, therefore a combination of an elastic solid and a Newtonian fluid (Figure 10.1(c));

Kelvin body. Like the Maxwell body, this combines elasticity and viscosity, but in parallel instead of in series (Figure 10.1(d));

Non-Newtonian fluid. A fluid whose rate of shear is not a linear function of the applied shear stress.

Many other models are possible, but these are the main ones that have been applied with some success in the earth sciences. In the discussion that follows, we present constitutive equations for most of the models. It should be emphasized that these are one-dimensional equations, comparable to Hooke's law of elasticity or Newton's law of viscosity. No attempt is made in this chapter to explore the full, three-dimensional constitutive equations of these complex materials.

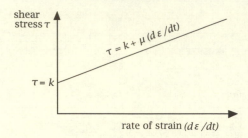

Fig. 10.3. Stress plotted against rate of strain for a Bingham plastic.

10.2.1 Ideal plastic

An ideal plastic is a material which will not deform at all until a critical shear stress is exceded. At higher shear stresses, it behaves as an ideal fluid (i.e., a fluid with zero viscosity). Because of this, the material will not transmit shear stresses that exceed the critical shear stress by more than an infinitesimal amount. It is sometimes called a Saint Venant material, after B. de Saint-Venant (1797–1886: he was not, in fact, a saint, but a French engineer). The stress–strain diagram of an ideal plastic is shown in Figure 10.2. Although it is clear that ice does have a finite viscosity, there are some aspects of ice flow that are better represented by an ideal plastic than by a Newtonian fluid (see next section).

A mechanical model for an ideal plastic is a weight resting on a rough surface: the weight will not move until friction is overcome, but after motion begins, it continues (at a rate not determined by friction) so long as the stress needed to overcome friction is exceeded.

10.2.2 Bingham plastic

This material may be considered to be a combination, in series, of an ideal plastic and a Newtonian fluid (Figure 10.1(b)). It does not deform at all until a critical shear stress is exceeded, but after that it deforms like a Newtonian fluid. A typical stress–rate-of-strain diagram is shown in Figure 10.3. The constitutive relation for a Bingham plastic is

$$\tau = k + \mu \frac{d\varepsilon}{dt} \tag{10.1}$$

Mud and lava show many flow phenomena that can best be explained using the Bingham plastic model. The model was originally developed in 1919 for paint by E. C. Bingham (1878–1945), an American engineer who is generally regarded as one of the founders of modern rheology. Paint has the interesting

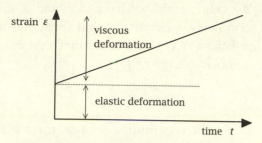

Fig. 10.4. Strain–time diagram for a Maxwell body when the applied stress is held constant.

property that it can be brushed onto a wall, like a viscous fluid, but that it does not run off the wall (provided that the paint film is thin enough) in the way that a true viscous fluid would. This is not because of surface tension but because the weight of the paint layer is not large enough to exceed its shear strength.

Note that, according to the models, neither an ideal plastic nor a Bingham plastic shows any elastic properties (we also ignored these properties in developing the Navier–Stokes equation for a viscous fluid). Real materials always do have some elastic properties, but these have a negligible effect on many of the phenomena exhibited by these materials.

10.2.3 Maxwell body

This is defined as a substance that shows both elastic ("solid") and viscous ("fluid") properties. It is named for James Clerk Maxwell (1831–79), one of the great physicists of the nineteenth century. It can be modeled by a spring and a dashpot (shock absorber) in series (Figure 10.1(c)).

Because one of these ideal materials is defined in terms of strain and the other in terms of strain rate, the constitutive equation for a Maxwell body is slightly more involved, but not complicated. The total strain ε observed in the specimen is the sum of elastic (solid) and viscous (fluid) deformations, $\varepsilon = \varepsilon_s + \varepsilon_f$. The rate of strain is the time derivative of ε:

$$\frac{d\varepsilon}{dt} = \frac{d\varepsilon_s}{dt} + \frac{d\varepsilon_f}{dt}$$

The elastic term for a uniaxial deviatoric stress σ is given by the time derivative of the constitutive relation for an elastic solid (in one dimension),

$$\frac{d\varepsilon_s}{dt} = \frac{1}{E}\frac{d\sigma}{dt}$$

where E is Young's modulus and we have assumed it does not vary with time. The viscous term is given exactly by that for a Newtonian fluid, (refer back to the equation before (9.14), and remember that ε is here a strain, not a rate of strain as it was in Chapter 9):

$$\frac{d\varepsilon_f}{dt} = \frac{\sigma}{2\mu}$$

Combining these, we get the constitutive relation for a Maxwell body:

$$\frac{d\varepsilon}{dt} = \frac{1}{E}\frac{d\sigma}{dt} + \frac{\sigma}{2\mu} \tag{10.2}$$

When an initial stress is applied to a Maxwell body, the time derivatives in Equation (10.2) dominate, and the immediate response is elastic. In the physical model, deformation consists entirely of extension of the spring. After applying the initial stress, two extreme conditions can be distinguished. In one, the applied stress is held constant, (so $d\sigma/dt$ in Equation (10.2) is zero), and the material behaves as a viscous fluid. This is the case shown in Figure 10.4. In the other extreme, the strain is constant (i.e., $d\varepsilon/dt = 0$), no further deformation is allowed to take place, and Equation (10.2) then gives the rate at which the applied stress is gradually relaxed as slow viscous flow replaces elastic deformation. In the simple physical model, the flow of fluid in the dashpot permits contraction of the spring, and therefore release of the tension originally applied.

A geological example of this type of behaviour is isostatic rebound. Removal of an ice sheet produces a rapid but small uplift, due to elastic expansion of the underlying crust, followed by a period of much slower but more extensive uplift, due to flow in the mantle. Smaller but similar effects are produced by the draining of large lakes, or (in reverse) by the filling of new reservoirs. In this case, we are not strictly dealing with a Maxwell body, because the elastic deformation takes place mainly in the crust and the flow mainly in the mantle. But the overall effect is similar. Many rocks, however, are capable of responding to stress by both elastic and viscous deformation.

10.2.4 Kelvin body

The Maxwell body combines an elastic body and a viscous fluid in series. It is also possible to combine these bodies in parallel (Figure 10.1(d)). The result is called a *Kelvin body*. Application of a stress to such a body produces no immediate strain, but the body gradually strains over time. The strain is gradual because the rate of strain is limited to that which occurs in the viscous fluid. In the physical model, the extension of the spring is limited

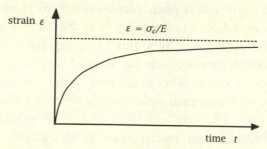

Fig. 10.5. Strain–time diagram for a Kelvin body when the applied stress is held constant.

by the rate at which the piston moves in the dashpot. Because the devices are in parallel, the applied stress is divided among the two materials and the constitutive relation is given by

$$\sigma = E\varepsilon + 2\mu\frac{d\varepsilon}{dt} \tag{10.3}$$

For a constant applied stress, the rate of strain in a Kelvin body decreases with time ($d^2\varepsilon/dt^2 < 0$, Figure 10.5); the proportion of the stress carried by the spring increases until all the stress acts on the spring, and the final strain is given by Hooke's law $\varepsilon = \sigma/E$.

Clays with a moderate water content behave over short time periods like Kelvin bodies. The mineral framework provides elasticity, but the clay reacts slowly to loads because of the low permeability, which limits the rate of water loss and therefore the rate of compaction (see Section 6.15). Only as the water escapes can the mineral framework be deformed elastically. If the load is removed at an early stage of compaction, the clay will slowly expand as water flows back into the pores. The elasticity of the mineral framework is not retained, however, over geologically long periods of time.

10.2.5 Non-Newtonian fluid

One obvious modification of the Newtonian fluid model is to suppose that the relation between stress and strain is nonlinear. The constitutive relation (for simple shear) can be written as

$$du/dy = a\tau^n \tag{10.4}$$

For a Newtonian fluid, $a = 1/\mu$ and $n = 1$. It is generally supposed that the strain is related to a power of the stress greater than one ($n > 1$). One way of looking at this is that the fluid has a viscosity which becomes lower as the rate of strain increases, a phenomenon observed in some real materials and called *shear thinning*. Another way is to dispense with the concept of viscosity and substitute a new modulus (*a* in Equation 10.4), which is the proportionality coefficient between the rate of strain and the appropriate power of the stress. Ice and other crystalline materials probably correspond more closely to non-Newtonian fluid models than to other models. It must be remembered that the "flow" of these crystalline materials is due to the operation of quite different mechanisms at the atomic level from those operating in liquids or gases.

Some non-Newtonian fluids show the opposite phenomenon: the viscosity increases as the rate of shear increases. Such behaviour is called *work hardening* or *dilatant* though the latter term has nothing to do with volume increase during shearing, which is the more common use of the term.

An example is provided by the shearing of a concentrated dispersion of sand. Bagnold (1956) showed that, at high rates of shearing and at concentrations higher than about 9 per cent by volume, the stress T was transmitted directly by collisions between grains, and was related to the rate of shearing du/dy, the size of the grains D, their density ρ_s, and their volume concentration. The form of the relationship can easily be determined from dimensional analysis (Chapter 3) and was confirmed experimentally by Bagnold:

$$T = C\rho_s D^2 \left(\frac{du}{dy}\right)^2 \tag{10.5}$$

where C is a numerical coefficient that depends on the grain concentration. This equation has the form of a constitutive equation for a non-Newtonian fluid: it predicts that there is always some strain, no matter how small the stress, and that the rate of strain is related to the square root of the stress. In other words, it is harder to shear the dispersion at high rates of shear than at low rates of shear. Bagnold's work has recently been elaborated by many other workers, who have tried to determine theoretically the full constitutive equations of a granular medium and to confirm their results experimentally or by computer simulation. Progress in this very difficult enterprise is summarized by Campbell (1990).

10.3 Flow of materials down slopes

To estimate the deformation of any natural material, we replace the surface stress terms in the equation of motion (Cauchy's first law, which is a statement of Newton's second law as applied to any continuum) by the constitutive relations for the material and solve for the deformation field. We will use this process throughout the rest of this chapter, but we consider only simple stress fields, to which our simplified constitutive equations can be applied. One of the simplest stress fields is simple shear (Chapter 7, Problem 1). It applies to steady uniform flow down a slope, and approximates many natural flows that take place in wide channels.

Let the depth of the flow be d (Figure 9.9). The shear stress acting on the plane at a distance y above the bed, and parallel to it, is due solely to the downslope component of the weight (per unit area) of the overlying material. The shear stress is given by

$$\tau = \gamma(d - y)\sin\alpha = \gamma_x(d - y) \tag{10.6}$$

where $\gamma\sin\alpha$ is the component of the unit gravity force acting downslope (the x-direction). To simplify the notation we will write $\gamma\sin\alpha$ in this section as γ_x. Equation (10.6) states that the stress variation with depth is linear in steady, uniform, gravity flow. At the base of the flow ($y = 0$), the stress is equal to $\gamma_x d$, and is commonly denoted τ_0. At the top of the flow (the free surface, $y = d$), the stress is equal to zero. This stress distribution is valid for the flow of any material.

We have limited ourselves to steady, uniform, two-dimensional planar flow. Though we assume steady flow, the equations can also be applied to unsteady flows provided that conditions change only slowly with time. Uniform conditions are closely approximated in the upper part of flows in many natural channels with irregular boundaries. Only near the bed or channel walls, or in strongly curved channels, does the stress distribution becomes substantially more complex.

In order to obtain a velocity distribution from Equation (10.6), we need a constitutive equation relating the shear (the rate of change of u in the y-direction) to stress. In this section, we use several different constitutive equations (collectively known as flow laws) to solve for the velocity distributions of different types of materials.

10.3.1 Newtonian fluid

In Chapter 9, we derived the velocity distribution for a Newtonian fluid by simplifying the full Navier–Stokes equation. Here we use a different approach. For this simple case, the constitutive equation simplifies to

$$\tau = \mu\frac{du}{dy}$$

and we can easily combine this with Equation (10.6) to yield

$$\gamma_x(d - y) = \mu\frac{du}{dy} \tag{10.7}$$

Integrating this, and using the no-slip condition ($u = 0$ at $y = 0$) to show that the constant of integration is zero, we obtain

$$u = \frac{\gamma_x}{2\mu}(2yd - y^2) \tag{10.8}$$

which is the same as Equation (9.40). In Chapter 9 we began with the general form of the equations of motion, derived using a very general form of the constitutive equations of a Newtonian fluid, and then simplified them. Here we have begun with the simplifications, which has enabled us to derive the stress distribution directly and combine it with a much simplified constitutive equation.

10.3.2 Non-Newtonian fluid

For a non-Newtonian fluid, the governing equation is a slight modification of Equation (10.7):

$$a[\gamma_x(d - y)]^n = \frac{du}{dy} \tag{10.9}$$

Integrating this equation is only a little more involved than integrating Equation (10.7). We make the further change of giving up the no-slip assumption. Instead we set $u = u_b$ at the bed. The result is

$$u = u_b + \frac{a}{n+1}(\gamma_x)^n[d^{n+1} + (d - y)^{n+1}] \tag{10.10}$$

It is convenient to write this equation in the form

$$\frac{u_{max} - u}{u_{max}} = \frac{\Gamma(d - y)^{n+1}}{u_b + \Gamma d^{n+1}} \tag{10.11}$$

where u_{max} is the maximum velocity (at $y = d$) and

$$\Gamma = \frac{a}{n+1}(\gamma_x)^n$$

For $u_b = 0$, Equation (10.11) simplifies to

$$\frac{u_{max} - u}{u_{max}} = \left(1 - \frac{y}{d}\right)^{n+1} \tag{10.12}$$

One reason for giving up the no-slip condition is that there is evidence that ice behaves as a non-Newtonian fluid that may nevertheless slip on its bed.

Note that, theoretically, this contradicts the basic constitutive equation for a non-Newtonian fluid (Equation 10.4) – but sometimes consistency is less important than usefulness.

As we have seen, a Bagnoldian "grain flow" is a type of non-Newtonian flow for which $n = 1/2$. In this case:

$$\frac{u_{max} - u}{u_{max}} = \left(1 - \frac{y}{d}\right)^{3/2} \tag{10.13}$$

10.3.3 Ideal and Bingham plastics

For an ideal plastic, all the flow takes place in an infinitesimally thin zone near the bed. The velocity distribution is uniform: the entire flow moves as a "rigid plug" and there is no internal deformation.

For a Bingham plastic, the equation of motion is

$$\gamma_x(d - y) = k + \mu\frac{du}{dy} \tag{10.14}$$

Using the no-slip condition, integration yields

$$u = \frac{\gamma_x}{\mu}\left(yd - \frac{y^2}{2}\right) - \frac{ky}{\mu}, \quad d - y \geq \frac{k}{\gamma_x} \tag{10.15}$$

The solution holds only in the lower part of the flow. Above the elevation given by $d - y = k/\gamma_x$, the shear stress falls below the yield strength k, and the material no longer deforms. This is, perhaps, clearer if we measure distance in the y-direction down from the free surface, which is accomplished by replacing y with $Y = d - y$. The result of this simple transformation is, after some rearranging,

$$u = \frac{\gamma_x}{2\mu}(d^2 - Y^2) - \frac{k(d - Y)}{\mu}, \quad Y \geq \frac{k}{\gamma_x} \tag{10.16}$$

Y is measured down from the free surface, so $u = 0$ at $Y = d$, the base of the flow. The profile holds only for the lower part of the flow ($\gamma_x Y \geq k$), where the shear stress exceeds the strength and viscous deformation can take place. Let the thickness of this zone be T. Above this there is a rigid plug in which there is no deformation. The thickness of the plug is k/γ_x, so the total thickness is $d = T + k/\gamma_x$. The velocity at the base of the plug is the maximum velocity u_{max} and is given by substituting $Y = k/\gamma_x$ into Equation (10.16). We eventually obtain

$$u_{max} = \frac{1}{\mu}\left(\frac{\gamma_x d^2}{2} + \frac{k^2}{2\gamma_x} - kd\right) \tag{10.17}$$

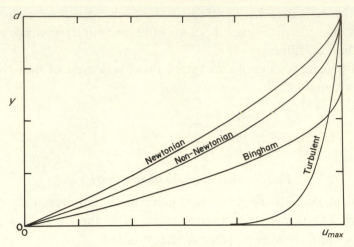

Fig. 10.6. Non-dimensional velocity distributions for two-dimensional flows of materials of differing rheology.

Substituting for d gives the simpler form

$$u_{\max} = \frac{\gamma_x T^2}{2\mu} \tag{10.18}$$

10.3.4 Summary

Examples of these velocity distributions, together with the velocity distribution for a turbulent flow (which will be derived in the next chapter) are shown in Figure 10.6. In all these case we have assumed that the "no-slip" condition holds at the lower boundary. Though this is known to be true for Newtonian and true non-Newtonian fluids, it is not always true for flows of granular debris or ice. Note the relatively small difference between the theoretical velocity distribution for a Newtonian fluid and a non-Newtonian grain flow. The difference would, of course, be larger for other non-Newtonian fluids, such as ice, which have a larger (or smaller) exponent in their constitutive equation.

10.4 Ice

Ice is a rock-like material that can be observed to flow in glaciers, and can also be readily studied in the laboratory. A great deal is now known about its behaviour. Glaciers have also caught the attention of some excellent theoreticians. Thus much is known about glacier flow. Useful short summaries

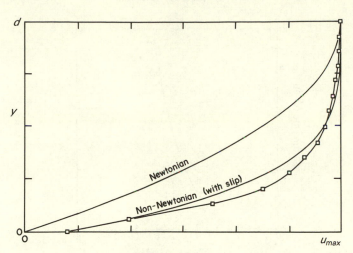

Fig. 10.7. Observed velocity profile in the Athabasca glacier: longitudinal section (after Paterson, 1981), compared with theoretical curves for a Newtonian and non-Newtonian fluid.

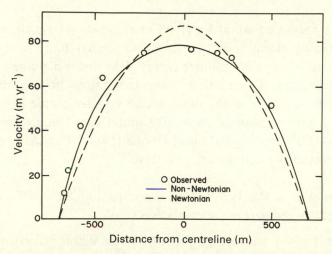

Fig. 10.8. Observed velocity distribution in the Saskatchewan glacier: surface observations. The theoretical distributions are also shown (after Paterson, 1981) for Newtonian flow (broken line) and non-Newtonian flow, with $n = 3$ (solid line).

are given by Kamb (1964) and Paterson (1981), and comprehensive treatises by Hutter (1983) and Lliboutry (1987).

The first hypothesis made about ice was that it flowed like a very viscous Newtonian fluid. The hypothesis was tested in two ways:

(1) Measured glacier movements were compared with those predicted from

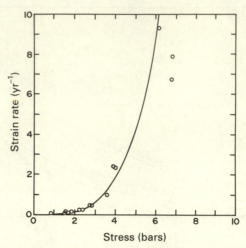

Fig. 10.9. Results of laboratory experiments on the deformation of ice. From Glen (1952).

the Navier–Stokes equation for flows in channels whose shape approximates those of glacial valleys, i.e., for somewhat more complex flows than the simple two-dimensional steady flow down a plane considered in the last chapter. Observations indicated systematic departures from the theory, for example in the shape of the velocity profile in both cross-section and surface plan (Figures 10.7 and 10.8). Details are given by Kamb (1964), Paterson (1981) and Hooke (1981). The main departures from theoretical predictions are as follows.

- There is slip at the bed, which accounts for 10%–90% of glacier movement in temperate valley glaciers.

- There is a lower velocity gradient (i.e., less deformation) away from the valley walls or floor than viscous flow theory predicts.

(2) Experiments were made in the laboratory on the deformation of single ice crystals and aggregates of crystals. Typical results are shown in Figure 10.9.

The results of both experiments and field observations show the following.

- Ice does seem to behave like a fluid under slow rates of strain: there is some flow no matter how small the stress.

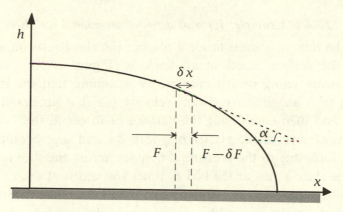

Fig. 10.10. Flow of a sheet of ideal plastic.

- Ice does not behave like a Newtonian fluid. At low rates of strain, the apparent viscosity of ice is much higher than at high rates of strain.

These observations suggest that ice may be treated as a non-Newtonian fluid. Experimental observations suggest that $n \approx 3$ and that at temperatures close to melting point the coefficient $a \approx 5 \times 10^{-15}$ kPa^{-3} s^{-1}. This value, for isotropic polycrystalline ice is about 100 times larger than for single crystals. The coefficient a in Equation (10.4) decreases as the temperature decreases, and is also affected by anisotropy and the impurities in glacier ice. Using this form of the constitutive equation in the equation for non-Newtonian flow (Equation 10.10), we get

$$u = u_b + \frac{a}{4}(\gamma_x)^3 [d^4 + (d-y)^4]$$

This solution is plotted in Figure 10.7, and a similar solution for flow between the walls of a channel of parabolic cross-section is shown in Figure 10.8.

Nye (1952, 1964) suggested that many aspects of glacier flow could be better modeled by assuming that glacier ice behaves as an ideal plastic rather than as a Newtonian fluid. The "true" model may be a non-Newtonian fluid, but the mathematics of such models are complicated, and a simpler model may be used if the predictions prove to be adequate. Nye found that quite satisfactory models of valley glaciers and ice sheets could be constructed by assuming that ice flowed as an ideal plastic with a shear strength of one bar (10^5 Pa). We shall discuss this below.

10.4.1 Example: form of a two-dimensional ice sheet

Consider the flow of a sheet of ideal plastic: the sheet rests on a horizontal base and the surface slopes at an angle α (Figure 10.10). We can find the shear stress acting on the base τ_0, by assuming that the ice is at rest or moving with an almost constant velocity (so that accelerations can be neglected) and then considering the balance of forces in the x-direction on a thin vertical slice of the glacier of width δx and unit breadth. The net horizontal force due to the pressure difference across the slice must then be equal to the shear stress at the bed τ_0 times the width of the slice δx. If we assume that the pressure is essentially hydrostatic (because of the relaxation of differential stresses within the ice by ductile deformation over long time scales), the pressure force per unit breadth acting on one side of a slice is found by integrating over the height of the slice. At position x with height h, the pressure force is

$$F = \int_0^h \gamma(h - z)dz$$

and the integration yields

$$F = \frac{\gamma}{2}h^2$$

At position $x + \delta x$,

$$F - \delta F = \frac{\gamma}{2}(h - \delta h)^2 \approx \frac{\gamma}{2}(h^2 - 2h\delta h)$$

The net pressure force is the difference between the quantities given in the two preceding equations, and acts towards the right in the diagram:

$$\delta F = \gamma h \delta h$$

Thus, the balance of forces is

$$-\tau_0 \delta x = \gamma h \delta h$$

Dividing by δx and letting it shrink to zero, we obtain

$$\tau_0 = -\gamma \frac{\partial h}{\partial x}h = -\gamma \tan \alpha h \tag{10.19}$$

This is the distribution of stress along the base of the ice sheet. For an ideal plastic, the applied stress cannot anywhere exceed the strength (the yield stress, ρ_0). If the shear stress at the base were greater than the strength, flow would take place until the stress was reduced to that value. If the stress at the base is less than the strength, no flow takes place and the ice sheet gradually builds up (by deposition on the surface) until the thickness and/or slope is such that the yield stress is produced. From the ideal plastic model, we conclude that $\tau_b = \tau_0$, which is a constant. For convenience, we define $h_0 = \tau_0/\gamma$, and rewrite Equation (10.19) in the simpler form

$$hdh = -h_0 dx \qquad (10.20)$$

Equation (10.20) can readily be integrated, giving:

$$h^2/2 = -h_0 x + C$$

If $x = L$ when $h = 0$, $C = h_0 L$ and solving for h gives

$$h = \sqrt{2h_0(L - x)} \qquad (10.21)$$

This is the equation of the profile of the ice sheet shown in Figure 10.10. The maximum height of the ice sheet is reached at the centre, where $x = 0$ and $h = \sqrt{2h_0 L}$. The form of the sheet does not depend on the rate of precipitation of snow on the surface, once the stress at the base has reached the yield stress; the model assumes that flow takes place very rapidly if the yield stress is exceeded. In real ice sheets it is enough that the accumulation rate is slow compared with the rate of flow of ice.

Elder (1976, pp. 149–50) applied the same model to the form of the Hawaiian island: he assumes the volcanic pile is so thick that the strength of the volcanic rock is just exceeded at the base. However, there are other, equally plausible models (e.g., Turcotte and Schubert, 1982, pp. 397–9).

This type of model applies to any substance with a finite yield stress. The plastic does not have to be ideal; it could be a Bingham plastic, or a pseudoplastic (in which the viscosity is not a constant), just so long as the rate of flow at the base, where the yield stress is exceeded, is rapid compared with the rate at which material is being added to the top of the sheet.

10.5 Debris

Debris flows are common in, but not restricted to, arid alluvial fans. Debris accumulates by weathering and hillslope processes in the steep-sided valleys that form the headwaters of many alluvial fans. Debris remains there until a rare storm, or a rapid snow melt event, provides sufficient water to saturate the debris and start the whole mass moving down the valley. The material, which consists of a mixture of mud, sand, and boulders (with a consistency of wet concrete) then flows in a channel over the surface of the alluvial fan. Debris flows are generally initiated on quite steep slopes, but can flow considerable distances and transport large boulders on slopes as low as 3–5 degrees. Boulders may be observed sticking out of the top of some flows, apparently "rafted along" by the flow, although measurements of the density of the boulders and the muddy debris show that the boulders are denser than the matrix which surrounds them. Direct observation of the flow surface suggests that in many cases flow is taking place in a quasi-laminar fashion, with strong velocity gradients at the margins of the flow. In other cases, flow seems to be turbulent. Speeds are generally of the order of metres per second.

Much larger, more catastrophic debris flows also take place. Some of these are produced by volcanic activity (as for example, in the Mount St Helens eruption). Debris flows composed largely of volcaniclastic materials are called *lahars*: frequently the debris contains more water, and therefore has a lower strength than the debris in arid watersheds, so lahars are not only larger but also more mobile and more hazardous.

Submarine debris flows are also common phenomena. A typical product as seen in the geological record is a *pebbly mudstone*, i.e., an unsorted mixture of cobbles and pebbles set in a muddy matrix.

Two different theoretical models have been applied to explain debris flows:

(1) Johnson (1970, Chapters 12–14) developed a model based on the assumption that the debris behaves as a Bingham plastic. Many of the features of muddy debris flows can be explained using this model. The model explains the transport of large boulders as due to a combination of buoyancy and matrix strength, which prevents settling of the boulders through the matrix. The strength of the debris also causes the flow to come to rest when the shear stress at the base is no longer sufficient to overcome the strength of the debris. At this point "mass deposition" takes place, i.e., the flow is itself transformed into the deposit. Estimates of the strength of debris in flows on arid region fans are about 10^3 Pa. Experiments (e.g., Major and Pierson, 1992) show that the strength, viscosity and other aspects of the rheology of muddy debris are very sensitive to the clay mineralogy and water content of the matrix.

Although a Bingham plastic model explains many properties of debris flows, there are other observations with which it is not consistent. Many debris flows are observed to contain very little cohesive mud, and it is difficult to understand how a mainly sandy matrix can retain the strength which is an essential part of the Bingham model. A further problem is that, during deposition, some debris flows are observed to "freeze" from the bottom up. Shear stress increases linearly below the flow surface and reaches a maximum at the base, so in the Bingham model the flow should freeze from the top down. It is likely that many debris flows are initiated as Coulomb-type failures along a discrete failure zone at depth (as described in Section 4.16) and this, too, is not consistent with the Bingham model. Finally, a Bingham model does not take account of momentum exchange via grain collisions within the flow, a factor that is likely to be important in rapid flows containing many large grains.

(2) Other authors (e.g., Takahashi, 1981) have developed an alternative model, based upon the theory of the flow of cohesionless grains developed originally by Bagnold, and more recently elaborated by Savage and others (for a summary and references see Campbell, 1990). This model (Equation 10.5) is essentially a non-Newtonian flow model. One problem with the model is explaining how debris continues to flow on slopes as low as 5 degrees, when the internal friction angle for "grain flows" is much larger than this.

If neither the Bingham plastic nor grain-flow models explain all properties of debris flows, what use do they have? First, a model that explains some aspects of flow behaviour can be an integral part of a complete model. Second, a model may be applicable in some, if not all, cases. That is, even if it is not general, a model may have some direct utility under particular circumstances. For example, although the Bingham model may not adequately explain the initiation or deposition of many debris flows, there are some whose behaviour during most of the flow is well described by the Bingham model. Hence, the model may provide useful information concerning flow dimensions, velocity, and travel distance (see Problem 1), provided that the properties of the flow fall within empirically defined bounds and the initiation and deposition conditions are otherwise understood.

In problems where the mechanical properties of the flow are complex and not fully understood, it is common that simple models explain only part of the observed behavior. Without a complete model, one must make use of direct observation to define the behavior of the flow and the particular conditions for which the model is approximately valid. Compare this with the application of a more complete model, such as that for Newtonian viscous flow. Empirical observations may be used to test this model, for example by using measurements of velocity to verify the equations for laminar flow (Sections 9.8, 9.9). Once verified, one may have confidence in applying this model to other Newtonian fluids and other geometric configurations. The Navier–Stokes equation, however, cannot be applied to all flows, and empirical information is still necessary to determine whether the fluid is Newtonian and the flow regime laminar rather than turbulent.

Some debris flows are observed to be turbulent, in which case neither of the models listed above can be applied to obtain velocity distributions or other characteristics of the flow. The critical criterion for initiation of turbulence in a Newtonian fluid is well known to be the Reynolds number (see next chapter). What are the criteria for Bingham plastics or non-Newtonian fluids?

In Chapter 9, we showed that the Reynolds number can be interpreted as the ratio between the inertia and viscous forces. But Bingham plastics not only have mass and viscosity, but also strength. So it is not surprising that experiments have shown that the initiation of turbulence in such materials depends not only on the Reynolds number, but also on the ratio between strength and viscous forces. From the constitutive equation for a Newtonian fluid, the viscous stress is proportional to $\mu U/d$, where U is the average speed, so the force is proportional to $\mu U d$. The strength k is a stress, so the strength force is proportional to kd^2. The ratio of the two gives the *Bingham number* :

$$B = \frac{kd}{\mu U} \qquad (10.22)$$

At small Bingham numbers (say < 0.1), strength would be much less important than viscosity, and we would expect the onset of turbulence to be controlled by the Reynolds number. At large values of the Bingham number it has been observed experimentally that the transition to turbulence takes place close to the relation $Re \approx 1000B$. This can be represented as

$$Re/B = \rho U^2/k \approx 1000$$

The ratio $\rho U^2/k$ is proportional to the inertia force divided by the strength force, and has been called the *Hampton number* (by Hiscott and Middleton, see Middleton and Southard, 1984, p. 358). Note that at high values of the Bingham number, the transition to turbulence seems to depend much more on strength than on viscosity.

Non-Newtonian fluids may also become turbulent, but the definition of an appropriate "generalized" Reynolds number is not a simple matter. Some quasi-empirical solutions are given by Skelland (1967, pp. 226–33). It seems to be clear that Bagnoldian grain flows can also become turbulent at large scales and speeds, but the appropriate criteria are not well understood.

10.6 Lava

The Bingham plastic model has also been applied to lava flows, notably by Johnson (1970) and Hulme (1974). For a recent review see Huppert

(1986). The following evidence suggests that lava does not flow as a simple Newtonian fluid.

(1) Lavas come to rest on relatively steep slopes, and flows have substantial thicknesses, and abrupt terminations. This suggests that lava has a shear strength, because the rates of cooling do not seem to be sufficient to explain these observations by access of strength on cooling (due to crystallization or solidification).

(2) Lava and debris flow morphologies show many resemblances, notably channels with a U shape, and with levéed borders. These features can be explained by a Bingham plastic model.

Details of the theory of lava flows can be found in the references at the end of the chapter. Hulme (1974) estimated that the shear strengths of terrestrial lavas varied from about 10^2 to 10^3 Pa for basaltic lavas to 10^5 Pa for lavas with a silica content as high as 60 per cent.

10.7 Rock rheology and sedimentary basins

In Section 8.10, we discussed sedimentary basins in terms of a simple elastic model. But there is evidence that rocks, even in the lithosphere, can respond to stress not only by elastic, but also by ductile, deformation. It is reasonable to hypothesize, therefore, that a Maxwell body should be used to model the lithosphere, rather than a simple elastic solid.

Suppose that we impose a sudden strain on a Maxwell body. For example, we might bend a rock stratum to make a broad fold. At first the deformation is entirely elastic, so if the applied stresses were relaxed the strain would disappear (the fold would straighten out). But after a period of time, the internal stresses gradually disappear as the material flows (as the rock deforms by creep, in the fold example). The final state is one of permanent (nonelastic) deformation and zero (or near-zero) stress. As rocks at relatively low pressures and temperatures do have elastic properties, but old folds retain almost none of the stresses which formed them, it is clear that rocks do behave in some ways like Maxwell bodies.

If there is no further strain after the initial deformation, the constitutive relation for a Maxwell body (Equation 10.2) reduces to

$$0 = \frac{\sigma}{2\mu} + \frac{1}{E}\frac{d\sigma}{dt}$$

This may be rearranged as

$$d\sigma/\sigma = -(E/2\mu)dt$$

and integrated to give

$$\ln \sigma = -(E/2\mu)t + C$$

Evaluating C for the initial conditions ($t = 0$, $C = C_o = \ln \sigma_o$) we get

$$\ln \sigma - \ln \sigma_o = -E/2\mu$$

or

$$\sigma = \sigma_o \exp[-(E/2\mu)t] \qquad (10.23)$$

This equation shows that the stress is an exponentially decreasing function of time. The *relaxation time t_o* of a Maxwell body is defined as

$$t_o = 2\mu/E \qquad (10.24)$$

After one relaxation time period, Equation (10.23) gives the result

$$\sigma = \sigma_o e^{-1} = 0.37\sigma_o$$

That is, after one relaxation time, the stress has been reduced to 37 per cent of its original value. The concept of a relaxation time is directly analogous to the concept of the half life of a radioactive isotope (though after a half life 50, not 63, per cent of the atoms have decayed).

Maxwell himself believed that all real fluids are actually Maxwell bodies, with finite relaxation times (by definition, a Newtonian fluid has zero relaxation time). Maxwell even calculated the relaxation time of an ideal gas, and found that it was 1.95×10^{-10} seconds (Reiner, 1960, p. 85). As this example illustrates, if our interest in the behavior of a Maxwell body is confined to its response to stresses over periods of time that are long compared with the relaxation time, then we can generally ignore the elastic component of its behavior and treat it as though it was a simple Newtonian fluid.

Let us see how this idea applies to the asthenosphere and the lithosphere. Using typical values for the viscosity and elasticity of the asthenosphere ($\mu = 4 \times 10^{19}$ Pa s and $E = 2 \times 10^{11}$ Pa), we get a value of the relaxation time of roughly 13 years. For the lithosphere, both viscosity and elasticity are larger, but the main increase is in the viscosity. In fact, the viscosity of the lithosphere is so high that it is very difficult to estimate accurately. Using $\mu = 10^{26}$ Pa s and $E = 8 \times 10^{10}$ Pa gives a relaxation time of 80 million years.

Isostatic phenomena and the development of sedimentary basins take times that are longer than thousands of years. Therefore the elastic properties of the asthenosphere can generally be ignored. The development of sedimentary basins generally takes times measured in tens of millions of years. By

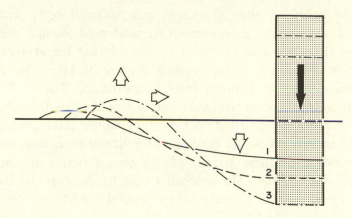

Fig. 10.11. Beaumont's model of a foreland basin. Profile 1 shows the elastic deformation resulting from emplacement of the load (shaded). The open arrows and profiles 2 and 3 show the later deformation resulting from viscous relaxation of the lithosphere. From Quinlan and Beaumont (1984, Fig. 3).

comparison, any viscous response of the mantle takes place very rapidly – so we can treat it as though it was a fluid of zero viscosity (as we did in Chapter 8). But, if we believe the above estimate of the viscosity of the lithosphere, the time for a viscous response of the lithosphere is similar to the time that it takes to develop sedimentary basins.

These considerations led Beaumont (1978, 1981; see also Quinlan and Beaumont, 1984) to develop a model for sedimentary basins based on the assumption that the lithosphere is a Maxwell body, overlying an inviscid asthenosphere. Since it is not possible to estimate the viscosity of the lithosphere reliably from other data, this model might be used to improve the guess given above.

The basic idea of Beaumont's model of foreland basins is that the basin is created in the first instance by imposing a sudden load on the lithosphere. The actual mechanism is probably major thrusting due to plate movements (which are very rapid movements if we are measuring time in millions of years). The first response of the lithosphere is elastic bending. As we saw in Chapter 8, this forms a sedimentary basin close to the load, and a more distant upwarp ("peripheral bulge"). The width of the basin is determined mainly by the thickness and rigidity of the lithosphere. The basin is then filled with sediment (imposing a further load) and the thrust load and upwarp are partly eroded (reducing the loads).

Beaumont found that, in the western Canada Cretaceous basin, purely elastic models yielded a sedimentary basin that was too broad compared with the real basin. If the lithosphere is a Maxwell body with viscous as well as elastic properties, Beaumont's model predicts that the basin is progressively deepened and narrowed by flow during the *relaxation phase* which follows the initial elastic bending (Figure 10.11). The peripheral bulge becomes higher and moves towards the basin. The model predicts that sediments deposited in the basin during this stage will be thicker and have a more restricted distribution than they would have according to a model based entirely on elastic bending of the lithosphere. Beaumont found that incorporation of these viscous effects gave a better agreement with the observed stratigraphy. He concluded that his best model indicated a lithosphere under western Canada with a flexural rigidity of 10^{25} N m and a relaxation time of 27.5 Ma.

Applying Equation (10.24) shows that this corresponds to a lithospheric viscosity of about 3.5×10^{25} Pa s, or about six to seven orders of magnitude higher than that of the asthenosphere. Not all workers who have applied elastic models to sedimentary basin development agree with Beaumont that a Maxwell model gives better agreement with the stratigraphic data than a simple elastic model. It is difficult to separate viscous relaxation effects from effects due to changes in the elastic constants as the lithosphere warms under sedimentary basins. For further discussion and references see Allen and Allen (1990, Chapter 4).

10.8 Review problems

(1) **Problem:** A debris flow 2 m thick moves down a slope inclined at an angle $\alpha = 5.7$ degrees, equivalent to $\sin \alpha = 0.1$. The bulk density of the flow was found to be 2400 kg m^{-3}. The debris approximated a Bingham material with a shear strength equal to 4×10^3 Pa, and a viscosity when shearing of 400 Pa s.

   (a) How thick was the rigid plug in this flow?

   (b) Neglecting inertia effects, and assuming that the slope slowly decreases without any change in the thickness of the flow, on what angle of slope will the flow come to a halt?

   (c) What was the approximate speed at which the rigid plug moved on the steeper slope?

   Answer:

   (a) The shear stress varies linearly with distance from the surface of the flow, Y:

$$\tau = \gamma Y \sin \alpha$$

The rigid plug is that part of the flow where $\tau < 4 \times 10^3$ Pa. Solving the equation for Y gives $Y = 1.7$ m.

(b) The flow will come to a halt when the shear stress at the bottom no longer exceeds the strength. We use the same equation as before, but set $Y = 2$ m and solve for $\sin \alpha$. The result is 0.085 (this angle is 4.9 degrees).

(c) The shear stress at the bed is $\tau_0 = \gamma_s d$. The strength is k so the stress available to produce viscous shear is $\tau_0 - k$. Since only an approximate answer is required, and the basal layer in which shearing takes place is thin ($T = 0.3$ m) we assume that the velocity gradient is linear. If the speed of the plug is U, then

$$\tau_0 - k = \mu(du/dY) \approx \mu(U/T)$$

Solving for U gives 0.53 m s^{-1}.

The complete analysis uses Equation (10.17), and the numerical result in this case is $U = 0.265$ m s^{-1}.

(2) **Problem:** Show that $d\epsilon/dt = d\dot{u}/dy$ for steady uniform flow down a slope, where ϵ is the total (engineering) component of shear strain, and $\dot{u}$ is the velocity component in the downslope direction.

Answer: This is a case of simple shear. The only strain is the shear strain corresponding to the shear stress τ_{yx}, where the flow is in the x-direction and y is the direction normal to the flow. Equation (7.16) gives the shear strain in terms of displacement gradients as:

$$\epsilon_{12} = \left(\frac{\partial u_2}{\partial x_1} + \frac{\partial u_1}{\partial x_2} \right)$$

For simple shear, $\partial u_2/\partial x_1 = 0$, so $\epsilon_{12} = \partial u_1/\partial x_2$. Differentiating with respect to time, we get:

$$\frac{d\epsilon}{dt} = \frac{d\epsilon_{12}}{dt} = \frac{d\dot{u}}{dy}$$

where we have replaced x_2 with y and the time derivative of the displacement u_1 is the x-direction *velocity* $\dot{u}$. (We have had to use u both for displacement and velocity, so here have used the dot notation to distinguish the two.)

(3) **Problem:**

(a) Derive an equation for the shear stress distribution in an inclined channel of semicircular cross-section. Assume radial symmetry of the flow.

(b) Use this to derive the equation for the velocity distribution of a Bingham plastic in the channel.

(c) (Data from Johnson, 1970, pp. 501–3.) In the laboratory, a debris flow is run through a channel of semicircular cross-section. The radius is 0.08 m, and the slope is 4.4 degrees. The density of the debris is 1700 kg m^{-3}. The central plug is observed to have a radius of 0.065 m and a speed of 0.21 m s^{-1}. What is the strength and viscosity of the debris?

Answer:

(a) This is derived using the same method we used for flow down a plane. For a flow with radial symmetry, it is convenient to use a circular coordinate system with its origin at the centre of the circle defined by the channel cross-section. Then the shear stress and velocity will be constant along semicircular surfaces of radius r. The downslope weight component of half cylinders of unit downslope length and radius r is $\gamma_x(\pi r^2/2)$, and the upslope component of resistance on the half cylindrical surface is $\pi r\tau$. The result of equating these two forces is

$$\tau = \gamma_x r/2$$

(b) We can now derive the velocity profile very easily. Given the stress distribution and the constitutive relation for a Bingham plastic we arrive at

$$\gamma_x r/2 = k + \mu(du/dr)$$

which is analogous to Equation (10.14). Integrating we obtain

$$u = \frac{\gamma_x}{4\mu}(R^2 - r^2) - \frac{k(R-r)}{\mu}, \qquad r \geq \frac{2k}{\gamma_x}$$

which is analogous to Equation (10.16).

(c) To find the strength, we use the result of (a), with $r = r_o$ at the lower boundary of the plug:

$$k = r_o\gamma_x/2 = 41.6 \text{ Pa}$$

We now use the result of (b), with $r = r_o$ at the lower boundary of the plug (this is the only place where we know both r and u):

$$\mu = \frac{\gamma_x}{4u}(R^2 - r_o^2) - \frac{k(R - r_o)}{u} = 0.34 \text{ Pa s}$$

10.9 Suggested reading

For an introduction to rheology:

Barnes, H. A., 1989. *An Introduction to Rheology.* New York, Elsevier, 199 pp. (QC189.5.B37)

Reiner, M., 1969. *Deformation, Strain and Flow: An Elementary Introduction to Rheology.* London, H. K. Lewis, third edition, 347 pp. (QC189.R3)

On the rheology of rocks see:

Lliboutry, L. A., 1987. *Very Slow Flows of Solids: Basics of Modeling in Geodynamics and Glaciology.* Boston, Martinus Nijhoff, 510 pp. (This work, by an expert on glacier flow, provides a systematic treatment of the rheology of rock and ice. Advanced, but does not assume prior knowledge of much mathematics. QE501.3.L57)

Ranalli, G., 1987. *Rheology of the Earth. Deformation and Flow Processes in Geophysics and Geodynamics.* Boston, Allen and Unwin, 366 pp. (Covers much the same ground as Lliboutry's book, is more concise, and assumes a stronger background in mathematics. QE501.3)

On mud and debris flows:

Johnson, A. M., 1970. *Physical Processes in Geology.* San Francisco, Freeman, Cooper, and Co., 577 pp. (Also discusses ice and lava. QE33.J58)

Middleton, G. V. and J. B. Southard, 1984. *Mechanics of Sediment Movement.* Soc. Econ. Paleontologists Mineralogists, Short Course No. 3, second edition, 401 pp. (QE471.2.M53)

For ice, besides Lliboutry (1987) see:

Hutter, K., 1982. Dynamics of glaciers and large ice masses. Ann. Rev. Fluid Mechanics, v.14, pp. 87–130.

Hutter, K., 1983. *Theoretical Glaciology.* Boston, Reidel, 510 pp. (Advanced treatment. GB2403.2H88)

Meier, M. F., 1960. Mode of flow of Saskatchewan Glacier, Alberta, Canada. US Geological Survey Professional Paper 351, 70 pp. (Descriptive and analytical study)

Paterson, W. S. B., 1981. *The Physics of Glaciers.* Oxford, Pergamon Press, second edition, 380 pp. (The best introductory text. GB2403.5.P37)

On lavas:

Hulme, G., 1974. The interpretation of lava flow morphology. Geophys. Jour. Royal Astronomical Soc., v.39, pp. 361–83.

Huppert, H. E., 1986. The intrusion of fluid mechanics into geology. Jour. Fluid Mechanics, v.173, pp. 557–94. (Review of recent applications of fluid mechanics to volcanology.)

On sedimentary basins:

Beaumont, C., 1978. The evolution of sedimentary basins on a viscoelastic lithosphere: theory and examples. Geophysical Jour. Roy. Astronomical Soc., v.55, pp. 471–97.

Beaumont, C., 1981. Foreland basins. Geophysical Jour. Roy. Astronomical Soc., v.65, pp. 291–329.

Quinlan, G. M. and C. Beaumont, 1984. Appalachian thrusting, lithospheric flexure, and the Paleozoic stratigraphy of the Eastern Interior of North America. Canadian Jour. Earth Sci., v.21, pp. 973–96.

11

Turbulence

11.1 Description of turbulence

In a classic series of experiments on flow in pipes, reported in 1883, the Irish engineer Osborne Reynolds (1842–1912) distinguished between two regimes of flow: (i) *laminar*, or streamline; and (ii) *turbulent*, and showed that the transition from one to another was determined by what is now known as the *Reynolds number*:

$$Re = UL\rho/\mu \qquad (11.1)$$

where ρ and μ are the density and viscosity of the fluid, U is the velocity, averaged over the pipe's cross-section, and L is a characteristic length (see Acheson, 1990, Chapter 9, for references and a more detailed discussion). For a pipe, the diameter D, or radius R, is the appropriate length. The term "turbulent" was not used by Reynolds but was introduced later by Lord Kelvin.

In laminar flow, the fluid appears to be moving in lines or layers, with one layer sliding smoothly past another. Diffusion from one layer to another is restricted to the molecular scale. Laminar flows are entirely predictable, or at least reproducible from one experimental run to the next.

In turbulent flow, irregular curved motions (eddies) are present, and mixing between different layers takes place at both molecular and much larger scales. Though the *average* properties of the flow (e.g., the discharge from a pipe) may be reproducible from one experimental run to the next, the detailed motions of individual fluid particles are not reproducible, and the average properties of the flow cannot be predicted from the general equation of motion (the Navier–Stokes equation) without making use of further empirical knowledge.

An experienced observer can often tell at once whether a flow is in the laminar or turbulent regime but it is not easy to give a precise definition or description of turbulence, and there was no clear distinction made between laminar and turbulent flows until the work of Reynolds. Lumley and Panofsky (1964) singled out the following features as diagnostic.

- Turbulence is **rotational**, and results in internal dissipation of kinetic energy to heat by means of a range (spectrum) of eddy sizes.
- Turbulence is **three-dimensional**, i.e., the eddies making up the flow have a more or less random orientation in space. Averaging over time reveals that many turbulent flows do display structure on a scale comparable to the scale of the whole flow, but the structure tends to be replaced by randomness at smaller scales.
- Turbulence, therefore, is **stochastic**, i.e., it can be fully described only in statistical terms.
- Turbulence is **diffusive**, i.e., a marked particle of fluid wanders away from the average direction of flow and, on the average, does not return.
- The time and length scales of turbulence are large compared with those of molecular diffusion. Therefore, **mixing** by turbulence is very effective at larger scales, but mixing at small scales still depends on molecular processes.

Because turbulence is essentially rotational, vortices and vorticity play an important role. We remind ourselves that a flow is rotational if the *vorticity* ω, where

$$\omega = \nabla \times \mathbf{u} = \text{curl}\,\mathbf{u} \qquad (11.2)$$

is not equal to zero. Not all rotational flows are turbulent (since there are perfectly predictable eddies developed at low Reynolds numbers, and most straight uniform laminar flows have vorticity), but all turbulence is rotational.

The *strength* of a vortex may be defined as the vorticity times the cross-sectional area. If we multiply this by the density of the fluid, the dimensions are $[ML^{-1}T^{-1}]$, corresponding to an energy per unit volume of the fluid. In a turbulent flow, in which there is shearing at all scales, vortices tend to be stretched by the shearing motion. Stretching reduces the cross-sectional area, and conservation of energy then requires an increase in the local vorticity. This in turn tends to lead to break-up of the vortex into smaller vortices:

> Big whirls have little whirls
> which feed on their velocity
> and little whirls have lesser whirls –
> and so on to viscosity...

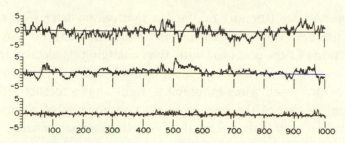

Fig. 11.1. Turbulent fluctuation of the x- (top), y- (middle), and z- (bottom) components of wind velocity 5.6 m above the ground (Panofsky and Dutton, 1984, p. 34). The units are seconds for the horizontal axes, and m s^{-1} for the vertical axes.

This little poem, a parody of one by Jonathan Swift, was composed by Lewis F. Richardson (1881–1953) and appeared in his pioneering book *Weather Prediction by Numerical Process* (published in 1922, long before computers made such methods practical). It nicely catches the essential physics of what is now called the *turbulent energy cascade*.

The simplest descriptive measures of turbulence are those which describe the statistics of the fluctuations in velocity components, observed at a point in a fluid (Figure 11.1). Let the components of instantaneous velocity in the x-, y-, and z-directions be denoted u, v, w and let the components in these directions averaged over time be $\bar{u}, \bar{v}, \bar{w}$: then we can use primes to define the *deviations* u', v', w' of the instantaneous from the time-averaged velocity components, e.g.,

$$u = \bar{u} + u' \tag{11.3}$$

A suitable measure of the average value of these turbulent velocity-component deviations is the root-mean-square, e.g.,

$$\text{rms } u' = \sqrt{\overline{(u')^2}} \tag{11.4}$$

Time averaging is indicated by the bar. Note that we cannot use a simple average of u', as this is equal to zero (by definition). The root-mean-square is simply the standard deviation of the instantaneous velocity. For a more complete statistical description, cross-correlations or covariances of two components are also required, e.g.,

$$\text{covar}(u', v') = \overline{u'v'} \tag{11.5}$$

In order to calculate these statistics, a velocity meter is required that is capable of measuring fluctuations of velocity in at least two different directions simultaneously, with a very rapid response time. Such an instrument is the *hot wire anemometer* (for air) or the *hot film anemometer* (for water). The basic principle here is that a fluid flowing past a heated thin wire transverse to the direction of flow cools the wire, thus changing its electric resistance. If the wire is parallel to the flow, the effect is much smaller. Thus an instrument consisting of two small wires mounted at right angles to each other permits measurement of two of the instantaneous components of a turbulent flow. More recently a new instrument, the *laser Doppler velocimeter*, which uses laser light reflected from dust particles in the flow, has become popular. Another instrument, which is much more robust and therefore more suited to use in the field, is the *electromagnetic flow meter*. For details of the theory and construction of these devices see Tritton (1988, Chapter 25) or the volume edited by Wendt (1974). The hot wire anemometer was developed some 20 years before reliable measurements could be made in water (these started in the 1960s and 1970s) so many of the classic investigations of turbulence were made in wind tunnels, and detailed studies of turbulence in natural environments such as rivers and nearshore marine environments are still rare.

One useful technique in the statistical description of turbulence, as in the description of other time series, is the calculation of an *energy spectrum*, which shows the contribution of each frequency of the turbulent eddies to the total mean-square velocity fluctuations (and therefore to the total kinetic energy of the turbulence). In other contexts such spectra are called *power* or *variance* spectra. Measured spectra (e.g., Figure 11.2) show that, though energy is carried by a whole range of frequencies, the energy contribution of a particular frequency is generally proportional to some negative power of that frequency. In other words, proportionately more of the energy is carried by the larger eddies. The shape of the energy spectrum can be explained as a "cascade". Energy is pumped into the system via large eddies generated by shearing and flow separation. Because of vortex stretching and shearing at their boundaries, the large eddies tend to break down into smaller eddies, and so on until the energy is dissipated, via viscosity, as heat.

Beginning in the 1960s, investigations of turbulence moved away from purely statistical description towards the experimental investigation of large-scale structures in turbulent shear flows, using "flow visualization" techniques such as high-speed cinematography, combined with the labeling of fluid particles by dye injection or the *hydrogen bubble* technique. In the latter, a thin platinum electrode oriented transverse to the flow is used to generate a line of tiny hydrogen bubbles. These bubbles are convected away from the wire by the flow, and they are so small that they rise only very slowly under buoyancy, and thus are carried with the flow and serve to mark fluid

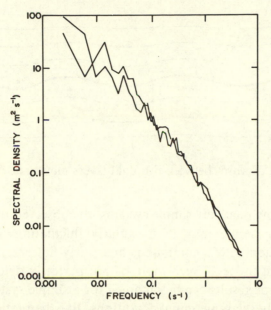

Fig. 11.2. Energy spectra of turbulence, for the *x*-component of wind velocity shown in Figure 11.1. Two spectra are shown, to indicate the statistical variation arising from repeated sampling of the same time series.

particles. Experiments using these visualization techniques reveal that there are quasi-periodic structures generated within turbulent shear flows, whose scale is large compared with the scale of the smaller, more truly random eddies. Because the larger structures are not *strictly* periodic, they are difficult to detect and measure using the conventional techniques of anemometry. A review of these studies is given by Cantwell (1981).

Finally, the latest trend in turbulence research is to distinguish between random turbulence and "chaos". Truly random phenomena are those where one event cannot be predicted from the event that preceded it. For example, in Equation (11.3), though it may be possible to predict the time-averaged velocity $\bar{u}$, it is generally assumed that it is not possible to predict the turbulent velocity component u', which is therefore described as a "random variable". Random variables are generally thought to be unpredictable because they are influenced by a multitude of causes not known to the observer. However, the existence of structure within turbulent flows, revealed by averaging over different time and length scales, shows that a simple random variable is not an adequate description of the turbulent motion.

Disturbance of a simple mechanical system generally leads to periodic motion – the classic examples are those of the pendulum and the vibrating

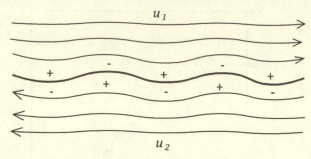

Fig. 11.3. Growth of waves between two fluid layers moving at different velocities.

string. But in some relatively simple systems it is observed that such periodic motion may degenerate into an irregularly fluctuating quasi-periodic or nonperiodic motion with a continuous and fairly flat energy spectrum, like that of a random process. Such motions are perfectly deterministic, in the sense that the same results can be obtained for a simple system by beginning again with exactly the same initial conditions. But the motion might appear to the observer to be not only highly irregular but also nonpredictable, hence it is termed *chaotic*. A further important characteristic of such chaotic motions is that very small changes in initial conditions sooner or later produce large differences in the observed motion. Therefore, although the motion is deterministic, its characteristics are such that it would rarely be possible to determine the early history of motion from observation of the later stages of development of the motion. So from the point of view of an observer who is trying either to predict future motion, or retrodict earlier motion (without knowing the exact initial conditions) the behavior of the system cannot be known far into the future or past, and might just as well be truly random.

Currently, researchers are actively investigating how far the chaotic behavior of such simple mechanical systems is relevant to the phenomenon of turbulence in fluids. The evidence is now very strong that the transition to turbulence in thermal convection (Bénard convection) is an example of deterministic chaos (see Chapter 12). At high Reynolds numbers, turbulence is closer to a truly random (high-dimensional) phenomenon than to dynamical chaos, though it may have originated as low-dimensional chaos. For brief reviews see Landahl and Mollo-Christensen (1986) and Tritton (1988, Chapter 24) and for an extended introduction see Bergé *et al.* (1986).

11.2 Origin of turbulence

Turbulence generally originates from the shearing of a fluid past a boundary. The reason why can be readily understood from Figure 11.3. Consider two fluid layers, with the same density and viscosity, but moving at different velocities u_1 and u_2. For simplicity, consider that the velocities are equal in magnitude (to u), but opposite in direction. If a small wavelike perturbation with an amplitude of δA is produced in the streamlines at the interface between the two flows, then along a streamtube close to the interface the width of the tube will vary by a factor proportional to δA. From continuity, the velocity will also vary (inversely to the streamtube width) by an amount $\pm \delta u$. From an application of Bernoulli's equation, it is clear that pressures above the crests of the waves, or below the troughs (labelled "$-$" in the figure), will be lower than those within the crests or troughs (labelled "$+$" in the figure). Therefore, the waves will continue to grow, unless there is some force that opposes these pressure differences. This force might be viscosity (or gravity, if the density of the lower layer were greater than that of the upper layer). Such waves are often called *Helmholtz*, or *Kelvin–Helmholtz waves*.

We can show that the transition to turbulence should depend on a Reynolds number by comparing the magnitude of the pressure tending to cause the wave to become unstable with that of the viscous stress tending to stabilize the waves. From Bernoulli's equation, the pressure difference along a single streamtube close to the interface is equal to

$$\frac{\rho}{2}[(u + \delta u)^2 - (u - \delta u)^2] = 2\rho u \delta u$$

From Newton's law of viscosity, the viscous stress on the interface is proportional to $\mu u / \delta A$. The ratio of these two forces per unit area is proportional to $\rho \delta u \delta A / \mu$, which is a Reynolds number. An increase in this number causes an increase in the disturbing pressure forces relative to the damping viscous forces.

A stability analysis for the case of two-dimensional waves damped by viscosity was first carried out in the 1930s, and verified experimentally by Schubauer and Skramstad (1947). The results are shown in Figure 11.4. As can be seen, the region of stability is defined not only by a Reynolds number, but also by a dimensionless measure of the frequency f of the disturbances affecting the boundary. Though this pioneer analysis was for the unrealistic case of two-dimensional disturbances, it helps us to understand two observations that can readily be made about the initiation of turbulence.

(1) There is generally a minimum critical value of the Reynolds number

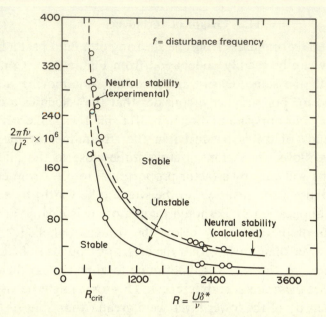

Fig. 11.4. Field of instability for a velocity discontinuity at a plane boundary. R is a Reynolds number based on the thickness δ^* of the laminar boundary layer. The ordinate is a dimensionless measure of the frequency f of the disturbance. After Schubauer and Skramstad (1947).

below which all flows are laminar (i.e., below which the flow is stable because all disturbances are damped out by viscosity, no matter what their frequency).

(2) Under special experimental conditions, laminar flow may persist to much higher Reynolds numbers.

In nature, a wide range of disturbance frequencies is generally present, so the minimum critical Reynolds number is also the maximum value for the persistence of laminar flow.

Recent experimental studies, well summarized in Tritton (1988) and Landahl and Mollo-Christensen (1986) have shown that the transition to turbulence is not abrupt. The first waves to develop are regular and predictable, but the regularity gradually breaks down as the Reynolds number is increased; chaotic flow patterns are developed in regions localized in space and time, thus such turbulence can be described as "intermittent". In flows shearing past a solid boundary, turbulence develops first at "spots" on the wall and then spreads out downstream of the spots. In flow past a cylinder, the flow is at first entirely symmetrical about the cylinder; next, it becomes asymmetrical and two regular vortices develop behind the cylinder; then the vortices are generated and shed alternately from one side and then the other of the cylinder (the *von Kármán vortex street*, named for the Hungarian engineer Theodor von Kármán, 1881–1963); and finally, the street of vortices breaks down to form a turbulent wake (Figure 9.13). At relatively low Reynolds numbers, the turbulent wake still exhibits a regular circulation, if the flow is averaged for a long enough time. At higher Reynolds numbers, it is increasingly difficult to detect any such structure in the wake. In this case, the turbulence can be described as "fully developed".

The flow past a cylinder illustrates an important aspect of the development of turbulence at a boundary: it is frequently related to the progressive development of a boundary layer, and the separation of flow from the boundary.

11.3 Boundary layers and flow separation

11.3.1 The boundary layer concept

The concept of a boundary layer was introduced in Chapter 9. The invention of this concept, its theoretical elaboration by Ludwig Prandtl (1875–1953), Geoffrey I. Taylor (1886–1975), and von Kármán, and the first experimental investigations all took place in the early twentieth century, and created the modern science of fluid mechanics and aerodynamics. Well-written early summary works that still preserve the flavour of those investigations are the volumes by Prandtl (1952), Prandtl and Tietjens (1934a,b), and Goldstein (1965). Goldstein (1969), Taylor (1970) and Oswatitsch and Wieghardt (1987) give some historical background. A fairly full discussion of how these ideas apply to sediment transport and natural boundary layers is given in Middleton and Southard (1984). What follows is a brief summary of some important points.

A boundary layer (BL) is defined as that part of the flow that is substantially affected by the presence of a solid boundary. The concept has meaning mainly at large Reynolds numbers. As an example, consider flow around a sphere or cylinder. At very low Reynolds numbers ($Re < 1$), it can be shown that the presence of the sphere changes a uniform flow approaching the sphere to a distance of the order of 100 times the diameter of the sphere (this is the reason why settling experiments at very low Reynolds numbers must be conducted in very wide tubes). At a distance of the order of 10 diameters, there are substantial differences between the real flow and the flow computed for an inviscid fluid. Therefore, at such low Reynolds numbers, there is no local part of the flow that may be designated the boundary layer.

The situation is very different at Reynolds numbers of the order of 100 or

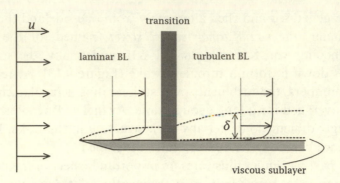

Fig. 11.5. Development of a boundary layer (BL) on a flat plate.

higher. The pattern of the flow upstream or cross-stream from the sphere is seen to be very like that of an inviscid fluid: viscous friction at the surface of the sphere affects the flow directly only in a thin region close to the boundary, and in the wake behind the sphere. The BL is very thin compared with the diameter of the sphere, reaching a maximum of only about one hundredth of the diameter. Nevertheless, what happens in this thin layer is responsible for the formation of the turbulent wake, and therefore for determining the drag on the sphere. Because the BL is so thin, it is possible to consider it as almost two-dimensional (i.e., as a thin flat sheet flow) within which the rate of shearing is very high. It therefore follows that the viscous forces are very large, and that flow within the BL tends to be "laminar" rather than turbulent.

11.3.2 Development of a boundary layer

We can understand the boundary layer better by considering flow past a thin plate, as shown by Figure 11.5. The BL starts to develop at the leading edge of the plate. As the fluid in contact with the plate is at rest, and the fluid immediately upstream is moving almost at the uniform approach velocity u, the BL is at first very thin and the velocity gradient across the BL is very large. The BL increases in thickness downflow, due to molecular diffusion of momentum near the boundary. Because the velocity gradient is so large, the rate of increase in thickness of the BL is at first very rapid, but becomes less rapid as the velocity gradient becomes less. Nevertheless the thickness of the BL increases in the flow direction. We could define a Reynolds number (Re_δ) based on the BL thickness δ, but in this case it is more useful to define one based on the distance x from the leading edge of the plate:

$$Re_x = ux/v \tag{11.6}$$

In 1908, H. Blasius, a student of Prandtl, developed an approximate theoretical solution for the thickness of the BL:

$$\delta/x \approx 5Re_x^{-1/2} \tag{11.7}$$

The growth of the laminar BL thickness is proportional to the square root of x. This growth continues until Re_x reaches a value of about 3×10^5. At this point the flow within the BL becomes turbulent, and the BL thickness increases more rapidly once again, because diffusion of momentum is effected much more readily by turbulence than by viscosity. Even when the BL becomes turbulent, however, there must always be a thin region next to the wall that is dominated by viscosity because the velocity must go to zero at the wall; this region is called the *viscous sublayer*. From experiment, the growth of the turbulent BL is given by

$$\delta/x \approx 0.16Re_x^{-1/7} \tag{11.8}$$

By substituting the definition of Re_x into Equation (11.8), we find that the growth of the turbulent BL is proportion to $x^{6/7}$; the turbulent BL grows more rapidly in the downstream direction than the laminar BL. If a value of Re_x near the transition to turbulence ($Re_x \approx 3 \times 10^5$) is substituted into Equations (11.7) and (11.8), we find that the turbulent BL is three to four times thicker near the transition point than the BL would have been if it had remained laminar.

11.3.3 Flow separation

If the pressure gradient in the flow is positive (and remember from Chapter 9 that this is an adverse pressure gradient because a downstream increase in pressure acts to decelerate the flow), the boundary layer flow may "separate" from the wall. By *separation*, we mean that the mean downstream current is separated from the wall by a region of slow, upstream-directed flow immediately next to the wall. Flow separation commonly occurs on a curved boundary, such as that shown in Figure 11.6.

As the diagram shows, the pressure at the outer margin of the BL is determined by the convergence or divergence of the streamlines. The BL is so thin that we may take the pressure within it to be equal to the pressure at its outer boundary (see Section 9.7). As the streamlines converge the velocity increases and, by Bernoulli's theorem, the pressure must decrease.

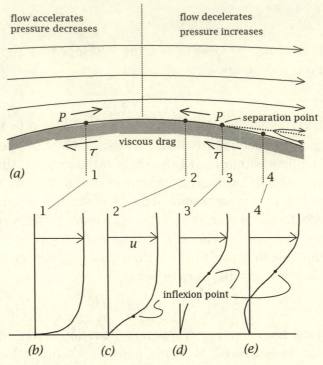

Fig. 11.6. Flow separation on a curved boundary. (*a*) Streamlines and pressures. (*b*)–(*e*) Velocity profiles at points shown on (*a*).

The reverse is true as the streamlines diverge. Outside the BL, viscous friction is negligible, so the acceleration and negative pressure gradient as the flow approaches the sphere are balanced by the deceleration and positive pressure gradient beyond it. In a completely inviscid flow, the flow exerts no net pressure force on the sphere.

Within the BL, however, the balance is upset by viscous friction. Where the streamlines (just outside the BL) converge, there is a negative pressure gradient which tends to accelerate the flow downstream, and thus to overcome the strong effect of viscous friction. Where the streamlines diverge, however, both the pressure gradient and viscous friction tend to decelerate the flow in the BL, and eventually the velocity close to the boundary is reduced to zero. Still further downstream, the flow is actually moving upstream close to the boundary.

The potential for flow separation is evident in the governing equation for flow in a laminar BL. If we adopt the standard BL assumptions (Section

9.7), the Navier–Stokes equation for steady flow in the x-direction becomes

$$\rho\left(u\frac{\partial u}{\partial x} + v\frac{\partial u}{\partial y}\right) = -\frac{\partial p}{\partial x} + \mu\frac{\partial^2 u}{\partial y^2} \tag{11.9}$$

At the wall, $u = v = 0$, so this equation reduces to

$$\frac{\partial p}{\partial x} = \mu\frac{\partial^2 u}{\partial y^2} \tag{11.10}$$

At the outer edge of the BL, the BL velocity profile must merge smoothly with that of the outer flow, for which $\partial u/\partial y = 0$. Because the velocity within the BL is less than that in the outer flow (so $\partial u/\partial y > 0$ in the BL), $\partial^2 u/\partial y^2$ must be negative in the outer region of the boundary layer. From Equation (11.10), we see that, if $\partial p/\partial x$ is negative, the curvature of the velocity profile at the wall is also negative, but if it is positive, $\partial^2 u/\partial y^2$ at the wall must also be positive. Because this curvature is always negative in the outer part of the BL, a positive pressure gradient requires an inflexion point in the velocity profile some distance from the wall (Figure 11.6(c)). If $\partial p/\partial x$ is sufficiently large, a point is reached where the shear stress and the velocity gradient at the wall is exactly zero (Figure 11.6(d)). At larger values of $\partial p/\partial x$ (Figure 11.6(e)), there is flow in the upstream direction along the wall, so flow must separate where the flow reversal takes place.

Downstream of the line of flow separation, there is a region of intense shear between fluid in the outer part of the BL, moving downstream, and flow in the inner part of the BL, moving upstream. Consequently the flow becomes unstable, and turbulence is generated.

Turbulence may be generated by shear without flow separation, but it tends to be suppressed by the presence of the solid boundary. Flow separation produces a zone of intense shear *away* from a solid boundary, and is therefore a powerful generator of turbulence. The eddies formed in this zone of shear rapidly grow in size and are carried (convected) downstream, thus producing a turbulent wake in the lee of the curved solid surface.

Turbulent wakes are characteristic of flow (at relatively high Reynolds numbers) past any kind of obstruction (e.g., grains settling through a fluid or resting on a bed with the flow passing over them). Wakes are also formed wherever there are sudden expansions in the flow, for example, in the lee of ripples or dunes, at places where a river channel suddenly widens, or at the margins of jets (as in the expanding flow at the mouths of deltaic distributary channels, or at volcanic vents).

Wakes are regions of reversed or low average velocities, so they tend to be regions of sediment deposition, rather than erosion. Over a ripple or dune,

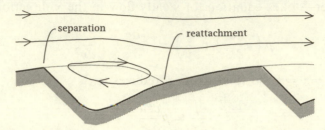

Fig. 11.7. Flow separation over a dune. Flow separates at the brink and reattaches on the stoss side, producing a reverse flow in the lee of the dune.

flow converges and accelerates up the back of the bedform, producing a zone of increasing shear stress, and therefore erosion of the bed due to increasing capacity to transport sediment (Figure 11.7). In the lee of the bedform brink, there is flow separation and expansion of flow. Grains that bounce or roll up the back are deposited at the brink (and eventually avalanche down the oversteepened lee slope). Some of the coarser grains taken into suspension on the back of the dune may be deposited in the wake, while the finer grains travel on in suspension. Similar processes take place, at larger scales, at channel expansions in rivers. Flow separation takes place at the margin of the channel, producing a large lateral eddy within which sediment may be deposited.

11.4 Reynolds stresses

Although the Navier–Stokes equation applies to both laminar and turbulent flow, we cannot solve it for the case of turbulent flow because each velocity and pressure term is a rapidly varying random function of time and space. In other words, there is no simple combination of the velocity vector v and pressure p that is a solution to the Navier–Stokes equation for turbulent flow. We can make substantial progress, however, if we substitute the time-averaged and fluctuating components for the instantaneous turbulent variables and then take the time average of the entire equation. There are good reasons to do this. First, it is often the time averaged properties of the flow which most interest us. Second, the time averaging process reveals which of the turbulent fluctuation terms are most important. To derive the time-averaged governing equations for turbulent flow, we replace each instantaneous velocity and pressure term in the governing equations with the sum of its mean and fluctuating component (Equation 11.3). Next, we take the time average of the entire equation. The time average $\bar{u}$ is defined by

$$\bar{u} = \frac{1}{T} \int_0^T u \, dt$$

where T, the averaging time, is long compared to the period of typical fluctuations, but short compared with the time over which the average flow changes. This is easy to write, but not always easy to achieve in real flows. For example, in large rivers, turbulent eddies may be hundreds of metres across, so to average out their effect requires times of the order of tens of minutes. In this time, the bed configuration may have changed due to migration of bedforms, or the discharge may have changed due to passage of a flood wave. The same problem, only worse, applies to many atmospheric and tidal flows.

The time averages of all fluctuating terms (u', v', w') are zero, by definition, and it may be shown that this is also true of their distance derivatives. Thus, the turbulent version of the continuity equation is no different from the laminar version, except that we now explicitly state that we are using time-averaged quantities:

$$\frac{\partial \bar{u}}{\partial x} + \frac{\partial \bar{v}}{\partial y} + \frac{\partial \bar{w}}{\partial z} = 0 \qquad (11.11)$$

To limit the number of terms we have to deal with, we will use the time-averaged BL equation, which is much shorter than the full Navier–Stokes equation and directly relevant to many natural flow problems. The time average of the full Navier–Stokes equation involves no additional important physical concepts, only more terms.

The x-component of the BL equation (Equation 11.9) is

$$\rho \left(u \frac{\partial u}{\partial x} + v \frac{\partial u}{\partial y} \right) = -\frac{\partial p}{\partial x} + \mu \frac{\partial^2 u}{\partial y^2}$$

Taking its time average yields

$$\rho \left(\bar{u} \frac{\partial \bar{u}}{\partial x} + \bar{v} \frac{\partial \bar{u}}{\partial y} \right) = -\frac{\partial \bar{p}}{\partial x} + \frac{\partial}{\partial y} \left(\mu \frac{\partial \bar{u}}{\partial y} - \rho \overline{u'v'} \right) \qquad (11.12)$$

We note that the time averages of products of fluctuating terms, such as $\overline{u'v'}$, are generally not zero.

Comparing the two equations, we can see that the main difference, other than the substitution of average values for instantaneous values, is the introduction of a new stress term. In the original BL equation, the stress term is the divergence of the viscous stress,

$$\frac{\partial \tau}{\partial y} = \mu \frac{\partial^2 u}{\partial y^2}$$

whereas in the time-averaged equation,

$$\frac{\partial \tau}{\partial y} = \frac{\partial}{\partial y} \left(\mu \frac{\partial \bar{u}}{\partial y} - \rho \overline{u'v'} \right) \tag{11.13}$$

In this equation, it appears that the viscous and turbulent stresses act in different directions but, in fact, they act in the same direction. This is so because when the average velocity increases in the positive y-direction (away from the boundary) a positive value of v' generally means that a fluid particle is moving to a region of higher $\bar{u}$, so that u' is negative, and the averaged cross-product $\overline{u'v'}$ is also negative. Thus

total stress = viscous stress + turbulent stress

The turbulent stress term is generally called the *Reynolds stress*, after Osborne Reynolds, who first made this analysis.

The turbulent stress is actually an inertia term left over from taking the time average of the left-hand side of Equation (11.9). It is included with the viscous stress because it functions physically and mathematically like a surface stress. In Equation (11.13), the turbulent stress term may be thought of as the mean vertical motion (due to v') of x-directed momentum (due to $\rho u'$). Just as viscosity in a gas can be explained as a transfer of momentum by movement of faster moving molecules into a layer of slower moving molecules, so the turbulent stress is caused by transfer of the momentum of fluid elements from faster-moving to slower-moving layers. In fact, this analogy was used by Prandtl in devising an early ("mixing length") theory of turbulence.

The analogous functions of the viscous and turbulent stresses can be made formally explicit by rewriting the stresses in Equation (11.13) as

$$\tau = (\mu + \eta) \frac{d\bar{u}}{dy} \tag{11.14}$$

Comparing this equation with Equation (11.13) shows that

$$\eta \frac{d\bar{u}}{dy} = -\rho(\overline{u'v'}) \tag{11.15}$$

η is termed the *dynamic eddy viscosity*. Dividing η by ρ gives the *kinematic eddy viscosity* ϵ, analogous to the kinematic (molecular) viscosity. The eddy viscosity is much larger than the molecular viscosity in turbulent flows, so the molecular term can be neglected, except close to the boundary. There the presence of the boundary inhibits turbulent fluctuations in the direction normal to it, so the turbulent stress disappears, and the very strong velocity gradient imposed by the no-slip condition produces a very large molecular stress.

It is often convenient to use a different way of writing the shear stress at the boundary, by defining the *shear velocity* $u_* = \sqrt{\tau_0/\rho}$. Then, close to the bed we may write

$$u_*^2 \approx \overline{u'v'}$$

That is, outside the viscous sublayer, the Reynolds stress is much larger than the viscous stress, and the shear velocity u_* is almost equal to the square root of the covariance of the turbulent velocity fluctuations. The correlation coefficient between u and v is given by

$$r^2 = \frac{\text{cov}(u',v')}{\text{rms}(u')\text{rms}(v')} \tag{11.16}$$

If we make the assumptions that in a turbulent BL this correlation is large (≈ 1) and that the two turbulent root-mean-squares are about the same order of magnitude, we have the further result that, close to the bed,

$$u_* \approx \text{rms}(v') \tag{11.17}$$

Some actual experimental data for boundary layers over smooth and rough boundaries are shown in Figure 11.8. The value of u_* is used to normalize the data. We see that, though $\text{rms}(u')$ is actually rather larger than $\text{rms}(v')$, $\text{rms}(v')$ is almost equal to u_* not only close to the boundary but also for most of the BL thickness. This is a useful result because we are often very interested in the intensity of turbulence near a boundary and it is much easier to measure the shear velocity u_* than the turbulence.

Writing the Reynolds stress as a viscosity does not make it easier to solve the BL equation (Equation 11.12), since we still need a method of predicting the eddy viscosity. Unlike the molecular viscosity, which is a property of the fluid, the eddy viscosity is a function of the flow (which in turn depends on the eddy viscosity). In practice, useful results can sometimes be obtained by assuming that the eddy viscosity is a constant (a technique still widely used in oceanography and meteorology). Approximate solutions can also be found by assuming that the distribution of the eddy viscosity, once determined for a particular geometry and Reynolds number, can be applied to other flow problems with similar properties.

In recent years, many "turbulence models" have been developed to predict

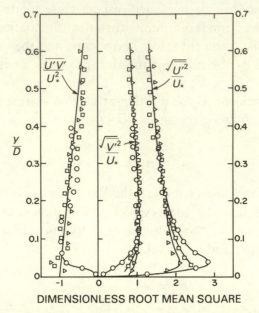

DIMENSIONLESS ROOT MEAN SQUARE

Fig. 11.8. The vertical variation in the dimensionless root-mean-square values of the velocity components u' and v' and their covariance in a boundary layer (after Grass, 1971). The circles are for a smooth boundary; the squares and triangles are for a rough boundary. Note the increase in the root-mean-square value of the vertical component v' close to the boundary, as the roughness increases.

the Reynolds stresses and to attempt to solve the turbulence "closure" problem, i.e., to obtain enough equations with known coefficients to solve the time-averaged Navier–Stokes equation. The models are designed for practical use (mostly in aeronautics), and make use of empirical coefficients based on experiment and engineering experience. One model, the so-called $k - \epsilon$ model, uses equations for the total turbulent energy k, and its rate of dissipation ϵ. The hydraulics applications have been reviewed by Rodi (1984).

11.5 Diffusion

In Chapter 6, we discussed diffusion in porous media flow. We return to the topic here because diffusion produced by turbulent eddies dominates the mixing in many natural systems. At the same time, we hope to provide a better understanding of the concept of momentum diffusion by considering the diffusion of a contaminant, or tracer, in turbulent flow.

11.5.1 Molecular diffusion

First, consider molecular diffusion in a static fluid. In 1885, the German biologist Adolf Fick (1829–1901) used a direct analogy with the flow of heat and electricity to deduce the relation governing the molecular diffusion of a material (*Fick's first law*):

$$J = -D\frac{\partial C}{\partial y} \tag{11.18}$$

where J is the mass flux per unit area (with dimensions $[ML^{-2}T^{-1}]$) of the tracer, D is the diffusion coefficient or *diffusivity* (with dimensions $[L^2T^{-1}]$), and C is the mass concentration of the tracer (with dimensions $[ML^{-3}]$). Equation (11.18) is directly analogous to Darcy's law (Equation 6.2) and Fourier's law for heat flux (Equation 12.2). Fick's law states that the mass flux of the tracer is linearly proportional to the gradient of its concentration. If the tracer mass is conserved (e.g., if there are no chemical reactions), continuity requires that the rate of change of tracer mass in an infinitesimal control volume must equal the flux across the control surface, which may be written in one dimension as

$$\frac{\partial C}{\partial t} = -\frac{\partial J}{\partial y} \tag{11.19}$$

Using Fick's first law to substitute for J gives

$$\frac{\partial C}{\partial t} = -\frac{\partial}{\partial y}\left(D\frac{\partial C}{\partial y}\right) \tag{11.20}$$

which is known as *Fick's second law*. If we assume that D does not vary with concentration (and therefore, with y), Fick's second law can be written for steady conditions as

$$\frac{d^2C}{dy^2} = 0 \tag{11.21}$$

which is the one-dimensional Laplace equation (Equation 6.41 and see Appendix D).

Fick's law applies equally well to molecular diffusion through a static fluid and diffusion across a laminar flow in a channel or duct. In both cases, the diffusion occurs via the random paths of individual fluid molecules. If this random motion occurs in the presence of a concentration gradient, there will be a net drift, or flux, of tracer down the concentration gradient.

A model that predicts Fickian diffusion can be constructed from assuming a simple "random-walk" model. Such a model (also called the "drunkard's walk") assumes that a single molecule follows a series of steps, each of which moves in a direction taken at random. Some observational support for such a model is provided by the "Brownian motion" of very small dust particles observed in liquids. The model not only leads to Fick's law of diffusion, but also predicts that tracer molecules advected by steady laminar flow from a point source will be distributed far downflow as a Gaussian (normal) distribution in the cross-flow direction. Suppose that at a distance x downflow, 95 per cent of the tracer molecules lie within $\pm\delta y$ of the meanflow direction, then, from the properties of the Gaussian distribution, $\delta y = 2\sigma_y$, where σ_y is the standard deviation of the tracer concentration in the y-direction.

We can also deduce that δy increases as the square root of x. Assuming that there is no velocity gradient in the y-direction, and that the rate of advection is much larger than the rate of diffusion in the x-direction, the amount of tracer within an incremental distance δx is independent of x. This means that the concentration gradient in the y-direction must decrease downflow at a rate inversely proportional to the width δy. If we further assume that the tracer is spreading out in all directions, the width is inversely proportional to the square root of the distance x. So the concentration gradient in the cross-flow direction is proportional to $\sqrt{x}$, and from Fick's law, the cross-flow flux must also be proportional to $\sqrt{x}$ – in other words, $\delta y \approx K\sqrt{x}$.

11.5.2 Turbulent diffusion

When the flow becomes turbulent, the rate of mixing increases dramatically. With turbulence, the diffusion process is not only much more rapid, it is also considerably more complex. For laminar flow, the diffusion coefficient D is a function of the fluid properties alone. In a turbulent flow, D also depends on properties of the flow, and, in particular, its velocity and depth, which control the rate of shearing within the flow. We state this formally by rewriting Fick's first law as

$$J = -(D_m + D_t)\frac{\partial C}{\partial y} \qquad (11.22)$$

where D_m and D_t are the molecular and turbulent diffusivities, respectively. In general, we expect D_t to be much larger than D_m.

The same type of reasoning that we used before to derive a random-walk model of molecular diffusion can be applied to phenomena in turbulent fluids, such as the spreading of a pollutant in the atmosphere as a "plume" downflow from a point source (e.g., a point near the top of a tall smoke

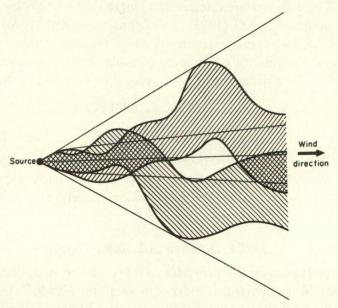

Fig. 11.9. Conical average form of a plume, due to large eddies perturbing smaller-scale parabolic plumes. (From Scorer 1968, Fig. 6 on p. 32).

stack). But there are several problems with this analogy. In deriving a Gaussian cross-flow distribution from a random-walk model, it is necessary to assume that the direction of each step is independent of the previous step. The corresponding assumption for turbulent velocity fluctuations is known to be false, because for short time intervals cross-flow velocities are known to be strongly correlated (e.g., $v(t)$ correlates strongly with $v(t + \delta t)$), and the correlation decreases as δt increases. The larger the eddy, the longer the correlation persists; in fact, the graph of the correlation coefficient against time can be used to define a mean size of the larger eddies (called the integral scale or *macroscale*). It is also known that, at large distances downflow, it is the large eddies that are most effective in causing cross-flow diffusion. So even though a simple Fickian model predicts that the width of a turbulent plume is proportional to $\sqrt{x}$, observation suggests that the rate of expansion is more nearly proportional to x, i.e., the plume has a conical, rather than a parabolic form (Figure 11.9).

Many other complications also affect real plumes: when gases emerge from a smokestack, they are usually hot and rise buoyantly, instead of being passively convected by the wind. There is generally some density (temperature) stratification, which makes turbulent exchange more difficult (or, if the stratification is unstable, easier) in the vertical than in the horizontal direction. These factors are discussed at length in the books by Scorer (1968, 1978), Pasquill and Smith (1983), and Lyons and Scott (1990). For neutral conditions (i.e., no density stratification) in the open country, Lyons and Scott suggest the following empirical formulae for the standard deviations (in metres) in the horizontal (y-) and vertical (z-) cross-flow directions:

$$\sigma_y = 0.08x(1 + 0.0001x)^{-1/2} \tag{11.23}$$

$$\sigma_z = 0.06x(1 + 0.0015x)^{-1/2} \tag{11.24}$$

The equations apply to distances between 100 m and 10 km. Note that, for small values of x, the equations predict a nearly linear expansion of the plume, but for large values of x they predict a nearly parabolic expansion.

11.5.3 Diffusion in boundary layers

From the discussion in the previous sections, we see that, because viscosity results from the molecular transport of momentum, Newton's law of viscosity is essentially a diffusion equation. We see this best by writing it in the form

$$\tau = \frac{\mu}{\rho} \frac{d(\rho \bar{u})}{dy}$$

$\rho \bar{u}$ is the fluid momentum per unit volume, τ is the flux of x-directed momentum in the y-direction and the kinematic viscosity μ/ρ is analogous to the molecular diffusion coefficient D_m.

In a turbulent BL, we can expand the equation to include the eddy viscosity. Referring back to Equation (11.14) shows that, when viscosity can be neglected, the turbulent diffusion coefficient for momentum is the kinematic eddy viscosity ϵ defined in Section 11.4. The term representing the turbulent momentum flux is the Reynolds stress $\rho(\overline{u'v'})$.

Returning now to the turbulent diffusion of a tracer, we need to relate the turbulent diffusion coefficient to the relevant properties of the flow. It is reasonable to suppose that the diffusion of a dissolved tracer takes place at about the same rate as the diffusion of the other molecules making up the fluid. In this case, the diffusion of the tracer should be nearly the same as the diffusion of momentum and $D_t = \beta \epsilon$, where β is a coefficient of order one.

We see, therefore, that the problem of turbulent diffusion can be reduced to the same problem we considered in the last section – how do we predict the eddy viscosity? Our discussion of diffusion in a turbulent plume shows that, even in the simplest type of turbulent flow, we cannot assume the eddy viscosity is independent of scale.

For more complex flows, it is generally still not possible to predict the eddy viscosity accurately, but there is an approximate solution for one case of great practical importance: the case of a turbulent boundary layer. In this case, we stand the problem on its head. The velocity distribution can be deduced by similarity arguments and the deduction is confirmed by experiments. We then deduce the eddy viscosity from the velocity distribution and use this to calculate the diffusion of a tracer carried by the flow.

11.6 Velocity distribution in a turbulent boundary layer

11.6.1 Viscous sublayer

Within a turbulent BL the diffusion of momentum by eddies is very efficient; therefore gradients in the mean velocity are relatively small. Yet the fluid in contact with the wall is at rest, so there must exist a region close to the wall where there is a very large velocity gradient. In this region viscous forces are very strong, and we can expect that the mean velocity gradient there is like that in a laminar BL. This region is called the *viscous sublayer* (Subsection 11.3.2). Its thickness δ is estimated experimentally to be

$$\delta \approx 12v/u_* \tag{11.25}$$

For most natural flows, this is very thin (less than 1 mm). Detailed flow visualization studies have shown that although flow in the sublayer is dominated by viscosity, it is not laminar, but fluctuates in a quasi-regular manner (Cantwell, 1981). Within the viscous sublayer, the velocity u should be a function only of the shear stress at the wall τ_0 (or alternatively the shear velocity u_*), the viscosity μ and the distance from the wall y. Dimensional analysis gives

$$\frac{\mu u}{\tau_0 y} = \text{constant}$$

Experiment (or theory) indicates the constant is unity, and the equation is generally written in the form

$$\frac{u}{u_*} = \frac{u_* y}{v} \tag{11.26}$$

In this equation, the two sides are both dimensionless. On the left-hand side, the mean velocity u is scaled by the shear velocity u_*. On the right-hand side, the distance from the wall y is scaled by v/u_*, which has the dimensions of length, and can be interpreted as a viscous length scale.

The meaning of such a length scale is made clearer by considering the effect on the velocity of small irregularities ("roughness elements") in the wall. If these roughness elements are entirely enclosed in the viscous sublayer, then the roughness has little effect on the rest of the flow and the wall is "hydraulically smooth". Suppose the roughness is characterized by a length k_S. Then we can define a *boundary Reynolds number* Re_* which is a ratio between this length and the viscous length scale:

$$Re_* = \frac{u_* k_s}{v} \qquad (11.27)$$

Experiments show that for $Re_* < 5$ the roughness has no direct effect on the velocity distribution outside the viscous sublayer. At the other extreme, for $Re_* > 60$, the viscous sublayer is so disrupted by the roughness elements that it effectively ceases to exist, and the velocity distribution does not depend on the viscosity (see below). So we can use the value of Re_* to define three different kinds of boundary: hydraulically smooth ($Re_* < 6$), transitional ($6 < Re_* < 60$), and hydraulically rough ($Re_* > 60$). Real roughness is a complex quantity, which can hardly be characterized by a single length scale. Luckily for earth scientists, the experiments (by Nikuradse, a student of Prandtl) that defined the roughness length scale k_s used the diameter of well-sorted natural sand grains.

11.6.2 Logarithmic layer

Next, we consider the velocity further from the wall. The velocity immediately next to the wall depends on the wall conditions represented by ku_*/v, so it will be difficult to derive a general equation for the velocity further from the wall. Instead, we seek a general relation by considering what factors control the *velocity gradient* $\partial u / \partial y$. This gradient does not necessarily depend on the particular velocities close to the wall. Now, we also argue that, if the flow is deep enough, there may exist a region within the flow that is at the same time far above the wall and far below the free surface (or the centre of the pipe, in pipe flow). If this is the case, we can argue that the velocity gradient in this region does not depend on any length scale other than y. In other words, in this zone $\partial u / \partial y$ does not vary strongly with k_s, v/u_*, or the flow depth, d. It turns out that we are often justified in making this assumption: many natural flows have depths that are at least two orders of magnitude greater than either k_s or δ.

If $\partial u / \partial y$ depends only on the wall shear stress (we will substitute u_* for τ_0) and the distance from the wall y, we have three variables in two fundamental dimensions, so the result is, again, a single dimensionless group:

$$\frac{u_*/y}{\partial u/\partial y} = \kappa \tag{11.28}$$

where the constant κ is called von Kármán's "universal" constant, and is observed from experiments to take a value in the range 0.38–0.41. Equation (11.28), sometimes called the "law of the wall", has been found experimentally to hold for many flow geometries (flow in pipes, in rectangular channels, in open channels, etc.). It has also been found to hold for both hydraulically rough and smooth boundaries, supporting our assumption that a single expression for a velocity gradient could satisfy both wall conditions. Equation (11.28) can be rearranged and immediately integrated, yielding

$$\frac{u}{u_*} = \frac{1}{\kappa} \ln y + C$$

or, moving the constant inside the logarithm,

$$\frac{u}{u_*} = \frac{1}{\kappa} \ln \left(\frac{y}{y_o} \right) \tag{11.29}$$

This version of Equation (11.28) is generally called the *logarithmic velocity distribution*. y_o is simply another form of the integration constant given by

$$y_o = e^{-\kappa C}$$

Inspection of Equation (11.29) shows that y_o is the height above the wall at which $u = 0$. Of course, we know that $u = 0$ at $y = 0$, not $y = y_o$, but this discrepancy is of no great concern because Equation (11.29) does not apply very close to the wall.

From our discussion of wall length scales, it follows that the actual velocity given by Equation (11.29) will depend on the relative magnitude of k_s and v/u_*, because these two parameters control the velocity profile near the bed. From experimental observations, for $k_s u_*/v < 5$, $y_o = v/9u_*$, and for $k_s u_*/v > 60$, $y_o = k_s/30$. For transitionally rough flow, the velocity depends on both wall length scales.

Finally, it turns out that Equation (11.29) applies with excellent accuracy throughout the entire boundary layer in pipes (excluding only the central part of the flow) and does a good job throughout the entire depth in open channel flow, except where the near-surface velocity is affected by the walls of the channel. These results are entirely empirical, so we have less confidence applying them to the entire BL than to the zone close to the bed, where we can argue that no length scale other than y is important. The results also apply only to relatively straight, uniform channels: we expect (and observe) significant departures from the logarithmic law if the channel is strongly curved, or if the depth of flow varies rapidly (as it does in some rivers, due to the presence of large bars or bed forms).

One valuable aspect of Equation (11.29) is that we can use the measured velocity distribution in a channel (and the known value of κ) to determine the eddy viscosity ϵ. We do this by combining the definition

$$\epsilon = \frac{\tau}{\rho(\partial \bar{u}/\partial y)} \tag{11.30}$$

derived from Equation (11.14) (neglecting viscosity), with the known linear distribution of shear stress in steady, uniform flow (Section 10.3),

$$\tau = \tau_0 \left(\frac{d - y}{d} \right) \tag{11.31}$$

where τ_0 is the stress at the bottom of the channel. Using Equation (11.28) to define $\partial \bar{u}/\partial y$ and Equation (11.31) to define τ, we can use Equation (11.30) to develop an expression for the turbulent kinematic eddy viscosity ϵ (see Review Problem 2).

$$\epsilon = u_* \left(\frac{d - y}{d} \right) \kappa y \tag{11.32}$$

This is the result mentioned at the end of Section 11.5. Of course, we had to assume the form of the velocity profile (Equation 11.28) in order to derive an expression for ϵ, whereas it would be preferable to determine ϵ independently and then calculate the velocity distribution. In addition, the velocity distribution we have used is strictly applicable only close to the wall and does not always apply throughout the flow. Nonetheless, wall-bounded shear flows are quite common and Equation (11.32), or a more complex empirical version that accounts better for the velocity distribution throughout the BL, provides a very useful approach to estimating the diffusion of momentum and tracers in a turbulent BL. We consider the turbulent diffusion of one tracer (sediment) in Review Problem 2.

Another common use of the logarithmic velocity distribution is to determine u_* (which may be difficult to measure in any other way) and the roughness height k_s. There are, however, some problems with this technique. Even in channels floored by sediment, k_s often differs from the diameter of the sediment, because the sediment is moulded by the flow into ripples or dunes, which act as additional roughness elements to the flow. If the dunes are large compared with the depth of the flow, the whole flow is constantly contracting and expanding, and the logarithmic profile may never become fully developed. In large flow systems, there are long-duration fluctutations in velocity produced by large turbulent eddies, so that measurements must be averaged over long times (tens of minutes) in order to obtain accurate average profiles. Experiments have also shown that a high concentration of coarse sediment moving in the flow produces departures in most of the flow from a simple logarithmic velocity distribution, which makes it difficult to apply Equation (11.29) to determine the shear velocity.

11.7 Review problems

(1) **Problem:** Suppose that a tracer is diffused from a region of high concentration (C_o at $y = 0$) toward a region (at $y = h$) where the concentration is kept very low (by constant replacement). Assume that the diffusion coefficient D is constant. What is the concentration profile?

Answer: This is a one-dimensional diffusion problem. We must integrate the Laplace equation (Equation 11.21) for the boundary conditions $C = C_o$ at $y = 0$ and $C = 0$ at $y = h$. The first integration gives

$$\frac{dC}{dy} = K_1$$

and the second gives

$$C = K_1 y + K_2$$

From the first boundary condition $K_2 = C_o$, and from the second

$$0 = K_1 h + C_o, \qquad K_1 = -C_o/h$$

So the solution is

$$C = C_o \left(1 - \frac{y}{h}\right)$$

(2) **Problem:**
  (a) Show how Equation (11.28) can be used to derive an equation for the turbulent diffusion coefficient D_t.
  (b) Assume that the material being diffused is suspended sediment, with a settling velocity w. Show that when there is a balance between settling and vertical diffusion the distribution of suspended sediment concentration C throughout the flow is given by:

$$\frac{C}{C_a} = \left[\left(\frac{d-y}{y}\right)\left(\frac{a}{d-a}\right)\right]^z$$

where $z = w/\beta \kappa u_*$ and $C = C_a$ at some reference level $y = a$.

Answer:

(a) From Equation (11.14), neglecting molecular viscosity,

$$\tau = \rho \epsilon \frac{d\bar{u}}{dy}$$

We can eliminate τ by using Equation (11.31):

$$\tau = \tau_0 \left(\frac{d-y}{d} \right)$$

Combining these two equations and converting from τ_0 to u_* gives

$$\epsilon = u_*^2 \left[\frac{d-y}{(d\bar{u}/dy)d} \right]$$

Now use Equation (11.28) to eliminate $d\bar{u}/dy$ $(= u_*/\kappa y)$,

$$\epsilon = u_* \left(\frac{d-y}{d} \right) \kappa y$$

By hypothesis $D_t = \beta \epsilon$ so

$$D_t = \beta u_* \left(\frac{d-y}{d} \right) \kappa y$$

(b) The rate of upward diffusion of sediment, through unit area, is given by

$$J_s = D_t \frac{dC}{dy}$$

The rate of downward settling of sediment is wC. Equating the two gives

$$\frac{dC}{C} = -\frac{wdy}{D_t}$$

Using the expression for D_t obtained in part (a) gives

$$\frac{dC}{C} = -\frac{(wd)dy}{\beta \kappa u_*(d-y)y}$$

Integrating this equation and evaluating the constant at $y = a$ gives the required result. The reference level is generally taken very close to the sediment bed, where the concentration C_a is measured or estimated theoretically.

(3) **Problem:** In Chapter 10, we presented constitutive equations for a variety of different types of substances. Write the Equation (11.28), applied to a turbulent BL, in the form of a constitutive equation. What kind of

rheological substance is a turbulent fluid? Why is this not generally true of all turbulent flows?

Answer: Equation (11.28) in this case is

$$\frac{d\bar{u}}{dy} = \frac{1}{\kappa y}\left(\frac{\tau_0}{\rho}\right)^{1/2}$$

Using the linear variation of τ with y (Equation 11.31) we can write this as

$$\frac{d\bar{u}}{dy} = \frac{1}{\kappa y}\left[\frac{d}{\rho(d-y)}\right]^{1/2}\tau^{1/2}$$

This is an equation relating the rate of shear to the square root of the shear stress, so there is an analogy with a non-Newtonian fluid. But it is not a real constitutive equation, because the rate of shear depends on the height above the bed y and there is no way of eliminating y from the equation. If we like, we can think of a turbulent fluid as a non-Newtonian fluid, with $n = 1/2$ and a value of a that varies nonlinearly with distance from the boundary (it approaches infinity near the boundary, decreases in the centre of the flow, and approaches infinity again near the free surface).

The equation cannot be applied to all turbulent flows because it is based on two equations that apply *only* to turbulent boundary layers. For different geometries or pressure distributions, the results might be quite different.

11.8 Suggested reading

Turbulence is a large subject, and only a few topics have been selected for discussion in this chapter. For further information the reader is referred to Middleton and Southard (1984) and the excellent text by Tennekes and Lumley (1972). Van Dyke (1982) has anthologized nearly 300 of the best photographs of fluid motion, many of which illustrate aspects of turbulence that cannot really be expressed in words or formulae.

Acheson, D. J., 1990. *Elementary Fluid Dynamics.* Oxford, Clarendon Press, 397 pp. (Vortex motion is analyzed in Chapter 5, boundary layers in Chapter 8, and turbulence in Chapter 9. TA357.A276)

Bergé, P., Y. Pomeau and C. Vidal, 1986. *Order Within Chaos. Towards a Deterministic Approach to Turbulence.* New York, John Wiley and Sons, 329 pp. (QA614.8.B4713)

Cantwell, B. J., 1981. Organized motion in turbulent flow. Ann. Rev. Fluid Mechanics, v.13, pp. 457–515.

Rodi, W., 1984. *Turbulence Models and Their Application in Hydraulics – A State of the Art Review*. The Netherlands, International Association of Hydraulic Research, second edition, 104 pp. (TA357.R63 1984)

Tennekes, H. and J. L. Lumley, 1972. *A First Course in Turbulence*. Cambridge MA, MIT Press, 300 pp. (QA913.T44)

Tritton, D. J., 1988. *Physical Fluid Dynamics*. Oxford, Clarendon Press, second edition, 519 pp. (QC151.T74)

Van Dyke, M., 1982. *An Album of Fluid Motion*. Stanford CA, The Parabolic Press, 176 pp. (TA357.V35)

Young, A. D., 1989. *Boundary Layers*. Washington DC, American Institute of Aeronautics and Astronautics, 269 pp. (TL574.B6. Y681)

12

Thermal convection

12.1 Introduction

Convection is the general term used in fluid mechanics for the transport of a fluid property by the mean motion of the fluid (the term *advection* is used by some authors). Dye that marks a fluid particle may be convected downstream, and so may the vorticity of the fluid. So too may be the temperature, or heat content.

In general, heat may be lost from (or gained by) a body by one of three mechanisms: (i) conduction, (ii) convection, and (iii) radiation. Gain of heat by radiation from the sun, and loss of heat by radiation from the atmosphere to space, are the dominant processes that determine the temperature of the surface of the earth. The interior of the earth is hot because of residual heat that was produced during the formation of the earth (mainly due to gravitational separation of the iron in the core from the silicates in the mantle) and because of heat produced by the decay of long-lived radioactive isotopes of potassium, uranium and thorium. Within solids or liquids, heat loss by radiation is negligible, because these materials are not transparent to infrared radiation. The heat is lost to the surface by conduction and convection.

Near the surface of the earth, the rigidity and viscosity of the lithosphere are too large to permit large-scale convection of the rock itself. Heat is lost to the surface mainly by conduction, and by the convection of pore fluids through permeable rocks, for example, at the mid-ocean ridges and in sedimentary basins. But within the mantle, most of the transfer of heat takes place by thermal convection, rather than by conduction. If this had not been the case, radioactive heat would have been generated faster than it could be conducted to the surface, and the mantle would have heated up until it either melted, or convection began (Stevenson and Turner, 1979). The heat flow in the ocean basins is such that if it were solely due to conduction, the temperature gradient would have to be so high that the mantle would melt at depths of about 100 km (Turcotte and Schubert, 1982, p.144).

Thermal convection is important not only within the mantle, but also within the atmosphere. Climate is determined partly by the balance between the radiation coming to the earth from the sun and the radiation leaving the earth for space, and partly by the way that heat is convected within the atmosphere by the general circulation: hot air rises from the equator and sinks at the tropics, and cold air sinks at the poles and rises again at the polar fronts (e.g., Strahler, 1971). Thus convection is a process of great importance to the earth sciences.

Before considering thermal convection itself, however, it is essential to understand the elements of heat conduction.

12.2 Conduction

The fundamental equation for the conduction of heat, and the mathematical techniques necessary to solve problems in heat flow were discovered by the French scientist and statesman, Joseph Fourier (1768–1830). *Fourier's law* states that the total heat flow $\partial Q/\partial t$ is proportional to the cross-sectional area A, normal to the direction of heat flow (say, the x-direction) and the temperature gradient in that direction $\partial T/\partial x$:

$$\frac{\partial Q}{\partial t} = -kA\frac{\partial T}{\partial x} \tag{12.1}$$

The coefficient of proportionality k, is called the thermal conductivity. The dimensions of k are $[MLT^{-3}\Theta^{-1}]$, where Θ is temperature: in SI units, they are generally given in terms of power, as watts per metre per kelvin. Near-surface rocks generally have values of about 2–3 W $m^{-1}K^{-1}$. The thermal conductivities of mudrocks are low (2 W $m^{-1}K^{-1}$), those of sandstones slightly higher (3 W $m^{-1}K^{-1}$). A few rocks (notably salt) have substantially higher values (6 W $m^{-1}K^{-1}$).

The heat flux (heat flow per unit area) is frequently written as q, so that Equation (12.1) becomes

$$q = -k\frac{\partial T}{\partial x} \tag{12.2}$$

In this form we see clearly that Fourier's law is a diffusion equation of the type discussed in Section 6.5 and Section 11.5. The heat flux near the surface of the earth is generally close to one heat flow unit (hfu), which is defined as 10^{-6} calories per square centimetre per second ($= 41.84$ m W m^{-2}). This

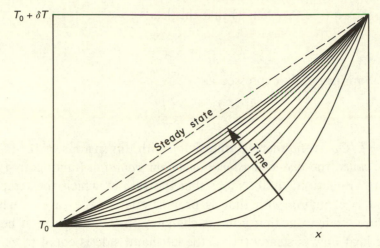

Fig. 12.1. Conduction of heat into a region of lower temperature (on the left) from a boundary (on the right) suddenly raised in temperature. Distance is normalized by the total thickness of the conducting region, and temperature by the temperature increment δT.

is a very small quantity compared to the solar heat flux, so it has no effect on the climate of the earth. It is insufficient to prevent the development of permafrost near the poles, though it does produce a measurable warming of the water in the lowest metre or two of the ocean basins.

If the heat flux is steady ($\partial T/\partial t = 0$), then Fourier's law of heat conduction can be combined with an equation of continuity for heat (assuming no internal heat production) to give Laplace's equation (Equation 6.41).

Often, geologists are more interested in the unsteady case. For example, how will the temperature vary with both time and distance from the intrusive contact when a thick sill of hot magma is emplaced in a flat-lying set of strata? This problem is an example of a one-dimensional heat flow problem. In its simplest form, suppose that one side of a uniform layer of material (of density ρ) is suddenly raised in temperature by an increment δT. Heat starts to diffuse into the region of lower temperature, but at first the temperature gradient is much larger close to the warmer side than further away from it (Figure 12.1). So long as a variable temperature gradient persists, more heat flows into a volume element from one side than leaves from the other: this heat is used to increase the temperature of the element, and the amount used per unit volume depends on the density and specific heat C. (C is defined as the heat necessary to raise the temperature of one kilogram by one kelvin. For the moment, we need not be concerned whether this is the specific heat at constant volume, C_v, or at constant pressure, C_p.) From simple continuity, it can be shown (see Problem 1) that this leads to

$$C\rho\frac{\partial T}{\partial t} = k\frac{\partial^2 T}{\partial x^2}$$ (12.3)

which is more generally expressed as

$$\frac{\partial T}{\partial t} = \kappa\frac{\partial^2 T}{\partial x^2}$$ (12.4)

where $\kappa = k/C\rho$ is the thermal diffusivity, with dimensions of $[L^2T^{-1}]$. This equation, called the *one-dimensional heat flow equation* (analogous to Fick's second law, Equation 11.20), states that the rate at which the temperature at a point is changing with time is proportional to the rate at which the temperature gradient at that point is changing in the direction of heat flow. So, if the heat flow is steady (i.e., if the left-hand side is equal to zero) then the temperature gradient must be a constant: the temperature must vary linearly in the direction of heat flow.

If the temperature gradient is not initially linear (as in the example of the intrusion of a sill), then as heat diffuses into a layer of thickness d the temperature gradient will eventually become almost linear and the heat flow almost uniform from one side of the layer to the other. The example shown in Figure 12.1 was computed numerically using the computer program EHEAT.

The *thermal relaxation time*, a characteristic time for heat to diffuse through the layer, may be obtained by dividing d^2 by κ:

$$t_T = d^2/\kappa$$ (12.5)

Figure 12.1 shows temperature gradients for ten increments of time, up to $t = 0.2t_T$. The dashed line is the linear, steady-state gradient: it is approached closely at times greater than $0.5t_T$.

In this text, we do not pursue the many interesting applications of heat conduction to the earth sciences, but introduce the topic only for its relevance to thermal convection. For a more complete discussion of heat flow problems of geological interest see Turcotte and Schubert (1982, Chapter 4).

12.3 Thermal instability

Thermal convection results from the fact that, when heated, most solids and fluids expand, and thus decrease in density. If heat is flowing upwards, this must be because a hotter, less dense layer lies below a colder, more dense layer. It is expected that this reverse density gradient will be gravitationally unstable and that convection will therefore occur. This may not be the case, however, if buoyancy is opposed by the strength of the material. Even if the layered material is a fluid, which has no strength, convection will not take place if conduction cools a particle of warmer fluid faster than it can rise by buoyancy. To understand the conditions under which this might be the case, we have to compare two characteristic times, the thermal relaxation time and the time it takes for the warmer fluid to rise up through the colder fluid.

We are only concerned with relative magnitudes, so we ignore numerical coefficients and use a single "characteristic length" d throughout this discussion. The expansion produced in a volume of size d by a temperature rise of δT is of order $\alpha \delta T d^3$, where α is the coefficient of thermal expansion, so the buoyancy force is of order

$$F_G \approx (g\rho\alpha\delta T)\,d^3 \qquad (12.6)$$

The force opposing the rise is due to viscosity and is given by Stokes' law (Equation 9.44):

$$F_V \approx \mu d v \qquad (12.7)$$

where v is the speed of rise.

If v is constant, the two forces must be equal. The time taken to rise a distance d is then given by

$$t_B = \frac{d}{v} \approx \frac{\mu}{(g\rho\alpha\delta T)\,d} \qquad (12.8)$$

For instability to persist, this time must be less than the thermal relaxation time due to conduction. The ratio of the two times is the *Rayleigh number* Ra:

$$Ra = \frac{t_T}{t_B} = \frac{(g\rho\alpha\delta T)\,d^3}{\mu\kappa} \qquad (12.9)$$

The Rayleigh number is named for Lord Rayleigh (John W. Strutt, 1842–1919) who first worked out a theory of thermal convection in 1916. It is expected that there will be a critical value of the Rayleigh number below which heat will be transmitted by conduction only, and any density instability will be damped out by viscosity. Above this value, the layer will become unstable and convection will begin. For a layer of uniform thickness and infinite width, the critical value may be calculated theoretically to be $Ra = 1707$ (Bergé and Dubois, 1984).

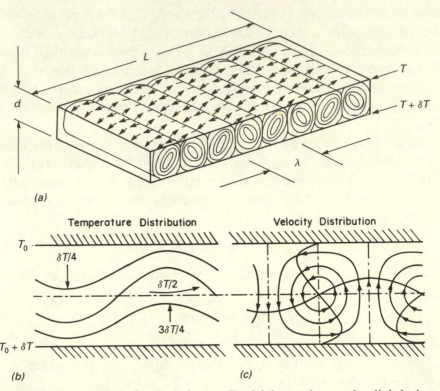

(a)

Temperature Distribution   Velocity Distribution

(b)   (c)

Fig. 12.2. (*a*) The pattern of convection at Rayleigh numbers only slightly larger than the critical value. For a vertical slice parallel to the long side of the rectangle, (*b*) gives the temperature distribution and (*c*) gives the velocity distribution. From Croquette (1989, Fig. 1).

12.4 Geometry of convection

The discussion given above indicates when convection will begin, but it gives no hint of what the pattern of flow will be. The boundary conditions generally specified are a horizontal layer of fluid with a uniform thickness d that is much smaller than the minimum horizontal dimension. The lower boundary is maintained at a higher temperature than the upper boundary, and gravity acts vertically. This type of convection is called Rayleigh–Bénard convection, after Lord Rayleigh and the French scientist, H. Bénard, who published the first experimental investigations in 1910 (it was discovered later that many of Bénard's results were strongly influenced by the surface tension of the fluid at the free surface – so they are not good examples of what is now called Bénard convection!). It has been found experimentally that the pattern is very sensitive to the geometry of the experimental enclosure, and also that it varies with Rayleigh number. At Rayleigh numbers just

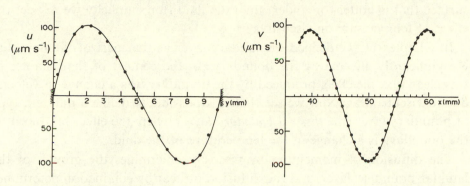

Fig. 12.3. Variation in the x- and y-components of velocity across a convection roll (after Bergé and Dubois, 1984). $Pr = 900$, $d = 1$ cm. u was measured along a vertical traverse passing through the centre of a roll and v was measured along a horizontal traverse at $y = d/2$.

above the critical value, in rectangular boxes whose horizontal dimensions are large relative to the vertical dimension, convection takes place in a series of regular, straight cylindrical rolls, whose horizontal dimension is almost equal to the vertical dimension d. The general pattern of motion is shown in Figure 12.2. The variation in velocity components in the horizontal and vertical directions, u and v, at a given level (for constant x or y, respectively) is almost sinusoidal, as shown in Figure 12.3:

$$u = u_{max} \sin(ay + \psi) \tag{12.10}$$

$$v = v_{max} \cos(ax + \psi) \tag{12.11}$$

where ψ is the phase lag, $a = 2\pi d/\lambda$ is the wavenumber, and λ is the wavelength, i.e., the distance between two ascending or descending limbs.

At higher Rayleigh numbers transverse oscillations of the convection rolls develop, and eventually the motion becomes turbulent. The type of convective motion depends not only on the Rayleigh number, but also on the *Prandtl number*,

$$Pr = v/\kappa \tag{12.12}$$

The Prandtl number is the ratio of two diffusion coefficients, the kinematic viscosity v which determines the diffusion of momentum by viscosity, and the thermal diffusivity κ which determines the diffusion of heat. The significance of the Prandtl number is best illustrated by considering the simultaneous development of a boundary layer in a viscous fluid moving past a solid surface that is hotter (or colder) than the fluid (for simplicity we neglect the effect of temperature on viscosity).

In Chapter 11, we defined a boundary layer as that part of the flow that is significantly affected by the boundary. In the context of that chapter, it was clear that the effect produced by the boundary was a change in velocity due to viscous drag. Now we see that it is possible to define a different type of boundary layer (the *thermal boundary layer*) where the effect produced by the boundary is a change in the temperature of the fluid.

The diffusion of momentum by viscosity determines the growth of the laminar boundary layer, and the diffusion of heat by conduction determines the growth of the thermal boundary layer. If the Prandtl number is unity, the two boundary layers grow at identical rates. If the Prandtl number is very large (as it is in the mantle, where it is of the order of 10^{23}) then the whole concept of a viscous boundary layer loses meaning – all parts of the convective motion are strongly affected by the boundary – but the thermal boundary layer remains well defined, because there is a strong temperature gradient close to the boundary. If the Prandtl number is relatively small, as it is in air (0.7), then heat diffuses more rapidly than momentum, and the thermal boundary layer is thicker than the viscous boundary layer. If the Prandtl number is greater than unity, as in water (7–8) or silicone oil (100–1000), the reverse is true.

At low Rayleigh numbers, thermal convection clearly shows a sequence of progressively more complex instabilities, leading ultimately to turbulence. Experiments on such instabilities are described by Bergé (1979) and Libchaber and Maurer (1980). Many of the experiments were carried out at very small scales (in cells less than one centimetre across) using silicone oil or liquid helium. For both liquids, it is relatively easy to change the Prandtl number: this is particularly the case for liquid helium, whose Prandtl number varies from about 0.4 to 0.8 as the temperature changes by a few degrees near absolute zero. These experiments have aroused much interest because they clearly show phenomena predicted by the theory of low-dimensional nonlinear dynamic chaos (see below).

Another dimensionless number often used in studies of heat convection is the ratio between the total heat flux, due to both conduction and convection, and the heat flux due only to conduction. This is the *Nusselt number Nu*:

$$Nu = qd/k\delta T \qquad\qquad (12.13)$$

Obviously, the more vigorous the convection, the larger the Nusselt number.

12.5 Equations of motion

More equations are needed to account for thermal convection than the Navier–Stokes equation, because account must be taken of the variation in density of the fluid with temperature. It is usual to make two simplifying assumptions.

(1) **Uniform viscosity.** This is certainly an over-simplification, for two reasons: (a) temperature generally has a strong effect on viscosity, even for a Newtonian fluid; and (b) the mantle almost certainly behaves as a non-Newtonian fluid, whose apparent viscosity is sensitive both to temperature and to variation in the rates of strain.

(2) **The Boussinesq approximation.** A change in temperature produces convection by changing the density, thus producing buoyancy forces even in a fluid that does not have a free surface. A change in density also produces a change in momentum, but the resulting change in the inertia forces is very small compared with the buoyancy forces. Thus, the change in the inertia forces may safely be neglected. The same assumption works well in dealing with flows in density-stratified fluids such as the atmosphere and ocean.

Using these assumptions the governing equations for two-dimensional convection are:

Continuity:

$$\frac{\partial u}{\partial x} + \frac{\partial v}{\partial y} = 0 \tag{12.14}$$

Navier–Stokes:

 x-direction:

$$\frac{\partial u}{\partial t} + u\frac{\partial u}{\partial x} + v\frac{\partial u}{\partial y} = -\frac{1}{\rho}\frac{\partial p}{\partial x} + v\left(\frac{\partial^2 u}{\partial x^2} + \frac{\partial^2 u}{\partial y^2}\right) \tag{12.15}$$

 y-direction:

$$\frac{\partial v}{\partial t} + u\frac{\partial v}{\partial x} + v\frac{\partial v}{\partial y} = -\frac{1}{\rho}\frac{\partial p}{\partial y} + v\left(\frac{\partial^2 v}{\partial x^2} + \frac{\partial^2 v}{\partial y^2}\right) + \frac{g\delta\rho}{\rho} \tag{12.16}$$

where $\delta\rho$ is the difference in density between the two layers.

Heat Flow:

$$\frac{\partial T}{\partial t} + u\frac{\partial T}{\partial x} + v\frac{\partial T}{\partial y} = \kappa\left(\frac{\partial^2 T}{\partial x^2} + \frac{\partial^2 T}{\partial y^2}\right) \tag{12.17}$$

The equation of continuity is unchanged, because of the Boussinesq approximation. The Navier–Stokes equations are modified only by the introduction of the buoyancy force $g(\delta\rho/\rho)$ in the vertical (y-) direction. The heat flow equation is a two-dimensional form of Equation (12.4), generalized to admit the possibility of convective changes, as well as local changes in temperature (see Section 9.3).

Just as the Navier–Stokes equation can be cast into dimensionless form, with the result that the Reynolds number appears naturally as a coefficient of the dimensionless viscosity force term (see Section 9.4), so a dimensionless form may be obtained for the governing equations for thermal convection (Croquette, 1989). In this case, the Rayleigh number appears as a coefficient on the right-hand side of the heat flow equation and the Prandtl number appears as a coefficient of all the terms on the right-hand side of the Navier–Stokes equation.

The critical Rayleigh number Ra_c at which convection begins can be derived theoretically from perturbing a linearized form of the governing equations given above. But the linear analysis does not give a realistic solution for the way the applied sinusoidal perturbation grows beyond the initial stage: in other words, it cannot be used to predict the flow pattern even at Prandtl numbers only just above the critical value.

The linear analysis fails because thermal convection is essentially a nonlinear phenomenon: it involves an important negative feedback, as follows. The cause of the vertical movement is the vertical gradient in density resulting from a temperature gradient. Vertical flow, however, tends to even out the vertical temperature and reduce the density gradient. Thus, an initially linear temperature gradient changes to a gradient which is steep near the boundaries (where the vertical component of velocity is small) to one which is almost uniform in the centre (Figure 12.4). At low values of the ratio $r = Ra/Ra_c$ this feedback leads to steady convective rolls, which become larger as r increases.

12.6 Lorenz equations

The first attempt at a nonlinear analysis of convection was that of Lorenz (1960, 1963, 1964). His 1963 paper is now famous as the key paper in nonlinear dynamical chaos; but it is also interesting as an attempt to study thermal convection. The Lorenz equations have generated a large literature: the following discussion follows Croquette (1989); see also Bergé *et al.* (1986) and Tritton (1988).

To obtain a linearized solution of the governing equations, the nonlinear

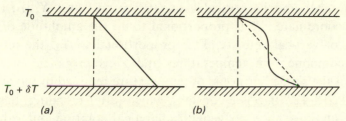

Fig. 12.4. Temperature gradient, averaged horizontally: (*a*) for conduction, and (*b*) across a convecting cell. From Croquette (1989, Fig. 1).

terms (such as the convective terms in the Navier–Stokes and heat flow equations) are simply discarded. The fundamental linear term in the temperature field (i.e., the variation of T in the x- and y-directions) is given by

$$T(x, y) = T_1 \sin(\pi y/d) \cos(ax) \tag{12.18}$$

and the linear term of the vertical component v of the velocity field is given by

$$v(x, y) = v_1 \sin(\pi y/d) \cos(ax) \tag{12.19}$$

These fields are simply generalizations of Equations (12.10) and (12.11). In nonlinear convection there are also terms involving multiples (harmonics) of the fundamental frequencies ($1/2d$ and $a/2\pi$): in theory, an infinite number might have to be considered. Lorenz chose to consider only the simplest possible nonlinear problem by retaining only one harmonic of the temperature field:

$$T'(x, y) = T_2 \sin(2\pi y/d) \tag{12.20}$$

This is a variation in temperature produced by the nonlinear convective term on the left-hand side of the heat flow equation. It corresponds essentially to the departures observed in Figure 12.4 from a linear temperature gradient: thus it is the most important effect of nonlinearity in convection.

Using these simplifications, Lorenz was able to reduce the governing equations to a set of three nonlinear ordinary differential equations:

$$\frac{dX}{dt} = -Pr(X - Y)$$
$$\frac{dY}{dt} = -Y + rX - ZX \tag{12.21}$$
$$\frac{dZ}{dt} = -b(Z - XY)$$

In these equations, Pr is the Prandtl number, and r is the ratio of the

Rayleigh number and its critical value (as before), but X, Y, and Z are not spatial coordinates. X is proportional to v_1, the amplitude of the linear component of vertical velocity, Y is proportional to T_1, the amplitude of the linear component of temperature, and Z is proportional to T_2, the amplitude of the single harmonic of temperature retained in the analysis (see Equations 12.18–12.20). b is a constant, which (after rescaling) has the value 8/3. The equations have no general analytical solution but can easily be integrated numerically.

It has been found that the Lorenz equations do correctly predict some aspects of thermal convection (e.g., heat flux) for a range of r values close to the critical value of unity. At large values of r ($r > 24$), the Lorenz equations predict a chaotic form of motion, in which the direction of convection reverses at apparently random intervals of time. Very small differences in the initial conditions produce very large differences in the solutions obtained at large times. It is these aspects of the equations that have generated the most interest. For such high Rayleigh numbers, it is clear that the Lorenz equations do not give a correct description of real Rayleigh-Bénard convection. They do describe certain other physical systems, however, such as a type of water wheel, and thermal convection in an upright circular tube (see Tritton, 1988, pp. 246–54).

The chaotic behaviour of the Lorenz equations results from their non-linearity, which arises directly from the convective term in the heat flow equation. At the very high Prandtl numbers characteristic of the earth's mantle, all the terms on the left-hand side of the Navier–Stokes and heat flow equations can be neglected, so the Lorenz approximation is not a suitable model of mantle convection. Stewart and Turcotte (1989) have considered the case of high Prandtl number: they extended Lorenz's original analysis by retaining as many as 40 nonlinear harmonics derived from the nonlinear terms on the right-hand side of the heat flow equation. They have found that these equations also have chaotic regimes. So it is quite possible that convective patterns in the mantle change in an apparently random fashion with time.

Lorenz's original analysis was made in order to throw some light on atmospheric convection. Many meteorologists now accept that nonlinear chaos may be one reason why it is not possible to predict the weather accurately for more than a few days: but it has not yet been fully demonstrated that this lack of predictability does arise from nonlinear chaos. The older explanation, that unpredictability arises from the very large number of different factors that may affect the weather, may still turn out to be correct. Though non-linear theory suggests that long-range prediction of some dynamic systems may be impossible, it has also developed techniques to *improve* the accuracy of short-term predictions.

12.7 Numerical simulation of convection

We have seen that it is possible to simplify the governing equations at very high Prandtl numbers. The behaviour of the full remaining set of equations may be simulated by numerical solution. The simplest case to investigate is two-dimensional convection in a rectangular "box" (which imposes artificial limits on the geometry of the convective rolls), but it is also possible to investigate convection within a spherical annulus, and to incorporate the effects of a rotating frame (Coriolis forces). The problem with all such simulations is that, because the flows are unsteady, it is necessary to use finite difference methods, and the demands on the capacity and speed of the computers are very large. This is also true of attempts to simulate the weather by using numerical models of the atmosphere (the so-called general circulation models).

Recent reviews of mantle convection have been given by Schubert (1992) and by Jarvis (1991); see also his contribution in the book edited by Peltier (1989). Some recent examples of "whole earth" simulations are given by Bercovici *et al.* (1989) and Machetel and Weber (1991). Most workers now believe that convection takes place not just in the asthenosphere but throughout the whole mantle, in spite of phase changes produced by the pressure gradient. The effect of temperature on viscosity has been incorporated in models studied by Christensen and Harder (1991).

12.8 Review problems

(1) **Problem:** Use the continuity equation for heat to derive the heat flow equation from Fourier's law.

Answer: Continuity for heat means that if there are no internal sources the heat flow into a small element must exceed the heat flow out, by the quantity required to raise the temperature of the element. The heat flows per unit area into and out of the element are given by Equation (12.2). Let the dimensions of the element be δx, δy, and δz. Using the same reasoning as we used in Section 6.6, we find that, for unit time the difference between the heat inflow and outflow in the x-direction is

$$\delta q(\delta y \delta z) = \frac{\partial q}{\partial x} \delta x (\delta y \delta z)$$

$$= \frac{\partial}{\partial x}\left(k\frac{\partial T}{\partial x}\right)\delta x \delta y \delta z$$

$$= k\frac{\partial^2 T}{\partial x^2}\delta x \delta y \delta z$$

Now, from the definition of heat capacity, a nett heat flow $\delta q(\delta y \delta z)$ will increase the temperature of the element at a rate $\partial T/\partial t$, where

$$\delta q(\delta y \delta z) = \rho \delta x \delta y \delta z C \frac{\partial T}{\partial t}$$

Equating these two expressions for the nett heat flow into the element,

$$\rho C \frac{\partial T}{\partial t} = k\frac{\partial^2 T}{\partial x^2}$$

which is the one-dimensional heat flow equation.

(2) **Problem:** Estimate the Rayleigh number for the asthenosphere. What temperature difference between the top and bottom is necessary to make it unstable?

Answer: To solve this problem we have to estimate several properties of the asthenosphere: its thickness ($d \approx 2 \times 10^5$ m), its density ($\rho \approx 3300$ kg m^{-3}), its viscosity ($\mu \approx 10^{19}$ Pa s), its thermal conductivity ($k \approx 3$ W m^{-1} K^{-1}), its heat capacity ($C \approx 10^3$ J kg^{-1} K^{-1}), and its coefficient of thermal expansion ($\alpha \approx 2 \times 10^{-5}$ K^{-1}). Putting these figures in the expression for the Rayleigh number gives

$$Ra \approx 570\delta T$$

A realistic estimate for the temperature difference between the top and base of the asthenosphere might be 100–200 K, so we can conclude that, using these data, the asthenosphere should be unstable. Extending the analysis to the whole mantle, or to the upper mantle (some 600+ kilometres thick), would also indicate instability.

12.9 Suggested reading

Bergé, P. and M. Dubois, 1984. Rayleigh–Bénard convection. Contemporary Physics, v.25, pp. 535–82.

Bergé, P., Y. Pomeau and C. Vidal, 1986. *Order within Chaos: Towards a Deterministic Approach to Turbulence.* New York, John Wiley and Sons, English edition, translated from the 1984 French edition by Laurette Tuckerman, 329 pp. (QA614.8.B4713)

Lorenz, E. N., 1963. Deterministic non-periodic flow. Jour. Atmospheric Sci., v.20, pp. 130–41 (The classic paper on chaos, and very readable).

Schubert, G., 1992. Numerical models of mantle convection. Ann. Rev. Fluid Mechanics, v.24, pp. 359–94.

Stevenson, D. J. and J. S. Turner, 1979. Fluid models of mantle convection. In M. W. McElhinny, ed., *The Earth, Its Origin, Evolution, and Structure*. New York, Academic Press, pp. 227–63. (QE35.G18)

Tritton, D. J., 1988. *Physical Fluid Dynamics*. Oxford, Clarendon Press, second edition, 519 pp. (Excellent extended treatment of thermal convection. QC151.T74)

Appendix A
List of symbols and vector notation

A.1 List of symbols

Dimensions are given where appropriate, in the mass [M], length [L], time [T], temperature [Θ], current [A] system. [O] means dimensionless. The corresponding SI units are the kilogram (kg), the metre (m), the second (s), the kelvin (K), and the ampere (A). Other common quantities and their SI units are as follows: force, the newton (1 N: 1 kg m s^{-2}); stress, the pascal (1 Pa = 1 kg m^{-1} s^{-2}); energy, the joule (1 J = 1 kg m^2 s^{-2}); temperature, degree Celsius (273.15 K = 0 °C).

Commonly used constants, dimensions or conversion factors include the following:

π	3.142	ratio of circumference to diameter
e	2.718	base of natural logs
g	9.81 m s^{-2}	acceleration due to gravity
G	6.673 $\times 10^{-11}$ N m^2 kg^{-2}	universal gravitational constant
R	6.37 $\times 10^6$ m	radius of earth
M	5.974 $\times 10^{24}$ kg	mass of earth

One calendar year is equal to 3.154 $\times 10^7$ seconds.

The algebraic symbols used for quantities in the text are given in the following list, which includes the dimensions of each quantity.

a	constant, cross-sectional area of tube [L^2], inverse of non-Newtonian viscosity
a	acceleration [LT^{-2}]
A	azimuth, point, area [L^2]

410

b	constant
B	Bingham number [O], point
c	constant, celerity [LT^{-1}], stiffness [$ML^{-1}T^{-2}$]
C	constant or coefficient, celerity [LT^{-1}],
	Chézy coefficient [$L^{1/2}T^{-1}$],
	cohesion [$ML^{-1}T^{-2}$], volume concentration [O],
	mass concentration [ML^{-3}]
C_D	drag coefficient [O]
C_L	lift coefficient [O]
C_v	coefficient of consolidation [L^2T^{-1}]
d	angle of plunge [O], dip [O], depth [L]
D	diameter [L], flexural rigidity [MLT^{-2}],
	diffusion coefficient [L^2T^{-1}]
e	elongation [O]
e_{ijk}	alternating symbol; see Table A.2
E	energy [ML^2T^{-2}], Young's modulus [$ML^{-1}T^{-2}$]
f	coefficient of friction [O], Coriolis parameter [O],
	function
f_b	boundary porosity [O]
F	formation factor [O], safety factor [O]
Fr	Froude number [O]
$\mathbf{F}$	force [MLT^{-2}]
$\mathbf{F}_D$	drag force [MLT^{-2}]
g	acceleration due to gravity [LT^{-2}]
G	universal gravitational constant [$M^{-1}L^3T^{-2}$],
	rigidity modulus [$ML^{-1}T^{-2}$]
$\mathbf{G}$	gravitational force [MLT^{-2}]
h	height [L], head [L]
H	height [L]
i	index, hydraulic gradient [O]
$\mathbf{i}$	unit vector
$\mathbf{I}$	angular momentum [ML^2T^{-1}]
I, II, III	invariants of stress (or strain) matrix

j	index
j	unit vector
J	mass flux due to diffusion $[ML^{-2}T^{-1}]$
k	constant, coefficient of earth pressure [O], bulk modulus $[ML^{-1}T^{-2}]$, permeability $[L^2]$, strength of plastic $[ML^{-1}T^{-2}]$, thermal conductivity $[MLT^{-3}\Theta^{-1}]$
K	hydraulic conductivity $[LT^{-1}]$
l	length $[L]$
ℓ	length $[L]$, direction cosine [O]
L	vector mean length $[L]$, wavelength $[L]$, length $[L]$
m	mass $[M]$, moment, direction cosine [O]
m_v	coefficient of compressibility $[M^{-1}LT^2]$
M	moment of inertia $[ML^2]$, Mach number [O], mass of earth $[M]$
M	moment of force $[ML^2T^{-2}]$
n	direction cosine [O], exponent
N	total number
Nu	Nusselt number [O]
N	normal force $[MLT^{-2}]$
p	pressure $[ML^{-1}T^{-2}]$
P	point, power $[ML^2T^{-3}]$, component of stress vector $[ML^{-1}T^{-2}]$, porosity [O]
Pr	Prandtl number [O]
P	pressure force $[MLT^{-2}]$
q	unit discharge $[LT^{-1}]$, unit heat flow $[MT^{-3}]$
q^*	pore velocity $[LT^{-1}]$
Q	discharge (volume rate of flow) $[L^3T^{-1}]$, heat $[ML^2T^{-2}]$
r	distance $[L]$, distance from origin $[L]$, radius $[L]$, electrical resistance $[ML^2T^{-3}A^{-2}]$
R	radius $[L]$, electrical resistivity $[ML^3T^{-3}A^{-2}]$
Re	Reynolds number [O]

s	distance in direction of motion [L], compliance [$M^{-1}LT^2$]
sf	scale factor [O]
S	slope [O], strength [$ML^{-1}T^{-2}$], specific surface area [L^{-1}]
t	time [T]
T	temperature [Θ], tortuosity [O], thickness [L], time [T]
$\mathbf{T}$	torque [ML^2T^{-2}], shear force [MLT^{-2}]
u	x-component of displacement [L] or velocity [LT^{-1}], neutral stress (pore pressure) [$ML^{-1}T^{-2}$]
u_*	shear velocity [LT^{-1}]
U	velocity averaged over depth [LT^{-1}]
v	y-component of displacement [L] or velocity [LT^{-1}]
V	sum of east–west components of azimuths [O], volume [L^3]
w	z-component of displacement [L] or velocity [LT^{-1}], settling velocity [LT^{-1}], work per unit mass [L^2T^{-2}]
W	sum of north–south components of azimuths [O], work [ML^2T^{-2}]
$\mathbf{W}$	weight [MLT^{-2}]
x	coordinate
$\bar{x}$	mean value
y	coordinate
z	coordinate
α	angle with x-axis [O], angle of slope [O], angle [O], coefficient of thermal expansion [Θ^{-1}], dispersivity [L]
$\boldsymbol{\alpha}$	angular acceleration [T^{-2}]
β	angle with y-axis [O], angular shear strain [O]
γ	specific weight [$ML^{-2}T^{-2}$], (total) shear strain [O]
δ	increment
δ_{ij}	Kronecker's delta

ϵ	engineering component of strain [O], kinematic eddy viscosity $[L^2T^{-1}]$
ε	elongation [O], strain [O], rate of strain $[T^{-1}]$ (in Chapter 9)
$\varepsilon_1,\ \varepsilon_2,\ \varepsilon_3$	principal strains [O]
η	dynamic eddy viscosity $[ML^{-1}T^{-1}]$
θ	angle with x-axis in cylindrical coordinates [O], angle with z-axis in spherical coordinates [O], co-latitude [O], dilatation or compression [O]
κ	von Karman's constant [O], thermal diffusivity $[L^2T^{-1}]$
λ	Lamé's constant $[ML^{-1}T^{-2}]$, second coefficient of viscosity $[ML^{-1}T^{-1}]$
Λ	matrix of direction cosines
μ	coefficient of dynamic viscosity $[ML^{-1}T^{-1}]$, Lamé's constant $[ML^{-1}T^{-2}]$, coefficient of friction [O]
ν	Poisson's ratio [O], kinematic viscosity $[L^2T^{-1}]$
Π	dimensionless product
ρ	density $[ML^{-3}]$
σ	stress, generally normal stress, $[ML^{-1}T^{-2}]$
$\sigma_1,\ \sigma_2,\ \sigma_3$	principal stresses $[ML^{-1}T^{-2}]$
τ	stress or strength, generally shear stress $[ML^{-1}T^{-2}]$
ϕ	angle of friction [O], fluid potential $[L^2T^{-2}]$, angle from x-axis in spherical coordinates [O], longitude [O], angle of curvature [O], rate of dilatation $[T^{-1}]$
ψ	phase lag [O], angle of friction [O] (in Chapter 6 only)
ω	angular rate of rotation $[T^{-1}]$, vorticity $[T^{-1}]$, infinitesimal rotation [O]
Ω	angular rate of rotation of earth $[T^{-1}]$
Ω	rotation tensor

A.2 Vector and index notation

Operation	Vector Notation	Index Notation (for ith component)
vector	$\mathbf{u}$	u_i
scalar product	$x = \mathbf{u} \cdot \mathbf{v}$	$x = u_j v_j$
vector product	$\mathbf{w} = \mathbf{u} \times \mathbf{v}$	$w_i = e_{ijk} u_j v_k$
gradient of scalar	grad $\phi = \nabla\phi$	$\dfrac{\partial\phi}{\partial x_i}$
divergence of vector	div $\mathbf{u} = \nabla \cdot \mathbf{u}$	$\dfrac{\partial u_j}{\partial x_j}$
curl of vector	curl $\mathbf{u} = \nabla \times \mathbf{u}$	$e_{ijk}\dfrac{\partial u_k}{\partial x_j}$
Laplacian	$\nabla^2\mathbf{u} = \nabla \cdot \nabla\mathbf{u}$	$\dfrac{\partial}{\partial x_j}\left(\dfrac{\partial u_i}{\partial x_j}\right) = \dfrac{\partial^2 u_i}{\partial x_j \partial x_j}$

e_{ijk} is the *alternating symbol* defined by

$e_{ijk} = 0$ if any two of i, j, k are equal

$\phantom{e_{ijk}} = 1$ if ijk is an even permutation

$\phantom{e_{ijk}} = -1$ if ijk is an odd permutation

An odd permutation is one that involves an odd number of transpositions, for example $123 \rightarrow 213$. An even permutation is one that involves an even number of transpositions, for example $123 \rightarrow 312$.

Appendix B
Properties of common fluids and rocks

B.1 Density and viscosity of fluids and rocks at surface temperature (20 °C) and surface pressure

Substance	Density (kg m^{-3})	Viscosity (Pa s)
methane	—	1.1×10^{-5}
air	1.2	1.8×10^{-5}
alcohol	785	1.197×10^{-3}
gasoline	660–90	3–6×10^{-4}
water	998	1.00×10^{-3}
seawater	1037	1.08×10^{-3}
mercury	13 550	1.552×10^{-3}
kerosene	810–20	2×10^{-3}
crude oils	850–950	0.01–0.1
glycerine	1262	1.495
80% in water	1209	0.0618
10% in water	1024	1.307×10^{-3}
motor oil (SAE 30)	≈ 870	0.5
liquid mud	1100	$0.003 - 1$
castor oil	969	1
honey	—	~ 10
Lyle's Golden Syrup	—	120
plasticine	1450	(nonlinear)
silicone putty	—	$10^4 - 10^5$
pitch	1070	10^6 (at 50 °C)–10^{10}
glacier ice	920	10^{13}
rock salt	2180	3×10^{15}
glass	2400–800	$10^{18} - 10^{21}$
Solenhofen limestone	2620	$> 10^{21}$

Note that an increase in temperature increases the viscosity of gases, but decreases the viscosity of liquids. Values for the viscosity of rocks are apparent viscosities at low strain rates; in reality, most rocks show non-Newtonian flow behaviour.

B.2 Variation in viscosity of fluids with temperature

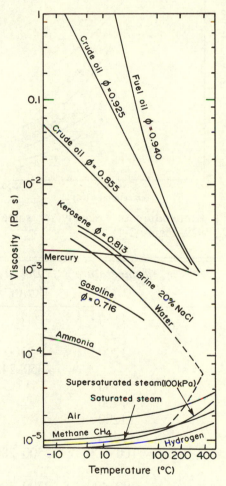

Fig. B.1. Variation with temperature of dynamic viscosity for some common fluids. ϕ is the density at 15 °C relative to that of water at 15 °C.

B.3 Density and viscosity at high temperatures

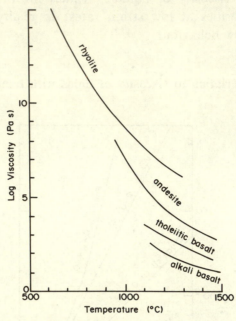

Fig. B.2. Variation with temperature of viscosity for some igneous rocks (after Williams and McBirney, 1979, Figure 2-4).

Substance	Temperature (°C)	Density (kg m^{-3})	Viscosity (Pa s)
basalt lava	1150		10^3–10^4
	1400		10
basalt magma	1200	2700	10^2–10^3
obsidian glass		2350–2450	10^8–10^9
silica glass	1100		4×10^{14}
	1400		5×10^9
olivine	1200		3×10^2
	1400		16
sedimentary rocks	100–200	2500–2800	
crust			
continental	300–400	2700–3000	
oceanic	100–200	2900	
lithosphere (below crust)	800–1200	3600	10^{22}–10^{23}
asthenosphere	1300–1400	3300	10^{19}–10^{20}
mantle (below asthenosphere)	1400–2400+	3400–5000+	10^{21}

B.4 Density, unconfined strength, and elastic moduli of some common rocks

Substance	Density (10³ kg m⁻³)	Uniaxial compressive strength (MPa)	Uniaxial tensile strength (MPa)	Young's modulus (GPa)	Poisson's ratio
shale	2.4–2.8	10–150	10	10–30	0.1–0.2
sandstone	2.1–2.8	30–90	1–15	10–60	0.2–0.3
limestone	2.2–2.7	40–120	5–10	10–80	0.25–0.3
dolomite	2.4–2.8	100	10	10–60	—
salt	2.18	—	—	0.03	0.05–0.1
granite	2.6–2.75	100–230	13–14	10–70	0.1–0.25
basalt	2.6–2.8	80–120	9	30–80	0.15–0.25
dunite	3.2–3.3	—	—	140–160	—

Densities of sedimentary rocks increase with depth of burial and may be predicted using empirical formulae such as those proposed by Sclater and Christie (1980). At high confining pressures, the strengths of rocks are approximated by Byerlee's law (see Chapter 4). A dash indicates that no data are known to the authors.

B.5 Density and strength of soils and sediments

Properties of uncompacted, uncemented soils and sediments vary widely, depending on their composition and stress history. Many properties correlate with grain size, which is used to give a rough classification of different materials in the following table. Silt sizes range from 0.002 mm to 0.062 mm, sand from 0.062 mm to 2 mm; grains coarser than 2 mm in size are referred to as gravel. The strength is related to the apparent cohesion c' and friction angle ϕ' by the Navier–Coulomb criterion:

$$s' = c' + \sigma_e \tan \phi'$$

where σ_e is the effective normal stress. For further discussion of soil cohesion and friction, see Chapter 4.

Material	Dry density ρ_b (kg m^{-3})	Apparent cohesion c' (kPa)	Apparent friction angle ϕ' (degrees)
sandy gravel	1440–2360	≈ 0	35–50
clean sand (uniform)	1350–1920	≈ 0	27–35
clean sand (poor sorting)	1380–2230	≈ 0	33–45
silty sand	1410–2060	≈ 0	27–34
cohesive soil (preconsolidation pressure < 1 MPa)	1300–1800	0–10	20–35 (peak) 8–35 (residual)

The above table permits the calculation of drained strengths (where pore pressure is hydrostatic). Undrained strengths C_u vary widely, depending on water content. They may be estimated very roughly from the following manual tests:

soft: easily moulded by fingers $C_u < 40$ kPa

firm: moulded by strong finger pressure   50 kPa $< C_u < 75$ kPa

stiff: can be indented by thumb   100 kPa $< C_u < 150$ kPa
        but not moulded by fingers

B.6 Sources of Data

Landolt, H. H., ed., 1982. *Physical Properties of Rocks*. Volume 1 of *Landolt–Börnstein Numerical Data and Functional Relationships in Science and Technology, Group V: Geophysics and Space Research*, Subvolume a, ed. V. Cermák *et al.*, 373 pp.; Subvolume b, ed. M. Beblo *et al.*, 604 pp., New York, Springer-Verlag.

Carmichael, R. S., 1989. *Practical Handbook of Physical Properties of Rocks and Minerals*. Boca Raton FL, CRC Press, 741 pp.

Carter, N. L., 1976. Steady state flow of rocks. Rev. Geophysics Space Physics, v.14, pp. 301–60.

Clark, S. P., ed., 1966. *Handbook of Physical Constants*. Geol. Soc. America Memoir 97, 587 pp. (Q199.N25)

Doake, C. S. M. and E. W. Wolff, 1985. Flow law for ice in polar ice sheets. Nature, v.314, pp. 255–7.

Emiliani, C., 1987. *Dictionary of the Physical Sciences: Terms, Formulas, Data*. Oxford University Press, 365 pp.

Emiliani, C., 1988. *The Scientific Companion: Exploring the Physical World with Facts, Figures, and Formulas*. New York, John Wiley and Sons, 287 pp.

Ferguson, C. C., 1983. Composite flow laws derived from high temperature experimental data on limestone and marble. Tectonophysics, v.95, pp. 253–66.

Forsythe, W. E., ed., 1959. *Smithsonian Physical Tables*. Washington DC, Smithsonian Institution, nineth revised edition, 827 pp.

Grim, R. E., 1962. *Applied Clay Mineralogy*. New York, McGraw-Hill, 422 pp. (TA455.C55G7).

Grolier, J., A. Fernandez, M. Hucher, and J. Riss, 1991. *Les Propriétés Physiques des Roches: Théories et Modèles*. Paris, Masson, 462 pp.

Gueguen, Y. and A. Nicolas, 1980. Deformation of mantle rocks. Ann. Rev. Earth Planetary Sci., v.8, pp. 119–44.

Heard, H. C., 1976. Comparison of the flow properties of rocks at crustal conditions. Phil. Trans. Royal Soc. London, v.A283, pp. 173–86.

Heard, H. C., I. Y. Borg, N. L. Carter and C. B. Rayleigh, eds., 1972. *Flow and Fracture of Rocks*. Amer. Geophysical Union, Geophysical Monograph Series 16.

Kaye, G. W. C. and T. H. Laby, eds., 1973. *Tables of Physical and Chemical Constants and Some Mathematical Functions*. New York, Longman, fourteenth edition, 386 pp.

McBirney, A. R. and T. Murase, 1984. Rheological properties of magmas. Ann. Rev. Earth Planetary Sci., v.12, pp. 337–57.

Poirier, J.-P., 1991. *Introduction to the Physics of the Earth's Interior*. Cambridge University Press, 264 pp.

Ryan, M. P. and J. Y. K. Blevins, 1987. The viscosity of synthetic and natural silicate melts and glasses at high temperatures and 1 bar (10^5 pascals) pressure and at higher pressures. US Geological Survey Bull. No. 1764, 563 pp.

Sclater, J. G. and P. A. F. Christie, 1980. Continental stretching: an explanation of the post-mid-Cretaceous subsidence of the central North Sea Basin. Jour. Geophys. Res., v.85, pp. 3711–39.

Touloukian, Y. S., W. R. Judd and R. F. Roy, 1989. *Physical Properties of Rocks and Minerals*. Volume II-2 of *CINDAS Data Series on Material Properties*, New York, Hemisphere, 548 pp.

Weijermars, R. and H. Schmeling, 1986. Scaling of Newtonian and non-Newtonian fluid dynamics without inertia for quantitative modelling of rock flow due to gravity (including the concept of rheological similarity). Phys. Earth Planetary Interiors, v.43, pp. 316–30.

Appendix C
Sets of linear equations

C.1 Introduction

Many commonly used algebraic and numerical methods were first developed in order to solve sets of linear equations. Consider the following simple example:

$$2x_1 + 4x_2 = 2 \qquad\qquad\text{(C.1)}$$

$$4x_1 + 11x_2 = 1 \qquad\qquad\text{(C.2)}$$

To solve this set of equations means finding the values of the unknown variables x_1 and x_2, that satisfy both of the equations. What does this mean graphically? Figure C.1 shows a plot of the two equations, and we can see that the lines representing the equations intersect at one point. At this point, the same values of x_1 and x_2 satisfy both equations, so this point, $(3, -1)$, is the solution we require.

How do we obtain this solution numerically? In this case, the procedure is

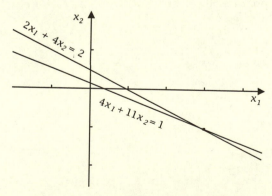

Fig. C.1. Intersection of straight lines.

422

simple: we eliminate one of the variables (say x_1) from the pair of equations, to give a solution for the other. To do this we multiply Equation (C.1) by 2 and then subtract this newly created equation from Equation (C.2).

$$
\begin{array}{r}
4x_1 + 11x_2 = 1 \\
- (4x_1 + \ 8x_2 = 4) \\
\hline
3x_2 = -3
\end{array}
$$

so $x_2 = -1$. Then the value of x_1 is easily obtained by substitution in either of the original equations. Note that it does not change the validity of the first equation to multiply through by a constant. More specifically, multiplying an equation by a constant does not change the values of the unknown variables.

From the graphical representation of the two equations, we can imagine that a solution does not always exist: it might be that the two lines would be parallel, in which case they would never intersect.

We now reformulate what we have just carried out for a general set of two equations:

$$a_{11}x_1 + a_{12}x_2 = b_1 \qquad (C.3)$$
$$a_{21}x_1 + a_{22}x_2 = b_2 \qquad (C.4)$$

A systematic way to eliminate x_1 from the second of these equations is to reduce the coefficients of x to unity, and subtract the first equation from the second. To do this we divide the first equation through by a_{11} and the second equation by a_{21}, and then subtract the first from the second. This leaves us with the following pair of equations:

$$x_1 + (a_{12}/a_{11})x_2 = b_1/a_{11}$$
$$(a_{22}/a_{21} - a_{12}/a_{11})x_2 = b_2/a_{21} - b_1/a_{11}$$

Now, to obtain a solution for x_2, we have to divide the second equation by $a_{22}/a_{21} - a_{12}/a_{11}$. If this expression is equal to zero, then no solution is possible. Note that a_{12}/a_{11} is the slope of the line defined by the first equation, and a_{22}/a_{21} is the slope of the line defined by the second. So we have confirmed algebraically that the condition for no solution is that the slopes of the two lines are equal.

C.2 Determinants

The two equations (C.1) and (C.2) can be represented by one matrix equation

$$\begin{bmatrix} 2 & 4 \\ 4 & 11 \end{bmatrix} \begin{bmatrix} x_1 \\ x_2 \end{bmatrix} = \begin{bmatrix} 2 \\ 1 \end{bmatrix} \tag{C.5}$$

whose general form is given by

$$\begin{bmatrix} a_{11} & a_{12} \\ a_{21} & a_{22} \end{bmatrix} \begin{bmatrix} x_1 \\ x_2 \end{bmatrix} = \begin{bmatrix} b_1 \\ b_2 \end{bmatrix} \tag{C.6}$$

The arrays of numbers or algebraic symbols are called *matrices* and may be written concisely as $[a_{ij}]$, $[x_j]$, $[b_i]$ or $\mathbf{A}$, $\mathbf{x}$, $\mathbf{b}$. Thus, we may write Equation (C.6) as $[a_{ij}][x_j] = [b_i]$ or

$$\mathbf{A}\mathbf{x} = \mathbf{b} \tag{C.7}$$

The algebra of matrices is defined so that linear equations can be written in this way (see Chapter 4, Subsection 4.17.4). The *determinant* of the 2×2 matrix $[a_{ij}]$ is defined as

$$|a_{ij}| = (a_{11}a_{22} - a_{12}a_{21}) \tag{C.8}$$
$$= (a_{22}/a_{21} - a_{12}/a_{11})a_{21}a_{11} \tag{C.9}$$

We see, therefore, that if the slopes of the two lines are equal, then the determinant of $\mathbf{A}$ is equal to zero. A matrix whose determinant is equal to zero is called *singular*, and it is a necessary and sufficient condition for the existence of a solution to a set of linear equations that the matrix $\mathbf{A}$ should be nonsingular (though we have not provided the full proof).

The generalization to three dimensions is:

$$|a_{ij}| = a_{11}a_{22}a_{33} + a_{12}a_{23}a_{31} + a_{13}a_{21}a_{32}$$
$$-a_{12}a_{21}a_{33} - a_{11}a_{23}a_{32} - a_{13}a_{22}a_{31} \tag{C.10}$$

Several different ways of remembering this formula have been proposed.

(1) In the expansion, the first subscripts are always 123, and the second subscripts are all possible permutations on 123, i.e. 123, 231, 312, 213, 132, and 321. The first three of these permutations involve either no exchange of indices (123) or two exchanges (e.g., 123 to 321 to 312). The remainder involve only one exchange of indices. Those terms that involve no or even exchanges are positive, and those that involve odd exchanges are negative.

One way to write this concisely is by using the *alternating symbol* e_{ijk}

which is defined to be equal to 1 if ijk is an even permutation of 123, -1 if ijk is an odd permutation of 123, and zero if any two of ijk are equal. Using this notation, and the Einstein summation convention (where a repeated index must be summed), a 3×3 determinant may be written

$$|a_{pq}| = e_{ijk}e_{rst}a_{ir}a_{js}a_{kt}/6 \tag{C.11}$$

(2) If we write out the array of indices in $[a_{ij}]$ twice, as follows:

$$
\begin{array}{cccccc}
11 & 12 & 13 & 11 & 12 & 13 \\
21 & 22 & 23 & 21 & 22 & 23 \\
31 & 32 & 33 & 31 & 32 & 33
\end{array}
$$

we can see that the various products in the formula are diagonals of the array: the positive terms are diagonals sloping down to the right, the negative terms are diagonals sloping down to the left. For example $a_{11}a_{22}a_{33}$ is positive, $a_{13}a_{22}a_{31}$ is negative. Note that we do not use the last column, to avoid duplication.

(3) The full 3×3 determinant can be written in terms of 2×2 determinants as follows:

$$|a_{ij}| = a_{11} \begin{vmatrix} a_{22} & a_{23} \\ a_{32} & a_{33} \end{vmatrix} + a_{12} \begin{vmatrix} a_{23} & a_{21} \\ a_{33} & a_{31} \end{vmatrix}$$

$$+ a_{13} \begin{vmatrix} a_{21} & a_{22} \\ a_{31} & a_{32} \end{vmatrix} \tag{C.12}$$

Determinants formed by eliminating one row and column of a larger determinant are known as *cofactors* of the element defined by the row and column removed. In the expansion given above, the first term is a_{11} multiplied by its cofactor, and the second is a_{12} multiplied by the negative of its cofactor.

Returning to the case of two linear equations, the solution for x_1 may be written

$$x_1 = \frac{b_1 a_{21} - b_2 a_{11}}{a_{11}a_{22} - a_{12}a_{21}} = \frac{\begin{vmatrix} b_1 & a_{11} \\ b_2 & a_{21} \end{vmatrix}}{\begin{vmatrix} a_{11} & a_{12} \\ a_{21} & a_{22} \end{vmatrix}} \tag{C.13}$$

that is, as the quotient (ratio) of two determinants.

Similarly it is possible to write a solution for a 3×3 matrix in terms of determinants (Cramer's rule, see Potter and Goldberg, 1987).

Note also that we can define a matrix $\mathbf{A}^{-1}$, called the *inverse* of the square matrix $\mathbf{A}$, which has the property:

$$\mathbf{A}\mathbf{A}^{-1} = \mathbf{A}^{-1}\mathbf{A} = \mathbf{I} \qquad\qquad (C.14)$$

where $\mathbf{I}$ is the unit matrix. Then the solution to the linear matrix equation (C.7), $\mathbf{A}\mathbf{x} = \mathbf{b}$, which is a matrix form of Equations (C.3) and (C.4), is formally given by multiplying each side of this equation by $\mathbf{A}^{-1}$:

$$\mathbf{x} = \mathbf{A}^{-1}\mathbf{b} \qquad\qquad (C.15)$$

Though this is a useful notation, it does not help much with the practical problem of how to solve sets of linear equations. The numerical solution of linear equations proves to be simpler than the numerical determination of the inverse matrix – though the same methods can be used for both.

In the procedure InvMat1, included in the program MATLIB, Cramer's rule is used to determine the inverse of 2×2 and 3×3 matrices. But, in general it is easier to solve linear equations, and invert matrices, using a different method, which will be discussed below.

C.3 Gaussian elimination

There are many different numerical methods in use for solving large sets of linear equations (or for determining the inverse of large matrices). For an excellent guide to these methods, with computer programs, see Press *et al.* (1989).

One of the simplest, but most effective, methods for solving systems of equations is an extension of the method used in Section C.1. Suppose the coefficients of a set of linear equations such as Equations (C.3) and (C.4) are given by a matrix $\mathbf{A}$. We proceed systematically to reduce the principal diagonal elements of $\mathbf{A}$ to unity and to eliminate the "lower half" of the matrix (i.e., the part below the principal diagonal, the elements a_{ij} for which $i > j$) by invoking procedures that set these elements equal to zero.

We have already seen how to do this for a 2×2 matrix. Next, we give an example of the systematic procedure for a set of three equations:

$$
\begin{array}{rrrr}
13x_1 & -8x_2 & -3x_3 = & 20 \\
-8x_1 & +10x_2 & -x_3 = & -5 \\
-3x_1 & -x_2 & +11x_3 = & 0
\end{array}
$$

• **Step 1:** write this set of coefficients as an array:

$$
\begin{array}{rrrr}
13 & -8 & -3 & 20 \\
-8 & 10 & -1 & -5 \\
-3 & -1 & 11 & 0
\end{array}
$$

To reduce the elements of the first column to 1 0 0:

  (i) divide the row 1 by $a_{11} = 13$;

 (ii) multiply row 1 by a_{21} and subtract it from row 2;

(iii) perform a similar operation for row 3.

The result of (i) is that row 1 becomes 1 -0.61 -0.23 1.5, and the result of (i), (ii), and (iii) together is to give the array

$$
\begin{array}{rrrr}
1 & -0.61 & -0.23 & 1.5 \\
0 & 5.1 & -2.8 & 7.3 \\
0 & -2.8 & 10.3 & 4.6
\end{array}
$$

For simplicity, only two significant figures are shown, but it is essential that the actual calculations be carried out more precisely. In fact, for large matrices, preserving adequate precision can be a challenging numerical problem in its own right.

- **Step 2:** now make $a_{22} = 1$ and $a_{32} = 0$, using two operations very similar to those in Step 1. That is, we divide row 2 by (the new) a_{22}, then multiply row 2 by a_{32} and subtract from row 3. The result is

$$
\begin{array}{rrrr}
1 & -0.61 & -0.23 & 1.5 \\
0 & 1 & -0.56 & 1.4 \\
0 & 0 & 8.7 & 8.7
\end{array}
$$

Note that there is no need to change the first row after the first step.

- **Step 3:** make $a_{33} = 1$ by dividing the third row by 8.7:

$$
\begin{array}{rrrr}
1 & -0.61 & -0.23 & 1.5 \\
0 & 1 & -0.56 & 1.4 \\
0 & 0 & 1 & 1
\end{array}
$$

This corresponds to the equations:

$$
\begin{aligned}
x_1 - 0.61x_2 - 0.23x_3 &= 1.5 \\
x_2 - 0.56x_3 &= 1.4 \\
x_3 &= 1
\end{aligned}
\tag{C.16}
$$

At this point we have a solution for x_3 and it would be easy to solve for x_1 and x_2 by substitution. We may take the systematic operation one step further, and obtain the solution by eliminating all nondiagonal elements of **A**, as follows.

- **Step 4:**

  (i) Eliminate a_{12} by multiplying row 2 by a_{12} and subtracting row 2 from row 1.

(ii) Eliminate a_{13} by multiplying row 3 by a_{13} and subtracting row 3 from row 1.

(iii) Eliminate a_{23} by multiplying row 3 by a_{23} and subtracting row 3 from row 2.

The result is

$$\begin{array}{cccc} 1 & 0 & 0 & 3 \\ 0 & 1 & 0 & 2 \\ 0 & 0 & 1 & 1 \end{array}$$

When we write this array of coefficients as equations we directly obtain $x_1 = 3$, $x_2 = 2$ and $x_3 = 1$. Note also that the first three columns are the unit matrix **I**.

C.4 Triangular matrices

If half of a matrix (above or below the principal diagonal) is composed of zeros it is described as triangular. Let us represent a triangular matrix as **U** if the lower half has the zeros, and **L** if the upper half has the zeros. Then Equations (C.16) can be written

$$\mathbf{Ux} = \mathbf{c}$$

If we define a lower triangular matrix **L** such that

$$\mathbf{LU} = \mathbf{A}$$

where $\mathbf{Ax} = \mathbf{b}$ is the original set of linear equations, then it turns out that

$$\mathbf{Lc} = \mathbf{b}$$

Such triangular matrices have many interesting properties, discussed in texts on linear algebra. One property useful in computation is

$$|\mathbf{U}| = |\mathbf{A}|$$

Because the lower half of **U** is composed of zeros, the determinant has a very simple form: it is simply the product of the diagonal elements of **U** (also called the *trace* of **U**). If **U** has been determined as part of the process of Gaussian elimination, it is very easy to determine the value of the determinant by the same method.

A computer implementation of the Gaussian elimination method is given in the program GAUJORD.

C.5 Suggested reading

Potter, M. C. and J. Goldberg, 1987. *Mathematical Methods.* Englewood Cliffs NJ, Prentice-Hall, second edition, 639 pp. (QA37.2P668)

Press, W. H., B. P. Flannery, S. A. Teukolsky and W. T. Vettering, 1989. *Numerical Recipes in Pascal: The Art of Scientific Computing.* Cambridge University Press, 759 pp. (QA76.73.P2N87)

Appendix D
Partial derivatives and differential equations

D.1 Differentiation

If we have a function of only one variable, $u = f(x)$, then its derivative is defined as

$$\frac{du}{dx} = \lim_{\delta x \to 0} \frac{f(x + \delta x) - f(x)}{\delta x} = f'(x) = f' \qquad \text{(D.1)}$$

If f is a function of more than one variable, $u = f(x, y)$, we correspondingly define two (or more) partial derivatives as:

$$\frac{\partial u}{\partial x} = \lim_{\delta x \to 0} \frac{f(x + \delta x) - f(x)}{\delta x} \bigg|_{y = \text{constant}} \qquad \text{(D.2)}$$

$$\frac{\partial u}{\partial y} = \lim_{\delta y \to 0} \frac{f(y + \delta y) - f(y)}{\delta y} \bigg|_{x = \text{constant}} \qquad \text{(D.3)}$$

Example: if $u = f(x, y) = xy^2$, then

$$\frac{\partial u}{\partial x} = \frac{\partial (xy^2)}{\partial x} \qquad \text{(D.4)}$$

Because y^2 is being held constant,

$$\frac{\partial (xy^2)}{\partial x} = y^2 \frac{\partial (x)}{\partial x} = y^2 \qquad \text{(D.5)}$$

Similarly

$$\frac{\partial u}{\partial y} = \frac{\partial (xy^2)}{\partial y} = x \frac{\partial (y^2)}{\partial y} = x(2y) \qquad \text{(D.6)}$$

Holding y constant, while varying x, is equivalent to taking a series of slices (each at a particular value of y) through the u-surface parallel to the x-axis. Similarly, the variation with y is seen by taking slices parallel to the y-axis. This is shown graphically in Figure D.1. A better way to visualize these

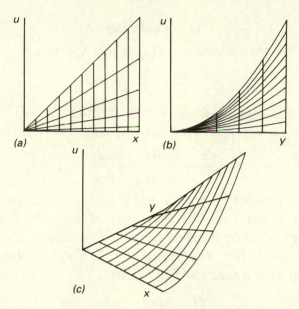

Fig. D.1. Graph of u: (a) against x, (b) against y. (c) Three-dimensional view: the surface u as a function of x and y.

surfaces is by running the computer program GRAPH3D, which produced these illustrations.

D.2 Higher derivatives

The concept of partial differentiation is easily extended to higher derivatives.

For $u = f(x, y)$, the two partial derivatives $\partial u / \partial x$ and $\partial u / \partial y$ may each be differentiated again with respect to either x or y, yielding

$$\frac{\partial^2 u}{\partial x^2} \quad \text{and} \quad \frac{\partial}{\partial y}\left(\frac{\partial u}{\partial x}\right), \qquad \frac{\partial^2 u}{\partial y^2} \quad \text{and} \quad \frac{\partial}{\partial x}\left(\frac{\partial u}{\partial y}\right)$$

In the example $u = xy^2$ we find

$$\frac{\partial^2 u}{\partial x^2} = 0, \quad \frac{\partial}{\partial y}\left(\frac{\partial u}{\partial x}\right) = 2y, \quad \frac{\partial^2 u}{\partial y^2} = 2x, \quad \frac{\partial}{\partial x}\left(\frac{\partial u}{\partial y}\right) = 2y$$

It can be shown that it is true in general that

$$\frac{\partial}{\partial y}\left(\frac{\partial u}{\partial x}\right) = \frac{\partial}{\partial x}\left(\frac{\partial u}{\partial y}\right) = \frac{\partial^2 u}{\partial x \partial y} \tag{D.7}$$

i.e., the order of partial differentiation makes no difference to the result.

D.3 Chain rule

The chain rule for one variable $u = f(x)$, where $x = x(t)$, is

$$\frac{du}{dt} = \frac{du}{dx}\frac{dx}{dt} \tag{D.8}$$

A comparable rule applies where u is a function of two variables. For example, suppose $u = f(x, y)$, where $x = x(t)$ and $y = y(t)$. Then

$$\frac{du}{dt} = \frac{\partial u}{\partial x}\frac{dx}{dt} + \frac{\partial u}{\partial y}\frac{dy}{dt} \tag{D.9}$$

Note that u is a function of both x and y so $\partial u/\partial x$ and $\partial u/\partial y$ are partial derivatives, but x and y are functions only of t, so dx/dt and dy/dt are ordinary derivatives. du/dt is called the *total derivative* of u.

In the expression $M dx + N dy$, where $M = f_1(x, y)$ and $N = f_2(x, y)$, there may or may not exist a function f such that

$$df = M dx + N dy \tag{D.10}$$

If df exists, we call it an *exact differential*. It can be shown that, if x and y are independent variables, a necessary and sufficient condition for the existence of an exact differential is that $\partial M/\partial y = \partial N/\partial x$.

Various other forms of the chain rule exist, for example, for the case where x and y are both functions of two other variables (say, s and t). See Potter and Goldberg (1987) for details.

D.4 Integration

For functions of a single variable, integration yields an arbitrary constant, whose value must be determined from initial or boundary conditions.

Integration of a partial differential equation is different, however, in that it yields not arbitrary constants, but arbitrary functions. For example, suppose $u = u(x, y) = xy$. Integrating with respect to x yields

$$v(x, y) = yx^2/2 + f(y) \tag{D.11}$$

and integrating again with respect to y yields:

$$w(x, y) = x^2 y^2/4 + g(x) + h(y) \tag{D.12}$$

The function $h(y)$ is obtained by integrating $f(y)$ with respect to y. Both h and g are arbitrary functions: any choice of function will give a possible solution to the differential equation. This can be seen by differentiating Equation (D.12): $g(x)$ disappears when we differentiate with respect to y, and $h(y)$ disappears when we differentiate with respect to x.

As in ordinary differential equations (DEs), so too in partial DEs the arbitrary functions are determined from the initial and boundary conditions. But, for partial DEs, this generally involves specifying u and its derivatives along certain regions of the xy-plane. If the boundaries themselves cannot easily be specified analytically, this obviously imposes further limitations on the possibility of obtaining analytical solutions.

D.5 Types of partial differential equation

A useful notation for partial DEs is to use a subscript to denote differentiation. For example, if $u = f(x, y)$, then

$$u_x = \partial u/\partial x \qquad \text{and} \qquad u_y = \partial u/\partial y$$

The general form of a partial DE with two independent variables x, y may be written

$$Lu(x, y) = f(x, y) \tag{D.13}$$

where L is a differential operator (e.g., $\partial/\partial x$, or a combination of such terms). In writing the equation this way, we have put all the terms involving differentiation on the left-hand side. If it turns out that $f(x, y) = 0$, then the equation is described as *homogeneous*. Otherwise the equation is *nonhomogeneous*.

Laplace's equation ($\phi_{xx} + \phi_{yy} = 0$ in two dimensions) is an example of a homogeneous partial DE. *Poisson's equation*, $\phi_{xx} + \phi_{yy} = f(x, y)$, is an example of a nonhomogeneous partial DE. Poisson's equation may arise in groundwater studies: $f(x, y)$ may represent a source or sink of groundwater originating from outside the aquifer under consideration.

A partial DE is described as linear if L has the both of the following properties:

(1) $L(u + w) = Lu + Lw$ and
(2) $L(cu) = cLu$

where in general u and w are each functions of both x and y, and c is a constant.

The property of *superposition* (addition) of solutions is an important property of linear partial DEs. It means that any two known solutions may be combined to produce a third solution.

The *order* of a partial DE is the order of the highest-order derivative in the equation. The *degree* (in u) is the highest power or number of cross-products of u and its derivatives. For example, if there are products of the type

$u(\partial u/\partial x)$ then the equation is of second degree in u. The general form of a second-order equation can be written

$$A\frac{\partial^2 u}{\partial x^2} + 2B\frac{\partial^2 u}{\partial x\partial y} + C\frac{\partial^2 u}{\partial y^2} + \text{lower-order terms} = 0$$

Second-order partial DEs are generally classified by the relative magnitudes of the coefficients of the various terms:

$$\text{hyperbolic,} \quad B^2 - AC > 0$$
$$\text{parabolic,} \quad B^2 - AC = 0$$
$$\text{elliptic,} \quad B^2 - AC < 0$$

The following are examples from the earth sciences (for simplicity, only the one-dimensional form of the equation is shown):

Hyperbolic: the wave equation

$$\frac{\partial^2 u}{\partial x^2} - \frac{1}{c^2}\frac{\partial^2 u}{\partial t^2} = 0$$

This equation describes seismic and sound waves in elastic media (Section 8.12).

Parabolic: the general diffusion equation

$$\frac{\partial^2 u}{\partial x^2} - \frac{1}{h^2}\frac{\partial u}{\partial t} = 0$$

The general diffusion equation arises from a simple diffusion equation, e.g., Darcy's law, or the equation for heat flow by conduction,

$$q = -k\frac{\partial T}{\partial x}$$

where T is the temperature. When an equation of this form is combined with a continuity equation, a general diffusion equation commonly arises. Besides the diffusion of heat (Section 12.2), similar equations arise from the flow of water in soil (under unsaturated conditions) and from the consolidation of porous rocks, which depends on the diffusion of pore fluid out of the rock (Section 6.15). The Navier–Stokes equation (Section 9.4) is also parabolic.

Elliptic: the Laplace equation

$$\frac{\partial^2 u}{\partial x^2} + \frac{\partial^2 u}{\partial y^2} = 0$$

The Laplace equation arises as a special case of the general diffusion equation for steady-state conditions (i.e., where u does not vary with

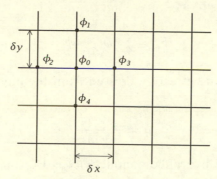

Fig. D.2. Definition of finite difference array.

time). Variants are the Poisson equation (with a non-zero constant on the right-hand side) and the Helmholtz equation (with an additional term in u), both of which may arise in the steady flow of groundwater (Chapter 6). The Laplace equation also describes steady irrotational flow and has been used extensively in fluid dynamics (Section 9.6).

See Potter and Goldberg (1987) or Moon and Spencer (1969) for further details about partial DEs.

D.6 Numerical solution of the Laplace equation or Poisson equation by relaxation

The relaxation technique is one type of iterative finite difference method that is well suited to solution of the Laplace equation. The technique was developed originally by Southwell in the 1930s, and first applied to groundwater problems by Shaw and Southwell (1941); see Freeze and Back (1983) for further details of early applications of numerical methods to hydrogeologic problems.

We begin by dividing the region of interest into a square grid defined by $\delta x = \delta y = \Delta$ (Figure D.2). Each grid point is generally called a *node*, and the values of $\phi(x, y)$ corresponding to the nodes are called the *elements* of the numerical array.

If we know the values of ϕ at the four nodes surrounding any unknown element, it might seem reasonable to estimate the unknown value simply by averaging the surrounding values. Now, if we proceed to change the values of the surrounding elements, we can estimate the correction needed to the unknown central element as

$$F_o = \sum_{i=1}^{4} \phi_i - 4\phi_o \qquad (D.14)$$

The same equation can be readily derived from the finite difference form of Darcy's law. This is

$$q = -\frac{K}{g}\frac{\delta\phi}{\delta x} \qquad (D.15)$$

Now assume that the hydraulic conductivity is isotropic. Then the rate of flow per unit area from, say, node 1 to node 0, is

$$q = -\frac{K}{g\Delta}(\phi_o - \phi_1) \qquad (D.16)$$

The total rate of flow into node 0 is therefore

$$Q = -\frac{K}{g\Delta}\left(\sum \phi_i - 4\phi_o\right) \qquad (D.17)$$

In the case where there are no sources or sinks at node 0, Q must be equal to zero.

The algorithm actually used in solving Laplace's equation is as follows:

(1) First establish the area of interest, and divide it into a net of nodes.
(2) Assign values of ϕ to those nodes where the values are known, or defined by the statement of the problem (i.e., make use of the initial or boundary conditions). If the value of a node is not known, assign to it the value zero.
(3) Proceed to recompute systematically all values that are permitted to change by the boundary condition, using Equation (D.14).
(4) Repeat the third step until the results converge on a steady solution, i.e., until the total value of all the differences computed is smaller than some previously defined limit.

We have assumed here that the initial and boundary conditions have been so chosen that convergence will take place. This will certainly be the case if the boundaries of the grid are defined by known (constant) values of ϕ. As an example, consider the grid shown in Figure D.3 (taken from Ousey, 1986). We are interested in the decrease (draw-down) of the hydraulic potential around a well. The well is situated in the centre of a square area: its potential is maintained at 5 units (by pumping) while we assume (unrealistically) that the potential at the boundaries of the square area are maintained at a constant potential of 10. Because of symmetry, it is unecessary to perform

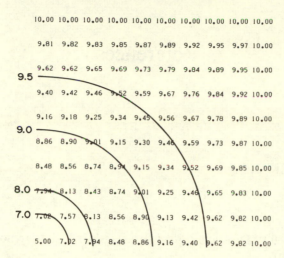

Fig. D.3. Flow towards a well in a confined aquifer (after Ousey, 1986).

the calculations for the whole square. The solution rapidly converges to the values shown in Figure D.3; contours may be drawn as shown. In realistic problems, involving much larger arrays, the rapidity of convergence can be greatly increased by using a slight modification of the original technique, called "over-relaxation". Details are given by Ousey (1986) and Williams (1987).

It can readily be seen how modern spreadsheet software can be adapted to apply this technique. Most professional spreadsheet packages can handle numerical arrays with several hundred rows and columns, and allow for recalculation of the array using a wide variety of user-defined formulae. In many cases, it will be easier and faster to make use of such spreadsheets than to obtain and use the software designed and marketed specifically to solve hydrogeological problems (see Olsthoorn, 1985, and Ousey, 1986; and, for other applications, see Arganbright, 1985).

References

Acheson, D. J., 1990. *Elementary Fluid Dynamics*. Oxford, Clarendon Press, 397 pp. (TA357.A276)

Allen, J. R. L., 1985. *Principles of Physical Sedimentology*. London, Allen and Unwin, 272 pp. (QE471.A5597)

Allen, P. A. and J. R. Allen, 1990. *Basin Analysis: Principles and Applications*. Oxford, Blackwell, 451 pp. (QE571.A45)

Angevine, C. L., P. L. Heller and C. Paola, 1990. *Quantitative Sedimentary Basin Modeling*. Amer. Assoc. Petroleum Geol. Continuing Education Course Notes No. 32, 132 pp. (QE472.A455)

Archie, G. E., 1942. The electrical resistivity log as an aid in determining some reservoir characteristics. Trans. Amer. Inst. Mining and Metallurgical Engineers, Petroleum Div., v.146, pp. 54–61.

Archie, G. E., 1950. Introduction to petrophysics of reservoir rocks. Amer. Assoc. Petroleum Geol. Bull., v.34, pp. 943–61.

Arganbright, D., 1985. *Mathematical Applications of Electronic Spreadsheets*. New York, McGraw-Hill, 165 pp. (QA76.95.A7)

Aris, R., 1962. *Vectors, Tensors, and the Basic Equations of Fluid Mechanics*. New York, Dover, 286 pp. (QA911.A69)

Aris, R., 1978. *Mathematical Modeling Techniques*. San Francisco, Pitman Advanced Publishing Program, 191 pp. (QA401.A68)

Atkinson, B. K., ed., 1987. *Fracture Mechanics of Rock*. New York, Academic Press, 534 pp. (QE431.6.F7)

Audet, D. M. and A. C. Fowler, 1992. A mathematical model of compaction in sedimentary basins. Geophys. Jour. International, v.110, pp. 577–90.

Bagnold, R. A., 1956. The flow of cohesionless grains in fluids. Phil. Trans. Roy. Soc. London, v.249 A, pp. 235–97.

Bagnold, R. A., 1960. Sediment discharge and stream power, a preliminary announcement. US Geological Survey Circular 421, 23 pp.

Bak, P., C. Tang and K. Wiesenfeld, 1988. Self-organized criticality. Physical Review, v.38 A, pp. 364–74 (see also Scientific American, January 1991, pp. 46–53).

Baker, W. E., P. S. Westine and F. T. Dodge, 1991. *Similarity Methods in Engineering Dynamics: Theory and Practice of Scale Modeling*. New York, Elsevier, revised edition, 384 pp. (TA177.B35)

438

Barnes, H. A., 1989. *An Introduction to Rheology.* New York, Elsevier, 199 pp. (QC189.5.B37)

Batchelor, G. K., 1970. *An Introduction to Fluid Dynamics.* Cambridge University Press, 615 pp. (QA911.B33)

Bayly, B., 1992. *Mechanics in Structural Geology.* New York, Springer-Verlag, 253 pp. (QE601.B36)

Bear, J. and A. Verruijt, 1987. *Modeling Groundwater Flow and Pollution.* Boston, Reidel, 414 pp. (TC176.B38)

Beaumont, C., 1978. The evolution of sedimentary basins on a viscoelastic lithosphere: theory and examples. Geophys. Jour. Royal Astronomical Soc., v.55, pp. 471–97.

Beaumont, C., 1981. Foreland basins. Geophys. Jour. Royal Astronomical Soc., v.65, pp. 291–329.

Beckinsale, R. P. and R. J. Chorley, 1991. *The History of the Study of Landforms. Vol. 3: Historical and Regional Geomorphology 1890–1950.* New York, Routledge, 496 pp. (QE505.C45 v.3)

Bercovici, D., G. Schubert and G. A. Glatzmaier, 1989. Three-dimensional spherical models of convection in the earth's mantle. Science, v.244, pp. 950–5.

Bergé, P., 1979. Experiments on hydrodynamic instabilities and the transition to turbulence. *Lecture Notes in Physics*, v.104. New York, Springer-Verlag, pp. 289–308.

Bergé, P. and M. Dubois, 1984. Rayleigh–Bénard convection. Contemporary Physics, v.25, pp. 535–82.

Bergé, P., Y. Pomeau, and C. Vidal, 1986. *Order Within Chaos: Towards a Deterministic Approach to Turbulence.* New York, John Wiley and Sons, 329 pp. (QA614.8.B4713)

Blatt, H., G. Middleton and R. Murray, 1980. *Origin of Sedimentary Rocks.* Englewood Cliffs NJ, Prentice-Hall, second edition, 782 pp. (QE471.B65)

Bolt, B. A., 1988. *Earthquakes.* New York, W.H. Freeman, 282 pp. (QE534.2.B64)

Borisenko, A. I. and I. E. E. Tarapov, 1968. *Vector and Tensor Analysis with Applications.* Englewood Cliffs NJ, Prentice-Hall, revised English edition, 257 pp. (QA261. B5513)

Brace, W. E., 1968. The mechanical effects of pore pressure on fracturing of rocks. Geol. Survey Canada, Paper 68–52, pp. 113–24.

Bridgman, P. W., 1921. *Dimensional Analysis.* Cambridge MA, Harvard University Press. Reprinted by Dover, 112 pp. (QC39.B85)

Bullen, K. E. and B. A. Bolt, 1985. *An Introduction to the Theory of Seismology.* Cambridge University Press, fourth edition, 499 pp. (QE534.2.B85)

Byerlee, J. D., 1978. Friction of rocks. Pure and Applied Geophysics, v.116, pp. 615–26.

Campbell, C. S., 1990. Rapid granular flows. Ann. Rev. Fluid Mechanics, v.22, pp. 57–92.

Cantwell, B. J., 1981. Organized motion in turbulent flow. Ann. Rev. Fluid Mechanics, v.13, pp. 457–515.

Capper, P. L and W. F. Cassie, 1976. *The Mechanics of Engineering Soils.* New York, John Wiley and Sons, sixth edition, 376 pp. (TA710.C26)

Carson, M. A., 1971. *The Mechanics of Erosion.* London, Pion, 174 pp. (QE471.C364)

Carson, M. A. and M. J. Kirkby, 1972. *Hillslope Form and Process.* Cambridge University Press, 475 pp. (GB406.C3)

Catchpole, J. P. and G. D. Fulford, 1973. Table 10-51. Dimensionless groups. In R. E. Bolz and G. L. Tuve, eds., *Tables for Applied Engineering Science.* Cleveland OH, CRC Press, pp. 1027–37. (TA151.C2)

Cathles, L. M. III, 1975. *The Viscosity of the Earth's Mantle.* Princeton University Press, 386 pp. (QE 511.C36)

Chapman, R. E., 1981. *Geology and Water: An Introduction to Fluid Mechanics for Geologists.* The Hague, Martinus Nijhoff, 228 pp. (QC809.F5C44)

Chayes, F., 1971. *Ratio Correlation: a Manual for Students of Petrology and Geochemistry.* Chicago, University of Chicago Press, 99 pp. (QE431.5.C47)

Cheeney, R. F., 1983. *Statistical Methods in Geology.* London, George Allen and Unwin, 169 pp. (QE33.2M3C48)

Cherry, J. A., R. W. Gillham and J. F. Pickens, 1975. Contaminant hydrogeology. Part 1: Physical processes. Geoscience Canada, v.2, pp. 76–84.

Christensen, U. and H. Harder, 1991. 3-D convection with variable viscosity. Geophys. Jour. International, v.104, pp. 213–26.

Church, M. and D. M. Mark, 1980. On size and scale in geomorphology. Progress in Physical Geography, v.4, pp. 342–91.

Cloetingh, S., 1988. Intraplate stresses: a new element in basin analysis. In K.L. Kleinspehn and C. Paola, eds., *New Perspectives in Basin Analysis,* New York, Springer-Verlag, pp. 205–30 (QE571.N39)

Cohen, I. B., 1987. *The Newtonian Revolution.* Norwalk CT, Burndy Library, 42 pp. (QC7. C662)

Cox, A. and R. B. Hart, 1986. *Plate Tectonics, How It Works.* Oxford, Blackwell, 392 pp. (QE511.4)

Croquette, V., 1989. Convective pattern dynamics at low Prandtl number: Part I. Contemporary Physics, v.30, No.2, pp. 113–33.

Cruden, D. M., 1976. Major rock slides in the Rockies. Canadian Geotechnical Journ., v.13, pp. 8–20.

Cruden, D. M. and J. Krahn, 1973. A reexamination of the geology of the Frank Slide. Canadian Geotechnical Jour., v.10, pp. 581–91.

Daily, J. W. and D. R. F. Harleman, 1966. *Fluid Dynamics.* Reading MA, Addison-Wesley, 454 pp. (QC151.D22)

Dally, J. W. and W. F. Riley, 1991. *Experimental Stress Analysis.* New York, McGraw-Hill, third edition, 639 pp. (TA407.D32)

De Marsily, G., 1986. *Quantitative Hydrogeology. Groundwater Hydrology for Engineers.* Translated from the French by G. de Marsily. New York, Academic Press, 440 pp. (GB1001.72.M37M3713)

Dixon, J. M. and J. M. Summers, 1985. Recent developments in centrifuge modelling of tectonic processes: equipment, model construction techniques and rheology of model materials. Jour. Structural Geol., v.7, pp. 83–102.

Domenico, P. A. and F. W. Schwartz, 1990. *Physical and Chemical Hydrogeology.* New York, John Wiley and Sons, 824 pp. (GB1003.2.D66)

Donnelly, R. J., 1991. Taylor–Couette flow: the early days. Physics Today, v.44, no.11 (Nov. 1991) pp. 32–9.

Dugas, R., 1955. *A History of Mechanics.* Translated by J. R. Maddox. New York, Dover, 671 pp. (QA802.D8613)

Elder, J., 1976. *The Bowels of the Earth.* Oxford University Press, 222 pp. (QE501.2E4)

Feynman, R. P., R. B. Leighton and M. Sands, 1963. *The Feynman Lectures on Physics. Vol. 1: Mainly Mechanics, Radiation, and Heat.* Reading MA, Addison-Wesley. (QC21.F43 v.1)

Fisher, N. I., T. Lewis and B. J. J. Embleton, 1987. *Statistical Analysis of Spherical Data.* Cambridge University Press, 329 pp. (QA276.F489)

Fjeldskaar, W. and L. Cathles, 1991. The present rate of uplift of Fennoscandia implies a low-viscosity asthenosphere. Terra Nova, v.3, pp. 393–400.

Foley, J. D. and A. Van Dam, 1990. *Computer Graphics: Principles and Practices.* Reading MA, Addison-Wesley, second edition, 1174 pp. (T385.C587)

Forsyth, D. W., 1979. Lithospheric flexure. Rev. Geophysics Space Physics, v.17, pp. 1109-14.

Forsyth, D. and S. Uyeda, 1975. On the Relative Importance of the Driving Forces of Plate Motion. Geophys. Jour. Roy. Astronomical Soc., v.43, pp. 163–200.

Freeze, R. A. and W. Back, eds., 1983. *Physical Hydrogeology.* Benchmark Papers in Geology, Vol.72. Stroudburg PA, Hutchinson Ross, 431 pp. (GB1003.2.P49)

Freeze, R. A. and J. A. Cherry, 1979. *Groundwater.* Englewood Cliffs NJ, Prentice-Hall, 604 pp. (GB1003.2.F73)

French, A. P., 1971. *Newtonian Mechanics.* New York, W. W. Norton and Co., 743 pp. (QC125.2.F74)

French, A. P. and M. G. Ebison, 1986. *Introduction to Classical Mechanics.* New York, Van Nostrand Reinhold, 310 pp. (QC125.2.F74)

Fung, Y.-C., 1977. *A First Course in Continuum Mechanics.* Englewood Cliffs NJ, Prentice-Hall, second edition, 340 pp. (QA808.2.F85)

Gellert, W., *et al.*, eds., 1977. *The VNR Concise Encyclopedia of Mathematics.* New York, Van Nostrand Reinhold, 760 pp. (QA40.V18)

Glen, J. W., 1952. Experiments on the deformation of ice. Jour. Glaciology, v.2, pp. 111–4.

Goldstein, H., 1980. *Classical Mechanics.* Reading MA, Addison-Wesley, second edition, 672 pp. (QA805.G6)

Goldstein, S., ed., 1965. *Modern Developments in Fluid Dynamics.* New York, Dover, 2 vols., 702 pp. (TL570.G55 v.1,v.2)

Goldstein, S., 1969. Fluid mechanics in the first half of this century. Ann. Rev. Fluid Mechanics, v.1, pp. 1–28.

Goodman, R. E., 1980. *Introduction to Rock Mechanics.* New York, John Wiley and Sons, 478 pp. (TA706.G65)

Gough, D. I. and W. I. Gough, 1987. Stress near the surface of the earth. Ann. Rev. Earth Planetary Sci., v.15, pp. 545–66.

Grass, A. J., 1971. Structural features of turbulent flow over smooth and rough boundaries. Jour. Fluid Mechanics, v.50, pp. 233–55.

Gratz, A. J., W. J. Nellis and N. A. Hinsey, 1992. Laboratory simulation of explosive volcanic loading and implications for the cause of the K/T boundary. Geophys. Research Letters, v.19, pp. 1391-4.

Greenwood, H. J., 1989. On models and modeling. Canadian Mineralogist, v.27, pp. 1–14.

Gretener, P. E., 1977. *Pore Pressure: Fundamentals, General Ramifications, and*

Implications for Structural Geology. Amer. Assoc. Petroleum Geol. Continuing Education Course Notes, No. 4, 87 pp. (QE601.2G74)

Gupta, H. K. and B. K. Rastogi, 1976. *Dams and Earthquakes.* New York, Elsevier, 299 pp. (QE539.G8)

Hampton, M. A., 1979. Buoyancy in debris flows. Jour. Sedimentary Petrology, v.49, pp. 753–8.

Hart, D. and A. Croft, 1988. *Modelling with Projectiles.* Chichester UK, Ellis Horwood, 151 pp. (QA862.P76.H37)

Harte, J., 1985. *Consider a Spherical Cow: a Course in Environmental Problem Solving.* Los Altos CA, William Kaufman, 283 pp. (QH541.15.M34H37)

Heard, H. C. and W. W. Rubey, 1966. Tectonic implications of gypsum dehydration. Geol. Soc. America Bull., v.77, pp. 741–60.

Heath, R. C., 1983. Basic ground-water hydrology. US Geol. Survey Water-Supply Paper 2220, 84 pp.

Hendry, A. W., 1977. *Elements of Experimental Stress Analysis.* Oxford, Pergamon Press, 193 pp. (TA407.H43)

Hesse, M. B., 1963. *Models and Analogies in Science.* Notre Dame IN, University of Notre Dame Press, 184 pp. (Q175.H56)

Hickman, S. H., 1991. Stress in the lithosphere and the strength of active faults. Rev. Geophysics, Supplement, April 1991, pp. 759–75.

Hodge, P. G. Jr., 1970. *Continuum Mechanics. An Introductory Text for Engineers.* New York, McGraw-Hill, 251 pp. (QA808.2.H6)

Hoffmann, B., 1966. *About Vectors.* New York, Dover, 134 pp. (QA161.H63)

Holsapple, K. A. and R. M. Schmidt, 1987. Point source solutions and coupling parameters in cratering mechanisms. Jour. Geophys. Research, v.92, No. B7, pp. 6350–76.

Holton, G. J., 1973. *Introduction to Concepts and Theories in Physical Science.* Revised with new material by S. G. Brush. Reading MA, Addison-Wesley, 589 pp. (QC23.H758)

Holtz, R. D. and W. D. Kovacs, 1981. *An Introduction to Geotechnical Engineering.* Englewood Cliffs NJ, Prentice-Hall, 733 pp. (TA710.H564)

Hooke, R. LeB., 1981. Flow law for polycrystalline ice in glaciers: comparison of theoretical predictions, laboratory data, and field measurements. Rev. Geophys. Space Physics, v.19, pp. 664–72.

Hubbert, M. K., 1937. Theory of scale models as applied to the study of geologic structures. Geol. Soc. America Bull., v.48, pp. 1459–520.

Hubbert, M. K., 1945. Strength of the Earth. Amer. Assoc. Petroleum Geol. Bull., v.29, pp. 1630–53.

Hubbert, M. K., 1969. *The Theory of Ground-water Motion and Related Papers.* New York, Hafner, 310 pp. (TC176.H83)

Hubbert, M. K. and W. W. Rubey, 1959, 1960, 1961. Role of fluid pressure in mechanics of overthrust faulting. Geol. Soc. America Bull., v.70, pp. 115–60; v.71, pp. 617–28; v.72, pp. 1587–94.

Hughes, W. F., 1979. *An Introduction to Viscous Flow.* New York, McGraw-Hill, 219 pp. (TA518.2.H83)

Huggett, R. J., 1985. *Earth Surface Systems.* New York, Springer-Verlag, 270 pp. (QE33.2.M3H75)

Hulme, G., 1974. The interpretation of lava flow morphology. Geophys. Jour. Royal Astronomical Soc., v.39, pp. 361–83.

Hunt, J. M., 1990. Generation and migration of petroleum from abnormally pressured fluid compartments. Amer. Assoc. Petroleum Geol. Bull., v.74, pp. 1–12 (and see discussion, v.75, pp. 326–38, 1991).

Huppert, H. E., 1986. The intrusion of fluid mechanics into geology. Jour. Fluid Mechanics, v.173, pp. 557–94.

Hutter, K., 1982. Dynamics of glaciers and large ice masses. Ann. Rev. Fluid Mechanics, v.14, pp. 87–130.

Hutter, K., 1983. *Theoretical Glaciology*. Boston, Reidel, 510 pp. (GB2403.2H88)

Istok, J., 1989. Groundwater Modeling by the Finite Element Method. Amer. Geophys. Union, Water Resources Monograph 13, 495 pp. (TC176.I79)

Iverson, R. M. and M. E. Reid, 1992. Gravity-driven groundwater flow and slope failure potential: 1. Elastic effective-stress model. Water Resources Research, v.28, pp. 925–38.

Jacob, C. E., 1950. Flow of Ground Water. Chapter V in H. Rouse, ed., *Engineering Hydraulics*. New York, John Wiley and Sons, pp. 321–86.

Jaeger, J. C., 1971. *Elasticity, Fracture, and Flow: with Engineering and Geological Applications*. New York, John Wiley and Sons, third edition, 268 pp. (QA931.J22)

Jaeger, J. C. and N. G. W. Cook, 1979. *Fundamentals of Rock Mechanics*. London, Methuen, third edition, 593 pp. (TA706.J32)

Jarvis, G. T., 1991. Two-dimensional numerical models of mantle convection. Advances in Geophysics, v.33, pp. 1–80.

Johnson, A. M., 1970. *Physical Processes in Geology*. San Francisco, Freeman, Cooper and Co., 577 pp. (QE33.J58)

Jordan, T. E. and P. B. Flemings, 1989. From geodynamic models to basin fill – a stratigraphic perspective. In T. A. Cross, ed., *Quantitative Dynamic Stratigraphy*, Englewood Cliffs NJ, Prentice-Hall, pp. 149–63. (QE651.Q26)

Jury, W. A., W. R. Gardner and W. H. Gardner, 1991. *Soil Physics*. New York, John Wiley and Sons, fifth edition, 328 pp. (S592.3.J86)

Kamb, B., 1964. Glacier geophysics. Science, v.146, pp. 353–65.

Kaula, W. M., 1972. Global Gravity and Tectonics. In E. C. Robertson, J. F. Hays and L. Knopoft, eds., *The Nature of the Solid Earth*. New York, McGraw-Hill, pp. 385–405. (QE515.N37)

Kieffer, S. W., 1984. Factors governing the structure of volcanic jets. In *Explosive Volcanism: Inception, Evolution, and Hazards*. Washington DC, National Academy Press, Studies in Geophysics, pp. 143–57.

Kieffer, S. W., 1989. Geological nozzles. Rev. Geophysics, v.27, pp. 3–38.

Kirchhoff, R. H., 1985. *Potential Flows: Computer Graphics Solutions*. New York, Marcel Dekker, 181 pp. (TA357.K57)

Kline, S. J., 1965. *Similitude and Approximation Theory*. New York, McGraw-Hill, 229 pp. (QA401.K65)

Lambe, T. W. and R. V. Whitman, 1969. *Soil Mechanics*. New York, John Wiley and Sons, 553 pp. (TA710.L245)

Landahl, M. T. and E. Mollo-Christensen, 1986. *Turbulence and Random Processes in Fluid Mechanics*. Cambridge University Press, 154 pp. (QA901.L278)

Langhaar, H. L., 1951. *Dimensional Analysis and Theory of Models*. New York, John Wiley and Sons, 166 pp. (QC39.L35)

Le Pichon, X., J. Francheteau and J. Bonnin, 1973. *Plate Tectonics*. New York, Elsevier, 300 pp. (QE511.4.L46)

Libchaber, A. and J. Maurer, 1980. Une expérience de Rayleigh–Bénard de géometrie reduite. Jour. Physique, Colloques C3, v.41, pp. 51–6.

Lighthill, M. J., 1986. *An Informal Introduction to Theoretical Fluid Mechanics*. Oxford, Clarendon Press, 260 pp. (QA911.L46)

Lliboutry, L. A., 1987. *Very Slow Flows of Solids: Basics of Modeling in Geodynamics and Glaciology*. Boston MA, Martinus Nijhoff, 510 pp. (QE501.3.L57)

Long, R. R., 1961. *Mechanics of Solids and Fluids*. Englewood Cliffs NJ, Prentice-Hall, 156 pp. (QA931.L84)

Lorenz, E. N., 1960. Maximum simplification of the dynamic equations. Tellus, v.12, pp. 243–254.

Lorenz, E. N., 1963. Deterministic non-periodic flow. Jour. Atmospheric Sci., v.20, pp. 130–41.

Lorenz, E. N., 1964. The problem of deducing the climate from the governing equations. Tellus, v.16, pp. 1–11.

Lumley, J. L. and H. A. Panofsky, 1964. *The Structure of Atmospheric Turbulence*. New York, John Wiley and Sons, 239 pp. (QC880.L84)

Lyons, T. J. and W. D. Scott, 1990. *Principles of Air Pollution Meteorology*. London, Belhaven Press, 224 p. (QC882.L96)

Macagno, E. O., 1971. Historico-critical review of dimensional analysis. Jour. Franklin Institute, v.292, pp. 391–402.

Mach, E., 1960. *The Science of Mechanics*. Translated by T. J. McCormack. LaSalle IL, Open Court, sixth edition, 634 pp. (QA802.M1313)

Machetel, P. and P. Weber, 1991. Intermittent convection in a model mantle with an endothermal phase change at 670 km. Nature, v.350, pp. 55–7.

Magara, K., 1975. Importance of aquathermal pressuring effect in Gulf Coast. Amer. Assoc. Petroleum Geol. Bull., v.59, pp. 2037–45.

Magara, K., 1978. *Compaction and Fluid Migration: Practical Petroleum Geology*. New York, Elsevier, 319 pp. (TN871.M328)

Major, J. J. and T. C. Pierson, 1992. Debris flow rheology: experimental analysis of fine-grained slurries. Water Resources Research, v.28, pp. 841–57.

Malvern, L. E., 1969. *Introduction to the Mechanics of a Continuous Medium*. Englewood Cliffs NJ, Prentice-Hall, 713 pp. (QA808.2.M3)

McEwen, A. S., 1989. Mobility of large rock avalanches: evidence from Valles Marineris, Mars. Geology, v.17, pp. 1111–4.

McGarr, A., 1980. Some constraints on levels of shear stress in the crust from observations and theory. Jour. Geophys. Research, v.85, pp. 6231–38.

McGarr, A. and N. C. Gay, 1978. State of stress in the earth's crust. Ann. Rev. Earth Planetary Sci., v.6, pp. 405–36.

McKenzie, D. P., 1969. Speculations on the Consequences and Causes of Plate Motions. Geophys. Jour. Roy. Astronomical Soc., v.18, pp. 1–32.

Means, W. D., 1976. *Stress and Strain. Basic Concepts of Continuum Mechanics for Geologists*. New York, Springer-Verlag, 339 pp. (QE604.M4)

Mehta, A., 1992. Real sand piles: dilatancy, hysteresis and cooperative dynamics. Physica A, v.186, pp. 121–53.

Meier, M. F., 1960. Mode of flow of Saskatchewan Glacier, Alberta, Canada. US Geological Survey Professional Paper 351, 70 pp.

Meissner, F. F., 1984. Petroleum geology of the Bakken Formation, Williston Basin, North Dakota and Montana. In G. Demaison and R. J. Murris, eds., *Petroleum Geochemistry and Basin Evolution*. Tulsa OK, Amer. Assoc. Petroleum Geol. Memoir 35, pp. 159–79. (TN870.5.P477)

Melosh, H. J., 1989. *Impact Cratering: A Geologic Process*. Oxford University Press, 245 pp. (QB603.C7M45)

Middleton, G. V. and J. B. Southard, 1984. *Mechanics of Sediment Movement*. Soc. Economic Paleontologists Mineralogists Short Course Notes 3, second edition, 401 pp. (QE471.2.M53)

Monicard, R. P., 1980. *Properties of Reservoir Rocks: Core Analysis*. Paris, Editions Technip, 168 pp. (TN870.5M65)

Moon, P. and D. E. Spencer, 1969. *Partial Differential Equations*. Lexington MA, D. C. Heath, 322 pp. (QA377.M6)

Morgenstern, N., 1967. Submarine slumping and the initiation of turbidity currents. In A. F. Richards, ed., *Marine Geotechnique*. Urbana IL, University of Illinois Press, pp. 189–220.

Mudford, B. S. and M. E. Best, 1989. Venture Gas Field, offshore Nova Scotia: Case study of overpressuring in region of low sedimentation rate. Amer. Assoc. Petroleum Geol. Bull., v.73, pp. 1383–96.

Nash, D., 1987. A comparative review of limit equilibrium methods of stability analysis. In M. G. Anderson and K. S. Richards, eds., *Slope Stability*. New York, John Wiley and Sons, pp. 11–75. (TA710.S553)

Neuman, S. P., and I. Neretnieks, eds., 1990. *Hydrogeology of Low Permeability Environments*. Hannover, Verlag Heinz Heise, 268 pp. (GB1197.7.N48)

Newman, W. M., 1979. *Principles of Interactive Computer Graphics*. New York, McGraw-Hill, second edition, 541 pp. (T385.N48)

Nur, A. and J. D. Byerlee, 1971. An exact effective stress law for elastic deformation of rock with fluids. Jour. Geophys. Research, v.76, pp. 6414–9.

Nye, J., 1952. The mechanics of glacier flow. Jour. Glaciology, v.2, pp. 82–93.

Nye, J., 1964. The flow of a glacier in a channel of rectangular, elliptical or parabolic cross-section. Jour. Glaciology, v.5, pp. 661–90.

Officer, C. B., 1974. *Introduction to Theoretical Geophysics*. New York, Springer-Verlag, 385 pp. (QC806.O24)

O'Hear, A., 1989. *An Introduction to the Philosophy of Science*. Oxford, Clarendon Press, 239 pp. (Q175.O454)

Olfe, D. B., 1987. *Fluid Mechanics Programs for the IBM PC*. New York, McGraw-Hill, 174 pp. + diskette. (QA901.O49)

Olsthoorn, T. N., 1985. The power of the electronic worksheet: modelling without special programs. Ground Water, v.23, pp. 381–90.

Oswatitsch, K. and K. Wieghardt, 1987. Ludwig Prandtl and his Kaiser-Wilhelm-Institut. Ann. Rev. Fluid Mechanics, v.19, pp. 1–26.

Ousey, J. R., Jr., 1986. Modelling steady-state groundwater flow using microcomputer spreadsheets. Jour. Geol. Education, v.34, pp. 305–11.

Panofsky, H. A. and J. A. Dutton, 1984. *Atmospheric Turbulence: Models and Methods for Engineering Applications*. New York, John Wiley and Sons, 397 pp. (QC880.4.T8P35)

Pasquill, F. and F. B. Smith, 1983. *Atmospheric Diffusion*. Chichester UK, Ellis Horwood, third edition, 437 pp. (QC880.4.D44P37)

Paterson, M. S., 1958. Experimental deformation and faulting in Wombeyan marble. Geol. Soc. America Bull., v.69, pp. 465–76.

Paterson, M. S., 1978. *Experimental Rock Deformation – the Brittle Field*. New York, Springer-Verlag, 254 pp. (QE604.P37)

Paterson, W. S. B., 1981. *The Physics of Glaciers*. Oxford, Pergamon Press, second edition, 380 pp. (GB2403.5.P37 1981)

Pedlosky, J., 1987. *Geophysical Fluid Dynamics*. New York, Springer-Verlag, second edition, 710 pp. (QC809.F5P43)

Peltier, W. R., ed., 1989. *Mantle Convection: Plate Tectonics and Global Dynamics*. New York, Gordon and Breach, 881 pp. (QE509.M2618)

Phillips, O. M., 1968. *The Heart of the Earth*. San Francisco, Freeman, Cooper and Co., 236 pp. (QE501.P48)

Phillips, O. M., 1991. *Flow and Reactions in Permeable Rocks*. Cambridge University Press, 285 pp. (TA357.P48)

Pippard, A. B., 1972. *Forces and Particles: An Outline of the Principles of Classical Physics*. New York, John Wiley and Sons, 321 pp. (QC21.2.P56)

Potter, M. C. and J. Goldberg, 1987. *Mathematical Methods*. Englewood Cliffs NJ, Prentice-Hall, second edition, 639 pp. (QA37.2P668)

Prandtl, L., 1952. *Essentials of Fluid Dynamics*. Glasgow, Blackie and Son, 452 pp. (QA911.P89)

Prandtl, L. and O. G. Tietjens, 1934a. *Fundamentals of Hydro- and Aeromechanics*. New York, Dover, 270 pp. (QA911.T5)

Prandtl, L. and O. G. Tietjens, 1934b. *Applied Hydro- and Aeromechanics*. New York, Dover, 311 pp.

Press, W. H., B. P. Flannery, S. A. Teukolsky and W. T. Vettering, 1989. *Numerical Recipes in Pascal: The Art of Scientific Computing*. Cambridge University Press, 759 pp. (QA76.73.P2N87)

Quinlan, G. M. and C. Beaumont, 1984. Appalachian thrusting, lithospheric flexure, and the Paleozoic stratigraphy of the Eastern Interior of North America. Canadian Jour. Earth Sci., v.21, pp. 973–96.

Ramberg, H., 1981. *Gravity, Deformation and the Earth's Crust*. New York, Academic Press, second edition, 452 pp. (QE601.R25)

Ramsay, J. G., 1967. *Folding and Fracturing of Rocks*. New York, McGraw-Hill, 568 pp. (QE606.R3)

Ramsay, J. G. and M. I. Huber, 1983. *The Technique of Modern Structural Geology: Vol. 1. Strain Analysis*. New York, Academic Press, 302 pp. (QE601.R254 1983, v.1)

Ranalli, G., 1987. *Rheology of the Earth. Deformation and Flow Processes in Geophysics and Geodynamics*. Boston, Allen and Unwin, 366 pp. (QE501.3)

Reiner, M., 1960. *Lectures on Theoretical Rheology*. Amsterdam, North Holland, third edition, 158 pp. (QC189.R35)

Reiner, M., 1969. *Deformation, Strain and Flow: An Elementary Introduction to Rheology*. London, H. K. Lewis, third edition, 347 pp. (QC189.R3)

Richardson, L. F., 1922. *Weather Prediction by Numerical Processes*. Cambridge University Press, 236 pp.

Richter, F. M., 1978. Mantle Convection Models. Ann. Rev. Earth Planetary Sci., v.6, pp. 9–19.

Richter, F. M. and D. McKenzie, 1978. Simple Plate Models of Mantle Convection. Jour. Geophysics, v.44, pp. 441–71.

Rodi, W., 1984. *Turbulence Models and Their Application in Hydraulics – A State of the Art Review.* The Netherlands, International Assoc. Hydraulic Research, second edition, 104 pp. (TA357.R63)

Roscoe, K. H., A. N. Schofield and C. P. Wroth, 1958. On the yielding of soils. Geotechnique, v.8, pp. 22–53.

Rouse, H. and S. Ince, 1957. *History of Hydraulics.* State University of Iowa, Iowa Institute of Hydraulic Research, 269 pp. (TC15.R86)

Rutter, E. H., 1986. On the nomenclature of mode of failure transitions in rocks. Tectonophysics, v.122, pp. 381–7.

Ryan, M. P., 1987. Neutral buoyancy and the mechanical evolution of magmatic systems. In B. O. Myers, ed., *Magmatic Processes, Physicochemical Principles.* Geochem. Soc. Special Publication 1, pp. 259–87.

Sabadini, R., K. Lambeck and E. Boschi, eds., 1991. *Glacial Isostasy, Sea-level, and Mantle Rheology.* Boston, Kluwer, 708 pp. (QE511.N272)

Sahay, B. and W. H. Fertl, 1988. *Origin and Evaluation of Formation Pressures.* Ahmedabad, India, Allied Publishers, 292 pp. (TN871.18.S243)

Scheidegger, A. E., 1975. *Physical Aspects of Natural Catastrophes.* New York, Elsevier, 289 pp. (GB70.S33)

Schmidt, G. W., 1973. Interstitial water composition and geochemistry of deep Gulf Coast shales and sandstones. Amer. Assoc. Petroleum Geol. Bull., v.57, pp. 321–37.

Scholz, C. H., 1990. *The Mechanics of Earthquakes and Faulting.* Cambridge University Press, 439 pp. (QE534.2.S37)

Schubauer, G. B. and H. K. Skramstad, 1947. Laminar boundary-layer oscillations and stability of laminar flow. Jour. Aeronautical Sci., v.14, pp. 69–78.

Schubert, G., 1992. Numerical models of mantle convection. Ann. Rev. Fluid Mechanics, v.24, pp. 359–94

Schwartz, F. W. *et al.*, 1990. *Ground Water Models: Scientific and Regulatory Applications.* Washington DC, National Academy Press, 303 pp. (TC176.G76)

Sclater, J. G. and P. A. F. Christie, 1980. Continental stretching: an explanation of the post-mid-Cretaceous subsidence of the central North Sea Basin. Jour. Geophys. Res., v.85, pp. 3711–39.

Scorer, R. S., 1968. *Air Pollution.* Oxford, Pergamon Press, 151 pp. (TD883.S36)

Scorer, R. S., 1978. *Environmental Aerodynamics.* Chichester UK, Ellis Horwood, 488 pp. (QC883.5.S36)

Sedov, L. I., 1959. *Similarity and Dimensional Methods in Mechanics.* New York, Academic Press, 363 pp. (QC39.S44)

Shaw, F. S. and R. V. Southwell, 1941. Relaxation methods applied to engineering problems, VII. Problems relating to the percolation of fluids through porous materials. Royal Soc. London Proc., v.178A, pp. 1–17.

Shaw, H. R. and D. A. Swanson, 1970. Eruption and flow rate of flood basalts. In E. H. Gilmore and D. Straling, eds., *Proc. Second Columbia River Basalt Symposium.* Cheney, Eastern Washington State College Press, pp. 271–99.

Simpson, D. W., 1986. Triggered earthquakes. Ann. Rev. Earth Planetary Sci., v.14, pp. 21–42.

Skelland, A. H. P., 1967. *Non-Newtonian Flow and Heat Transfer*. New York, John Wiley and Sons, 469 pp. (TA357.S5)

Smith, P. and R. C. Smith, 1990. *Mechanics*. New York, John Wiley and Sons, second edition, 321 pp. (QA805.S673)

Sommerfeld, A., 1964. *Mechanics of Deformable Bodies*. New York, Academic Press, 396 pp. (QC20.S69 v.2)

Southard, J. B., L. A. Boguchwal and R. D. Romea, 1980. Test of scale modelling of sediment transport in steady unidirectional flow. Earth Surface Processes, v.5, pp. 17–23.

Spencer, A. J. M., 1980. *Continuum Mechanics*. New York, John Wiley and Sons, 183 pp. (QA808.2S63)

Stacey, F. D., 1992. *Physics of the Earth*. Brisbane, Australia, Brookfield Press, third edition, 513 pp. (QC806.S65)

Statham, I., 1977. *Earth Surface Sediment Transport*. Oxford, Clarendon Press, 184 pp. (QE471.2.S73)

Stevenson, D. J. and J. S. Turner, 1979. Fluid models of mantle convection. In M. W. McElhinny, ed., *The Earth, Its Origin, Evolution, and Structure*. New York, Academic Press, pp. 227–63. (QE35.G18)

Stewart, C. A. and D. L. Turcotte, 1989. The route to chaos in thermal convection at infinite Prandtle number: 1. Some trajectories and bifurcations. Jour. Geophys. Research, v.94, no. B10, pp. 13 707–17.

Stommel, H. M., 1989. *An Introduction to the Coriolis Force*. New York, Columbia University Press, 297 pp. (QC880.4.C65S76)

Strahler, A. N., 1958. Dimensional analysis applied to fluvially eroded landforms. Geol. Soc. America Bull., v.69, pp. 279–300.

Strahler, A. N., 1971. *The Earth Sciences*. New York, Harper and Row, second edition, 824 pp. (QE501.S83)

Suppe, J., 1985. *Principles of Structural Geology*. Englewood Cliffs NJ, Prentice-Hall, 537 pp. (QE601.S94)

Sutton, O.G., 1957. *Mathematics in Action*. London, Bell, second edition, 236 pp. Reprinted by Dover in 1984. (QA461.S96)

Swanson, D. A., T. L. Wright and R. T. Nelz, 1975. Linear vent systems and estimated rates of magma production and eruption for the Yakima basalt on the Columbia plateau. Amer. Jour. Sci., v.275, pp. 877–905.

Takahashi, T., 1981. Debris flow. Ann. Rev. Fluid Mechanics, v.13, pp. 57–77.

Taylor, G. I., 1970. Some early ideas about turbulence. Jour. Fluid Mechanics, v.41, pp. 3–11. (Nos. 1–2 of this volume are devoted to a symposium on turbulence.)

Tennekes, H. and J. L. Lumley, 1972. *A First Course in Turbulence*. Cambridge MA, MIT Press, 300 pp. (QA913.T44)

Terzaghi, K., 1943. *Theoretical Soil Mechanics*. New York, John Wiley and Sons, 510 pp. (TA710.T4)

Terzaghi, K., 1960. *From Theory to Practice in Soil Mechanics*. New York, John Wiley and Sons, 425 pp. (TA710.T33F)

Terzaghi, K. and R. B. Peck, 1967. *Soil Mechanics in Engineering Practice*. New York, John Wiley and Sons, second edition, 729 pp. (TA710.T33)

Thorn, C. E., 1988. *An Introduction to Theoretical Geomorphology*. Boston, Unwin Hyman, 247 pp. (GB401.5.T54)

Timoshenko, S. P., 1953. *History of Strength of Materials*. New York, McGraw-Hill, 452 pp. (TA405.T58H)

Toth, J., 1963. A theoretical analysis of groundwater flow in small drainage basins. Jour. Geophys. Research, v.68, pp. 4795–812.

Tritton, D. J., 1988. *Physical Fluid Dynamics*. Oxford, Clarendon Press, second edition, 519 pp. (QC151.T74)

Truesdell, C., 1966. *Six Lectures on Modern Natural Philosophy*. New York, Springer-Verlag, 117 pp. (QA802.T75)

Truesdell, C., 1968. *Essays in the History of Mechanics*. New York, Springer-Verlag, 384 pp. (QC122.T7)

Turcotte, D. L., 1979, Flexure. Advances in Geophysics, v.21, pp. 51–86.

Turcotte, D. L. and G. Schubert, 1982. *Geodynamics: Applications of Continuum Physics to Geological Problems*. New York, John Wiley and Sons, 450 pp. (QE501.T83)

Uyeda, S., 1978. *The New View of the Earth: Moving Continents and Moving Oceans*. San Francisco, W. H. Freeman, 217 pp. (QE511.5.U9313)

Van Dyke, M. D., 1982. *An Album of Fluid Motion*. Stanford CA, The Parabolic Press, 176 pp. (TA357.V35)

Verhoogen, J. *et al.*, 1970. *The Earth. An Introduction to Physical Geology*. New York, Holt, Rinehart and Winston, 748 pp. (QE501.V38)

Verruijt, A., 1982. *Theory of Groundwater Flow*. London, Macmillan, second edition, 144 pp. (TC176.V36)

Vogel, S., 1981. *Life in Moving Fluids. The Physical Biology of Flow*. Princeton NJ, Princeton University Press, 352 pp. (QH505.V63)

Von Engelhardt, W. and J. Zimmermann, 1988. *Theory of Earth Science*. Translated by Lenore Fischer. Cambridge University Press, 381 pp. (QE33.E5313)

Walker, G. P. L., 1989. Gravitational (density) controls on volcanism, magma chambers, and intrusions. Australian Jour. Earth Sci., v.36, pp. 149–65.

Walker, R. G., 1990. Facies modeling and sequence stratigraphy. Jour. Sedimentary Petrology, v.60, pp. 777–86.

Wang, H. F. and M. P. Anderson, 1982. *Introduction to Groundwater Modelling, Finite Difference and Finite Element Methods*. San Francisco, W. H. Freeman, 237 pp. (TC176.W36)

Watts, A. B. and J. R. Cochran, 1974. Gravity anomalies and flexure of the lithosphere along the Hawaiian-Emperor seamount chain. Geophys. Jour. Roy. Astronomical Soc., v.38, pp. 119–41.

Waythomas, C. F. and G. P. Williams, 1988. Sediment yield and spurious correlation – towards a better portrayal of the annual suspended-sediment load of rivers. Geomorphology, v.1, pp. 309–16.

Weijermars, R. and H. Schmeling, 1986. Scaling of Newtonian and non-Newtonian fluid dynamics without inertia for quantitative modelling of rock flow due to gravity (including the concept of rheological similarity). Physics Earth Planetary Interiors, v.43, pp. 316–30.

Wendt, R. E. Jr., ed., 1974. *Flow, Its Measurement and Control in Science and Industry, Vol. 1, Part 2, Flow Measuring Devices*. Pittsburgh, Instrument Soc. America (TC177.S84)

Westfall, R. S., 1971. *Force in Newton's Physics: the Science of Dynamics in the Seventeenth Century*. New York, American Elsevier, 579 pp. (QA802.W48)

Westfall, R. S., 1977. *The Construction of Modern Science: Mechanisms and Mechanics*. Cambridge University Press, 171 pp. (QA802. W47)

Westfall, R. S., 1980. *Never At Rest: A Biography of Isaac Newton*. Cambridge University Press, 908 pp. (QC16.N7W35)

White, F. M., 1986. *Fluid Mechanics*. New York, McGraw-Hill, 732 pp. (TA357.W48)

Whitaker, S., 1968. *Introduction to Fluid Mechanics*. Englewood Cliffs NJ, Prentice-Hall, 457 pp. (QA911.W38)

Williams, G., 1987. An introduction to relaxation methods. Byte, January 1987, pp. 111–24.

Williams, H. and A. R. McBirney, 1979. *Volcanology*. San Francisco, Freeman, Cooper, 397 pp. (QE522.W53)

Willis, J. C. and N. L. Coleman, 1969. Unification of data on sediment transport in flumes by similarity principles. Water Resources Research, v.5, pp. 1330–6.

Wilson, L., 1972. Explosive volcanic eruptions, II. The atmospheric trajectories of pyroclasts. Geophys. Jour. Royal Astronomical Soc., v.30, pp. 381–92.

Wilson, L., 1980. Relationships between pressure, volatile content and ejecta velocity in three types of volcanic explosions. Jour. Volcanology Geothermal Research, v.8, pp. 297–313.

Wilson, L., H. Pinkerton, and R. MacDonald, 1987. Physical processes in volcanic eruptions. Ann. Rev. Earth Planetary Sci., v.15, pp. 73–95.

Wood, D. M., 1990. *Soil Behaviour and Critical State Soil Mechanics*. Cambridge University Press, 462 pp. (TA710.W598)

Yalin, M. S., 1971. *Theory of Hydraulic Models*. London, Macmilian, 266 pp. (TC163.Y35)

Yalin, M. S. and J. W. Kamphuis, 1971. Theory of dimensions and spurious correlation. Jour. Hydraulic Research, v.9, pp. 249–65.

Young, A. D., 1989. *Boundary Layers*. Washington DC, Amer. Institute of Aeronautics and Astronautics, 269 pp. (TL574.B6.Y681)

Zoback, M. L., 1992. First- and second-order patterns of stress in the lithosphere: the World Stress Map Project. Jour. Geophys. Research, v.97, pp. 11 703–28.

Index of names

451

Subject index